Natural Polymers for Drug Delivery

We dedicate this book to:

Our Parents

Amit, Aditya & Nandika

Raghavi, Gaayathri & Virinchi

Natural Polymers for Drug Delivery

Edited by

Harsha Kharkwal

Amity University, India

and

Srinivas Janaswamy

South Dakota State University, USA

CABI

CABI is a trading name of CAB International

CABI
Nosworthy Way
Wallingford
Oxfordshire OX10 8DE
UK

Tel: +44 (0)1491 832111
Fax: +44 (0)1491 833508
E-mail: info@cabi.org
Website: www.cabi.org

CABI
745 Atlantic Avenue
8th Floor
Boston, MA 02111
USA

Tel: +1 (0) 617 682 9015
E-mail: cabi-nao@cabi.org

A catalogue record for this book is available from the British Library, London, UK.

Library of Congress Cataloging-in-Publication Data

Names: Kharkwal, Harsha, editor. | Janaswamy, Srinivas, editor. | C.A.B. International, issuing body.
Title: Natural polymers for drug delivery / edited by Harsha Kharkwal and Srinivas Janaswamy.
Description: Oxfordshire, UK ; Boston, MA : CABI, [2016] | Includes bibliographical references and index.
Identifiers: LCCN 2016013451 | ISBN 9781780644479 (alk. paper) | ISBN 9781786390745 (epub)
Subjects: | MESH: Polymers--therapeutic use | Drug Delivery Systems--methods
Classification: LCC RS199.5 | NLM QT 37.5.P7 | DDC 615/.6--dc23 LC record available at http://lccn.loc.gov/2016013451

ISBN-13: 978 1 78064 447 9

Commissioning editor: Sreepat Jain and Rachael Russell
Associate editor: Alexandra Lainsbury
Production editor: Tim Kapp

Typeset by SPi, Pondicherry, India
Printed and bound in the UK by CPI Group (UK) Ltd, Croydon, CR0 4YY

Contents

List of Contributors

I. Barman, Department of Mechanical Engineering, Johns Hopkins University, Baltimore, Maryland 21218, USA and Department of Oncology, Johns Hopkins University, Baltimore, Maryland 21287, USA.

A. Bose, Particle Design Research Group and Non-Destructive Biomedical and Pharmaceutical Research Centre, iPROMISE, Universiti Teknologi MARA, Puncak Alam, 42300, Selangor, Malaysia.

E. Cevher, Department of Pharmaceutical Technology, Faculty of Pharmacy, Istanbul University, 34116 Istanbul, Turkey.

N. Dhiman, Amity Institute of Pharmacy, Amity University Uttar Pradesh, Noida 201303, India.

S.K. Jain, Hamdard Institute of Medical Sciences and Research, Hamdard University, Delhi, India.

S. Janaswamy, Department of Dairy and Food Science, Berg Agricultural Hall, Room 108, South Dakota State University, Brookings, SD 57007, USA.

M. Joshi, Amity Institute of Nanotechnology, Amity University Uttar Pradesh, Noida 201303, India.

D.P. Katare, Amity Institute of Biotechnology, Amity University Uttar Pradesh, Noida 201303, India.

H. Kharkwal, Amity Center for Carbohydrate Research and Amity Institute of Phytomedicine and Phytochemistry, Amity University Uttar Pradesh, Noida 201303, India.

B. Malhotra, Amity Institute of Biotechnology and Amity Center for Carbohydrate Research, Amity University Uttar Pradesh, Noida 201303, India.

S. Mishra, Amity Institute of Biotechnology, Amity University Uttar Pradesh, Noida 201303, India.

R. Pandey, Department of Pediatrics, University of Connecticut Health, Farmington, Connecticut, USA.

P. Panthari, Amity Institute of Phytochemistry and Phytomedicine, Amity University Uttar Pradesh, Noida 201303, India.

A.Y. Peköz, Department of Pharmaceutical Technology, Faculty of Pharmacy, Istanbul University, 34116 Istanbul, Turkey.

A. Pradhan, Global Health, College of Public Health, University of South Florida, GH-IDRB Labs 3720 Spectrum Blvd, Suite 304 Tampa, Florida 33612, USA and Queensborough Community College, City University of New York, 222-05 56th Avenue, Bayside, New York 11364, USA.

R. Prasad, Department of Mechanical Engineering, Johns Hopkins University, Baltimore, Maryland 21218, USA; and Amity Institute of Microbial Technology, Amity University Uttar Pradesh, Noida 201303, India.

A.D. Sezer, Department of Pharmaceutical Biotechnology, Faculty of Pharmacy, Marmara University, 34668 Istanbul, Turkey.

A. Srivastava, Biological Sciences and Geology, Queensborough Community College, 222-05 56th Avenue, Bayside, New York 11364, USA.

A.K. Tyagi, Department of Experimental Therapeutics, Division of Cancer Medicine, The University of Texas MD Anderson Cancer Center, 1901 East Road, Houston, Texas 77054, USA.

A. Varma, Amity Institute of Microbial Technology, Amity University Uttar Pradesh, Noida 201303, India.

M.P. Yadav, Sustainable Biofuels and Co-Products Research Unit, Eastern Regional Research Center, ARS, USDA, 600 East Mermaid Lane, Wyndmoor, Pennsylvania19038, USA.

T.W. Wong, Particle Design Research Group & Non-Destructive Biomedical and Pharmaceutical Research Centre, iPROMISE, Universiti Teknologi MARA, Puncak Alam, 42300, Selangor, Malaysia.

1 Natural Polymers for Drug Delivery: An Introduction

Harsha Kharkwal[1,*] Bhanu Malhotra[2] and Srinivas Janaswamy[3]
[1]Amity Center for Carbohydrate Research and Amity Institute of Phytomedicine and Phytochemistry, Amity University, Noida, India; [2]Amity Institute of Biotechnology and Amity Center for Carbohydrate Research, Amity University, Noida, India; [3]Department of Dairy and Food Science, South Dakota State University, South Dakota, USA

Abstract
Natural polymers are macromolecules composed of repeating structural units joined by covalent bonds. Carbohydrates, proteins and muscle fibres are known examples and have potential as drug delivery systems. A typical delivery system aims at slow and tissue-specific release, and as natural polymers exhibit biodegradability and biocompatibility they are well suited for this purpose. Natural polymers are also utilized as excipients and over the years, new advances in the treatment of diseases using the approach of site specific drug delivery by the utilization of polymers have emerged with several promises. This chapter highlights some available examples with an emphasis on their potent applications and properties in the drug domain.

Introduction

A polymer is a macromolecule with repeating monomeric structural units joined covalently. Carbohydrates, proteins and muscle fibres are common types of polymers. Carbohydrates are polyhydroxy aldehydes or ketones, and could be further classified as monosaccharides, disaccharides and polysaccharides. A polysaccharide consists of more than 20 repeating monomeric units. Polysaccharides can be homopolysaccharides if they contain only one repeating monomeric unit (e.g. cellulose, glycogen, starch and chitin), or heteropolysaccharides if two or more different kinds of monomers are present (e.g. peptidoglycan bacterial cell walls and glycosaminoglycans) (Pérez and Mulloy, 2005). These natural systems can be modified chemically to create biocompatible and biodegradable non-toxic entities, and have readily gained popularity in the pharmaceutical industry as drug delivery agents (Harborne, 1987). Plant-based polymers have also been investigated for this purpose. In addition, various liquid ophthalmic suspensions, buccal films, film-coating agents and microspheres have been proven to be effective (Pandey and Khuller, 2004; Chamarthy and Pinal, 2008; Alonso-Sande *et al.*, 2009).

The history of using silicone rubber as a carrier (Folkman and Long, 1964) set the stage for the design and development of prolonged drug delivery systems, and since then the use of polymers in drug therapy has advanced significantly. Several scientific journals highlight the use of polymers as drug vehicles, and Table 1.1 gives an historical perspective with citations of published

*Corresponding author. E-mail: hkharkwal@amity.edu

Table 1.1. Top polymer related reviews cited in 'Advanced Drug Delivery Reviews' according to Web of Science core collection in 2014. (Adapted, with permission, from Merkle, 2015.)

Subject	Year	Rank	Citations	Reference
Block copolymer micelles	2001	3	1620	Kataoka *et al.*, 2001
Biodegradable nanoparticles	2003	4	1230	Panyam and Labhasetwar, 2003
Environment-sensitive hydrogels	2001	5	1221	Qiu and Park, 2001
Nanoparticles	2002	6	1213	Brigger *et al.*, 2002
Hydrogels	2002	7	1083	Hoffman, 2002
Peptide and protein PEGylation	2002	9	803	Roberts *et al.*, 2002
Dendrimers	2005	10	788	Svenson and Tomalia, 2005
Drug release from HPMC delivery systems	2001	11	772	Siepmann, 2001
Thermo- and pH-responsive polymers	2006	12	771	Schmaljohann, 2006
Nanoparticle targeting	2004	13	712	Brannon-Peppas and Blanchette, 2004
Thermosensitive hydrogels	2002	14	695	Jeong *et al.*, 2002

HPMC, hydroxypropyl methylcellulose; PEG, polyethyleneglycol.

articles in *Advanced Drug Delivery Reviews*, according to the Web of Science core collection in 2014.

The need for natural polymers

Research has focused on the beneficial properties of natural polymers, especially towards delivering toxic therapeutic agents to the target tissue. The use of natural polymers and their derivatives not only enhances the drug availability at the target tissues, but is also regarded as a safe means of delivery. Some of the special characteristics of natural polymers that are attractive are their:

- Biodegradability – they pose no harmful environmental effect and are 100% biodegradable.
- Lack of toxicity – they are non-toxic.
- Economy – they are inexpensive and large quantities can easily be obtained.
- Safety – their natural availability bestows the required safety without any harmful side effects.
- Availability – they are widely distributed globally; for example, cellulose can easily be extracted in large quantities (Prajapati *et al.*, 2013).

Some of the disadvantages include the chances of microbial contamination when exposed to the external environment, uncontrolled hydration rate because of differences in availability and the presence of different species.

Classification of Natural Polymers

Natural polymers, mainly polysaccharides, are obtained from various sources including plants, microbes, algae and fungae. Some are neutral and others, such as the carboxylate or sulfate groups, possess a negative charge. Chitosan is the only cationic polysaccharide currently known (Fig. 1.1).

- Plant origin – starch, hemicellulose, cellulose, agar, glucomannan, pectin, guar gum, locust bean gum, gum acacia, gum tragacanth and psyllium
- Microbial origin – curdlan, gellan, xanthan
- Algal origin – alginate, carrageenan
- Fungal origin – chitin, pullulan, scleroglucan

Drug Delivery Applications of Polysaccharides

Polysaccharides are used as coating agents, polymer matrices, tablets formulations, and emulsifying and gelling agents (Prajapati *et al.*, 2013).

Tablet adjuvant formulations

Polysaccharides have been used in tablet formulation due to their inherent adhesive nature.

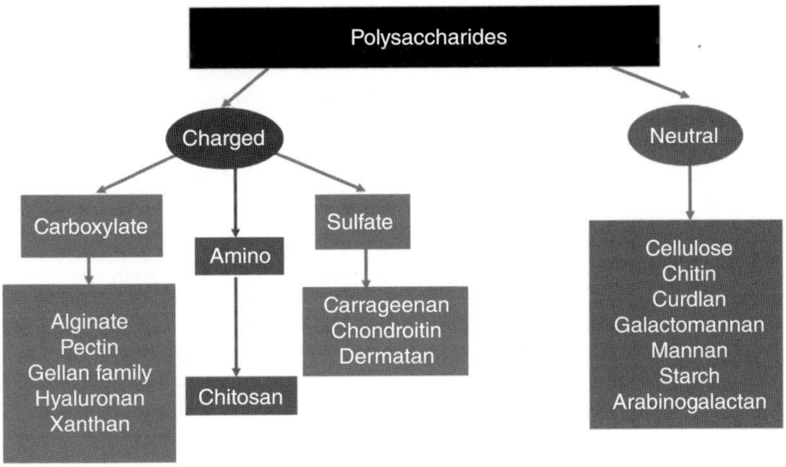

Fig. 1.1. Classification of polysaccharides used in drug delivery.

They adsorb large amount of water and swell, so acting as disintegrants. They also provide cohesiveness to the powder formulations and can easily be incorporated into tablets or granules (e.g. guar gum and acacia).

Mucoadhesive agents

Their main purpose is to control release of the drugs over a stipulated time. Furthermore, they can be retained in the intestinal lining and stomach for longer durations, enhancing drug absorption (e.g. karaya gum and sodium alginate).

Emulsifying and suspending agents

Natural polymers provide stability to emulsions because of their interfacial absorption. They can also form films with high tensile strength and resist coalescence among the droplets (e.g. xanthan gum and acacia gum).

Gelling agents

Mucilage and gums form gels either alone or in combinations with other gums. The gelation is due to inter- and intra-molecular associations among the chains leading to three-dimensional networks that can, in turn, trap large amounts of water. These formulations can be prepared through physical methods, for example changing pH and temperature, as well as by chemical treatments through adding suitable reagents. Examples include carrageenan and locust bean gum.

Coating agents

Certain natural polymers have the intrinsic ability to act as coating agents that protect the drugs from degradation and allow release in a controlled manner (e.g. pectin and sodium alginate).

Sustaining agents in dosage form

Matrix tablets are the most prominent oral drug delivery systems because of their sustained release and easy formulation properties (e.g. locust bean gum and karaya gum).

The main purpose of developing drug delivery systems casting polymers is to abolish any toxic product accumulation inside the body. This is quite feasible as natural polysaccharides do not generate any unusual products inside the body. Instead, they are eliminated easily as carbohydrate units during the regular metabolic processes, and so the polysaccharide disappears after serving its purpose. The biodegradation proceeds with bond breakage within the monomers leading to erosion of the bulk polymer

(Peppas, 1984). Various routes cause polymer degradation:

- hydrolysis;
- photolysis;
- its solubilising nature;
- brittleness;
- biodegradation;
- thermo-degradation; and
- structural weakening.

Polymer Drug Release Mechanism

The therapeutic agents attached to the polymers can be released at a controlled rate from the polymeric matrices via different mechanisms. The delivery of a drug over a specified time period to the tissues exploits various properties of polymers. One prominent example is that of stimuli-sensitive polymers releasing the drug only when there is a change in pH or temperature (Kaur *et al.*, 2014).

Degradation

Certain biodegradable polymers degrade inside the body under normal physiological and biological processes. They can also be designed to break under hydrolysing conditions, which results in smaller and manageable chain lengths without any side effects.

Diffusion

A reservoir device is often used, where the drug is located in the core of the tablet, capsule or polymeric network with a shell surrounding it. The shell might be composed of some type of polymer that will dictate the rate of diffusion of the drug from the core. With this mechanism, water will diffuse into the core and dissolve the drug inside, which will then diffuse out. Swelling or degradation of the shell can occur, depending on the polymer. Two different types of diffusion systems exist:

1. Only dissolved drug within the core. The drug load decreases over time as it diffuses out of the core.

2. Initial drug concentration within the core is higher than the aqueous solubility concentration. As the dissolved drug diffuses out more, an amount of drug is dissolved within the core, and the drug load will be constant for a longer period of time.

Swelling

Swelling is another type of controlling phenomenon involved in drug delivery. The matrix former has the capacity to swell and control the drug release rate. When polymer swelling leads to an increase in the length of diffusion pathways, the system volume increases, lowering the drug concentration gradient. This results in a slower release of the drug into the bulk system. In contrast, swelling of the polymer can enhance the molecular mobility, leading to faster release.

Overall, immense progress has been made in diffusion-controlled systems and solvent-activated formulations of drug release. Also, through the use of hydrogels and various other polymeric carrier systems, it is now possible to establish a very safe passage for the therapeutic drug to the target regions, and more importantly to inhospitable physiological regions. Polymeric substances having a controlled molecular architecture can be specifically engineered to provide response to the external stimulus. It has been shown that the therapeutic agents conjugated to the polymer show relatively improved drug release kinetics by preventing carrier accumulation. Polymer drug conjugates also help to improve the circulatory half-life for the cytoplasmic delivery of therapeutics.

Natural Polymers in Drug Delivery

Hierarchical evolution of present-day drug delivery systems began with the use of polymeric carriers that evoked spatiotemporal release of drugs in the implanted reservoir systems. Undoubtedly, conventional drug delivery systems have made great contributions towards treating disease. However, the growing need for special, accurate and potent biological therapeutic delivery procedures that target specific drug delivery protocols demands novel delivery systems and mechanisms. Recent advances also highlight the need for feedback control of the drug delivery systems (Heller, 2005). There are a number of hurdles to overcome in targeting specific delivery and so implementing intelligent delivery systems is a feasible approach. These may allow not only the development of proper routes for intracellular drug transport but also specific targeting and recognition through feedback control systems (Langer and Peppas, 2003). Many natural and synthetic systems are being examined, and some are described below.

Collagen

Collagen is the most abundant protein of the animal kingdom, and is found in the extracellular matrix of connective tissue. It has a characteristic triple helical structure of repeating glycine–proline–hydroxyproline repeats. As we write, around 19 different collagen systems have been employed in pharmaceutical and medical applications, of which 80–90% belong to type I, type II and type III. Collagen possesses outstanding biocompatibility, few immunity problems and good biodegradability (Harkness, 1961). Collagen formulations are made with the combination of liposomes, in which therapeutic agents are encapsulated in the liposomes and then fused with collagen, making a scaffold or gel that not only prolongs the drug release rate but also increases its therapeutic efficiency. Collagen pellets and tablets used in Japan are known as monolithic devices, and are specialized rods approximately 1 mm diameter and 15 mm length. These are injected using a syringe in plunger for the local delivery of minocycline and lysozyme to treat periodontitis. These pellets have also been used to deliver IL-2 in vivo via a mini-pellet subcutaneous injection (Alonso-Sande *et al.*, 2009).

Another example of the use of collagen in the pharmaceutical industry is the development of the collagen corneal shield, a type of ophthalmic lens. Collagen, being a structural protein, provides support and protection to the eye and treats a variety of eye conditions after corneal transplantation and surgery. Collagen shields are made from procaine and bovine collagen with different dissolving times. For example, Bausch & Lomb Pharmaceuticals has developed collagen contact lenses called BioCora collagen shields. These have potential in delivering corticosteroid and other conjunctivital antibiotics to the eye. Certain water soluble antibiotics and steroids like Vancomycin, Gentamycin, pilocarpine and Amphotericin-B are being used along with shields of collagen to minimize the rubbing of eyelids. Collagen shields marketed preparations include: MediLenso, Biocora, Irvine, ProshieldO and Chiron. They provide a structural scaffold and behave as short-term bandage protections, permitting the oxygen transfer required for metabolism of the eye. These shields dissolve in the cornea, providing good lubrication to the eye.

Rosin

Rosin, obtained from *Pinus palustris* Miller and *Pinus linnae*, is a non-volatile natural polymer with phenomenal biocompatibilities and biodegradation capabilities (Berkland *et al.*, 2002). It is a tricyclic diterpene containing carboxylic acids and non-acidic components. Rosin has applications in printing inks, paper-sizing agents and chewing gum, to name a few. It is also a renewable chemical for polymer synthesis and a good film-forming agent, along with its derivatives, and thus is used for prolonged release of drugs as well as for enteric coating and transdermal drug delivery (Sheorey and Dorle, 1991a). Resin displays high compatibility with a variety of drugs having varied molecular weights and water solubility (Felder *et al.*, 2003). Rosin has also been evaluated for its encapsulating properties in different pharamaceutical preparations (Sheorey and Dorle, 1991b). Microspheres of rosin with glycerol esters could be developed for controlled drug release. The release rate is found to depend on the size and morphology and the polymer degradation rate. Resin polymers maintain the drug concentration in the target tissue within the permissible therapeutic range, and implants can be prepared from them as they are completely degradable inside the body and do not require removal.

Rosin also has some special properties and behaves as an anti-tumour agent. It possesses similar actions of inflammation to Poly (DL-lactic-co-glycolic acid) (PLGA) (Liu *et al.*, 2003). Rosin polymers show enhanced emulsifying ability and good skin permeability with homogeneity and spreadability for transdermal drug delivery. Transdermal patches with enhanced pharmacodynamic performance and pharmacokinetics could be accomplished by combining rosin with polyvinyl pyrrolidone (Bohme, 2002).

Chitosan

Chitosan, a polymer of glucosamine and N-acetyl glucosamine, is obtained by deacetylation of chitin from the exoskeleton of crustaceans. It is a cationic polymer, biocompatible, biodegradable and non-toxic. It is used extensively in powder formulations, gels, emulsions and tablets. It is not only a good excipient but also an exceptional platform for parenteral delivery. Chitosan is also a good antimicrobial agent and could mask flavours. Low molecular weight chitosan exhibits reduced toxicity on Caco-2 cells. In association with ovalbumin gels, chitosan is used for cosmetic and pharmaceutical applications, and along with non-ionic surfactants is a good bioadhesive agent at different physiological pH regimes. Another important application of chitosan is as a vaginal delivery system for metronidazole. Introduction of thiol groups in the preparation certainly improves its mucoadhesion properties, which in turn enhance bioadhesion by increasing the residence time of drugs in the mucosal lining of the vagina (Uhrich *et al.*, 1999).

Starch

Starch is synthesized by plants and stored as an energy reserve. After cellulose, it is the second most abundant carbohydrate in the plant kingdom. World starch production in the year 2000, based on estimates from the European Union (EU) Commission and the United States Department of Agriculture (USDA), was 48.5 million tons (www.starch.dk/ISI/market/index.asp), of which, 39.4 million tons were from maize, 4.1 million tons from wheat and 2.6 million tons from potatoes. The rest is comprised of cassava, rice and other sources. It is a heterogeneous polymer of α-D-glucose units linked by α-(1, 4)-bonds and α-(1, 6)-linkages. Starch is found in two forms: amylose and amylopectin. Amylose is a linear polymer of several hundred α-(1,4)-linked glucose moieties and amylopectin a branched molecule composed of α-(1,4)-linked glucose moieties along with branching through α-(1,6) linkages. Starch and its derivatives have gained a niche in the pharmaceutical applications as a filler, diluent, glidant, disintegrant, and binder. It is economical and readily available. Also, starch is being used as an excipient in the extended release preparations due to its ease of enzymatic degradation and low compactibility. Enzymatic hydrolysis further improves its excipient potential (Hong *et al.*, 2014). Pregelatinized starch is routinely used as a controlled release matrix. Starch tablet formulations follow zero-order kinetics, but the release rates can be fine-tuned, for example by altering compaction force and tablet geometry. Important research has been carried out into developing synthetic derivatives of starch for various drug delivery systems, and these have been formulated.

Gelatin

Gelatin is a water-soluble polymer produced as a denaturation of collagen. It possesses good biodegradability and low antigenicity and could be used effectively in pharmaceutical applications. It is a protein and can be manipulated by appropriate changes in the isoelectric point for drug delivery applications. Its charge changes from positive to negative in the normal body physiological pH range. Gelatin is not only a good material for cell culture but is also used in tissue engineering. Gelatin-based systems are also used in controlled-release tablet formulations carrying appropriate therapeutic agents. Liposome-loaded active compounds are also made using PEG-gelatin gel systems which act as good scaffolds for prolonged drug release profiles (Foox and Zilberman, 2015).

A number of derivatives of natural polymers have been exploited worldwide for more than two decades. Derivatization not only enhances the physicochemical properties of these polymers but also couples their use with some synthetic polymers in drug delivery strategies. One example is the achievement of resistance by plasmid DNA in cancer treatment, by incorporating micelles into the polyion complex (PIC) micelles, which protect it from nuclease digestion (Kataoka *et al.*, 1999). The use of natural polymers and their derivatives have laid an important platform for modern drug delivery strategies.

Hibiscus mucilage

Hibiscus rosa-sinensis (Family Malvaceae), commonly known as the China rose, is a landscape shrub growing to 7–12 feet (2–4 m). It has glossy dark-green leaves which are medium textured, and produced throughout the year (King, 1999).

Diclofenac sodium formulation has been prepared in the form of tablets using *H. rosa-sinensis* leaves, and subsequently mucilage was developed. Its release retardant activity has been studied in all the sustained release formulations. A number of physicochemical properties of this mucilage have been investigated in diclofenac sodium formulations. The resulting matrix formulated tablets showed a better uniformity of weight, friability and hardness. The swelling of these formulations and their characteristic property of release rate *in vitro* showed that the mucilage from dried leaves of *H. rosa-sinensis* can potentially be used in sustained drug release applications. The mucilage followed zero-order reaction kinetics (Jani and Shah, 2008).

Aloe mucilage

Aloe vera leaves and Burm.f. (*A. barbadensis*) have also been investigated by isolating the exudates of adjacent cells of the vascular bundles and from the central parenchyma leaf tissue of *A. vera*. The exudates are bitter and yellow in colour, and contain glycosides of dihydroxyanthraquinone (Vázquez *et al.*, 1996). The *A. vera* parenchyma contains lipids, amino acids, proteins, vitamins, enzymes and various carbohydrates. Arabinan, glucogalactomannan, arabinorhamnogalactan, galactan and glucuronic acids are the main constituents (Choi and Chung, 2003). The combination of povidone and mucilage of dried *A. barbadensis* can be used effectively for sustained release formulations (Hamza and Aburahma, 2010).

Fenugreek mucilage

Fenugreek, *Trigonella foenum-graecum*, is a herb belonging to the Fabaceae family. A high percentage of mucilage is present in the fenugreek seeds along with a natural gummy agent in the seed coatings. It does not dissolve in water and forms a tacky mass upon exposure to fluids and becomes slick with over exposure (Shakuntala *et al.*, 2011). The isolation of the husk from the seeds is done by reducing the size and suspending the seeds in chloroform before decanting them. Chloroform extraction removes the oily part which can then be dried (Avachat *et al.*, 2007). Fenugreek at 66% w/w was superior in retarding drug release compared to hypomellose (Nokhodchi *et al.*, 2008).

Guar gum

Guar gum is obtained from the endosperm of the leguminous plant *Cyamopsis tetragonolobus*. Its extraction is through drying the pods in sunlight and their manual separation from the seeds. The commercial gum extraction involves mechanical roasting, sieving, differential attrition and polishing. The seeds are then broken and the endosperm releases the germ. The seed breaks into the two halves of the endosperm, which are referred to as the guar splits. These refined splits are coated with fine fibrous material layer forming the husk which can be removed from the endosperm by polishing. Different processing techniques are employed to obtain finished powders of refined guar splits. Guar gum consists of 1, 4-linked mannose units with 1, 6-galactose as side groups. It is used as a cosmetic thickener and to prevent ice crystals in ice creams. *In vitro* studies on the tablets made with xanthan gum, pectin and guar gum revealed that furosemide could be released at pH 7.2 in a sustained manner for 15 h (Inpharma Weekly, 1992). More importantly, guar gum tablets show a higher swelling index compared to xanthan gum and pectin.

Conclusions and Future Prospects

Currently, natural polymers have received considerable attention due to their value in environmental protection, as well as in the maintenance of human health. These polymers and their derivatives, coupled with biodegradable polymers, are also used in active packaging, fibre reinforcements and tissue engineering. In addition, they also have applications in mucosal, colonic and targeted protein/peptide, gene/vaccine, anticancer and drug delivery. In the pharmaceutical and medical arenas, these systems have received considerable interest as intelligent materials such as artificial tissues. Thorough understanding of the relationships between structure, property and performance will contribute significantly to the advancement of modern science and technology, and there is promise in the future of more sophisticated and better understood natural polymer systems.

Acknowledgements

We thank Dr Ashok K. Chauhan, Founder President, Ritanand Balved Educational Foundation, for support. The guidance provided by Shri. Atul Chauhan, Chancellor, Amity University Uttar Pradesh, and Prof. Dr (Mrs) Balvinder Shukla, Vice Chancellor, Amity University Uttar Pradesh is greatly appreciated. The publications cited in this chapter provided numerous insights, and we are grateful to their eminent authors.

References

Alonso-Sande, M., Teijeiro-Osorio, D., Remuñán-López, C. and Alonso, M. (2009) Glucomannan, a promising polysaccharide for biopharmaceutical purposes. *European Journal of Pharmaceutics and Biopharmaceutics* 72(2), 453–462.

Avachat, A., Gujar, K., Kotwal, V. and Patil, S. (2007) Isolation and evaluation of fenugreek seed husk as a granulating agent. *Indian Journal of Pharmaceutical Sciences* 69(5), 676.

Berkland, C., King, M., Cox, A., Kim, K. and Pack, D. (2002) Precise control of PLG microsphere size provides enhanced control of drug release rate. *Journal of Controlled Release* 82(1), 137–147.

Bohme, K. (2002). Buprenorphine in a transdermal therapeutic system – A new option. *Clinical Rheumatology* 21(S1), S13–S16.

Brannon-Peppas, L. and Blanchette, J. (2004) Nanoparticle and targeted systems for cancer therapy. *Advanced Drug Delivery Reviews* 56(11), 1649–1659.

Brigger, I., Dubernet, C. and Couvreur, P. (2002) Nanoparticles in cancer therapy and diagnosis. *Advanced Drug Delivery Reviews* 54(5), 631–651.

Chamarthy, S. and Pinal, R. (2008) Plasticizer concentration and the performance of a diffusion-controlled polymeric drug delivery system. *Colloids and Surfaces A: Physicochemical and Engineering Aspects* 331(1–2), 25–30.

Choi, S. and Chung, M. (2003) A review on the relationship between *Aloe vera* components and their biologic effects. *Seminars in Integrative Medicine* 1(1), 53–62.

Felder, C., Blanco-Príeto, M., Heizmann, J., Merkle, H. and Gander, B. (2003) Ultrasonic atomization and subsequent polymer desolvation for peptide and protein microencapsulation into biodegradable polyesters. *Journal of Microencapsulation* 20(5), 553–567.

Folkman, J. and Long, D. (1964) The use of silicone rubber as a carrier for prolonged drug therapy. *Journal of Surgical Research* 4(3), 139–142.

Foox, M. and Zilberman, M. (2015) Drug delivery from gelatin-based systems. *Expert Opinion on Drug Delivery* 12(9), 1547–1563.

Hamza, Y. and Aburahma, M. (2010) Design and in vitro evaluation of novel sustained-release matrix tablets for lornoxicam based on the combination of hydrophilic matrix formers and basic pH-modifiers. *Pharmaceutical Development and Technology* 15(2), 139–153.

Harborne, J.B. (1987) *The Wealth of India, Raw Materials*. Volume 1a, revised edn. CSIR, New Delhi.

Harkness, R. (1961) Biological functions of collagen. *Biological Reviews* 36(4), 399–455.

Heller, A. (2005) Integrated medical feedback systems for drug delivery. *AIChE Journal* 51(4), 1054–1066.

Hoffman, A. (2002) Hydrogels for biomedical applications. *Advanced Drug Delivery Reviews* 54(1), 3–12.

Hong, Y., Liu, G. and Gu, Z. (2014) Recent advances of starch-based excipients used in extended-release tablets: a review. *Drug Delivery*, 23(1), 12–20.

Inpharma Weekly (1992) Gummed up with guar gum. *Inpharma Weekly* 849(1), p.25. Available at: rd.springer.com/article/10.2165%2F00128413-199208490-00057 (accessed 27 April 2016).

Jani, G. and Shah, D. (2008) Evaluation of mucilage of *Hibiscus rosasinensis* Linn. as rate controlling matrix for sustained release of diclofenac. *Drug Development and Industrial Pharmacy* 34(8), 807–816.

Jeong, B., Kim, S. and Bae, Y. (2002) Thermosensitive sol–gel reversible hydrogels. *Advanced Drug Delivery Reviews* 54(1), 37–51.

Kataoka, K., Harada, A. and Nagasaki, Y. (2001) Block copolymer micelles for drug delivery: design, characterization and biological significance. *Advanced Drug Delivery Reviews* 47(1), 113–131.

Kataoka, K., Harada, A., Wakebayashi, D. and Nagasaki, Y. (1999) Polyion complex micelles with reactive aldehyde groups on their surface from plasmid DNA and end-functionalized charged block copolymers. *Macromolecules* 32(20), 6892–6894.

Kaur, A., Kaur, A., Kaur, V.P., Kaur, M. and Murthy, R.S. (2014) Polymeric drug delivery approaches for colon targeting: a review. *Drug Delivery Letters* 4(1), 38–48.

King, S. (1999) Medicinal plants of the world: chemical constituents, traditional and modern medicinal uses by Ivan A. Ross (U.S. Food and Drug Administration) reviewed in *Journal of Natural Products* 62(1), 203–204.

Langer, R. and Peppas, N. (2003) Advances in biomaterials, drug delivery, and bionanotechnology. *AIChE Journal* 49(12), 2990–3006.

Liu, J., Xiao, Y. and Allen, C. (2003) Polymer-drug compatibility: a guide to the development of delivery systems for the anticancer agent, ellipticine. *Journal of Pharmaceutical Sciences* 93(1), 132–143.

Merkle, H.P. (2015) Drug delivery's quest for polymers: Where are the frontiers? *European Journal of Pharmaceutics and Biopharmaceutics* 97(B), 293–303.

Nokhodchi, A., Nazemiyeh, H., Khodaparast, A., Sorkh-Shahan, T., Valizadeh, H. and Ford, J. (2008) An in vitro evaluation of fenugreek mucilage as a potential excipient for oral controlled-release matrix tablet. *Drug Development and Industrial Pharmacy* 34(3), 323–329.

Pandey, R. and Khuller, G. (2004) Polymer based drug delivery systems for Mycobacterial infections. *Current Drug Delivery* 1(3), 195–201.

Panyam, J. and Labhasetwar, V. (2003) Biodegradable nanoparticles for drug and gene delivery to cells and tissue. *Advanced Drug Delivery Reviews* 55(3), 329–347.

Peppas, N. (1984) Controlled drug delivery. Vol. I: Basic concepts reviewed in *Journal of Controlled Release* 1(1), 84–85.

Pérez, S. and Mulloy, B. (2005) Prospects for glycoinformatics. *Current Opinion in Structural Biology* 15(5), 517–524.

Prajapati, V., Jani, G., Moradiya, N. and Randeria, N. (2013) Pharmaceutical applications of various natural gums, mucilages and their modified forms. *Carbohydrate Polymers* 92(2), 1685–1699.

Qiu, Y. and Park, K. (2001) Environment-sensitive hydrogels for drug delivery. *Advanced Drug Delivery Reviews* 53(3), 321–339.

Roberts, M., Bentley, M. and Harris, J. (2002) Chemistry for peptide and protein PEGylation. *Advanced Drug Delivery Reviews* 54(4), 459–476.

Schmaljohann, D. (2006) Thermo- and pH-responsive polymers in drug delivery. *Advanced Drug Delivery Reviews* 58(15), 1655–1670.

Shakuntala, S., Pura Naik, J., Jeyarani, T., Madhava Naidu, M. and Srinivas, P. (2011) Characterisation of germinated fenugreek (*Trigonella foenum-graecum* L.) seed fractions. *International Journal of Food Science & Technology* 46(11), 2337–2343.

Sheorey, D. and Dorle, A. (1991a) Release kinetics of drugs from rosin-glycerol ester microcapsules prepared by solvent evaporation technique. *Journal of Microencapsulation* 8(2), 243–246.

Sheorey, D. and Dorle, A. (1991b) Effect of solvents on the characteristics of rosin walled microcapsules prepared by a solvent evaporation technique. *Journal of Microencapsulation* 8(1), 71–78.

Siepmann, J. (2001) Modeling of drug release from delivery systems based on hydroxypropyl methylcellulose (HPMC). *Advanced Drug Delivery Reviews* 48(2–3), 139–157.

Svenson, S. and Tomalia, D. (2005) Dendrimers in biomedical applications—reflections on the field. *Advanced Drug Delivery Reviews* 57(15), 2106–2129.

Uhrich, K., Cannizzaro, S., Langer, R. and Shakesheff, K. (1999) Polymeric systems for controlled drug release. *Chemical Reviews* 99(11), 3181–3198.

Vázquez, B., Avila, G., Segura, D. and Escalante, B. (1996) Antiinflammatory activity of extracts from *Aloe vera* gel. *Journal of Ethnopharmacology* 55(1), 69–75.

2 Cellulose-based Polymeric Systems in Drug Delivery

Bhanu Malhotra[1], Harsha Kharkwal[2,*] and Madhav P. Yadav[3]
[1]Amity Institute of Biotechnology and Amity Center for Carbohydrate Research, Amity University, Noida, India; [2]Amity Center for Carbohydrate Research and Amity Institute of Phytomedicine and Phytochemistry, Amity University Uttar Pradesh, Noida, India; [3]Sustainable Biofuels and Co-Products Research Unit, USDA, Wyndmoor, Pennsylvania, USA

Abstract

The pharmaceutical industry requires the development of biodegradable, biocompatible, non-toxic, site-specific drug delivery polymers which can be easily coupled with drugs to be delivered orally, topically, locally or parenterally. The use of the most abundant biopolymer, cellulose, along with its derivatives, is intended to develop sustainable controlled release dosage forms by their easy fabrication into hydrophilic matrices. This amalgamation of the use of natural polymers and their derivatives with the drugs has proved to be one of the most effective means of delivering complex drugs without any side effects to the target sites in the body. Both cellulose esters and ethers are well researched and exploited as coatings in various tablet formulations. In this chapter we provide a detailed discussion of the use of cellulose as a powerful biopolymeric material in drug delivery to various sites in the body; we also discuss the intervention of nanotechnology to develop cellulose nanofibrils as powerful moieties for therapeutic drug delivery.

Introduction

The pharmaceutical industry develops new drug delivery formulations and procures, compounds, stores and dispenses them. The main aim is to deliver new and efficient cost-effective drug delivery strategies which are safe and non-toxic. As we have already seen in the previous chapter of this book (Chapter 1), various biopolymers can be exploited differently to be used in drug site specific drug delivery procedures (Buurma *et al.*, 2003; Thompson, 2006). Cellulose, along with its derivatives, has been exploited for decades in drug delivery formulations and its use is most commonly reported as a viscosity enhancer in topical formulations, suspending agents in oral formulations.

Structure of Cellulose and Properties

Cellulose is a naturally occurring biopolymer mainly present in the cell walls of many organisms ranging from prokaryotes to eukaryotes. The cell walls of bacteria, cyanobacteria, fungi, green algae, amoebae, ferns, mosses, angiosperms and gymnosperms are composed of cellulose (Kennedy and Knill, 2001; Bochek, 2003). Cellulose, as shown in Fig. 2.1, is made up of anhydro-D-glucopyranose units linked together via 1,4 linkages. Cellulose is

*Corresponding author. E-mail: hkharkwal@amity.edu

primarily insoluble in water and in most solvents because of strong inter- and intra-molecular hydrogen bonding between the chains (Hinterstoisser and Salmén, 2000). Cellulose has wide-ranging applications, such as in netting, coatings, paper, composites and upholstery, in spite of its poor solubility. The hydroxy groups of 1,4 linked glucan residues of cellulose are placed at C2 (secondary), C3 (equatorial), and C6 (primary) positions. Side group CH_2OH is found in trans-gauche arrangement relative to the C4–C5 and O5–C5 bonds. Due to the supramolecular structure of cellulose its solid state is presented in crystalline as well as amorphous forms. The degree of crystallinity (DP) in cellulose can vary from 40% to 60% according to the pretreatment and its origin. Table 2.1 shows the DP in various origins.

Cellulose morphology also has an enormous effect on its reactivity. The hydroxyl moieties of cellulose of amorphous origin are highly reactive and accessible as compared to the cellulose of crystalline origin where it is completely inaccessible due to inter-chain hydrogen bonding (Marchessault, 1994). Plant cellulose is found in two native forms: cellulose I and cellulose II, which are both crystalline (Kuga and Malcolm Brown, 1988). Cellulose II is formed from cellulose I after its treatment with aqueous NaOH (Kolpak and Blackwell, 1975; Langan *et al.*, 1999; Saxena, 2005). There are four different crystalline polymorphs of cellulose: I, II, III and IV; of these four, cellulose II is the most stable. Treatment of I and II with liquid ammonia results in cellulose III (Sarko *et al.*, 1976; Chanzy *et al.*, 1987; Wada *et al.*, 2001). Heating cellulose III results in cellulose IV (Buleon and Chanzy, 1980). Figure 2.2 shows a flowchart of the conversion of cellulose into its various polymorphs (Klemm *et al.*, 2005).

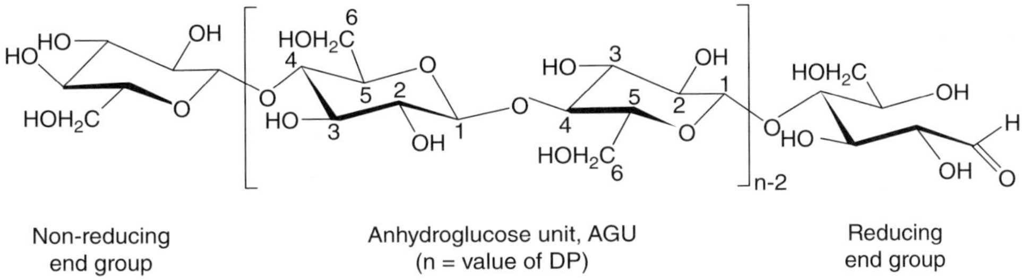

Fig. 2.1. Structure of cellulose containing anhydroglucose units joined via β-1,4 linkage. (Reproduced, with permission, from Marques-Marinho and Vianna-Soares, 2013.)

Table 2.1. Average DP of cellulose obtained from different sources. (Reproduced, with permission, from Kamel *et al.*, 2008.)

Source	Wood	Valonia	Cotton	*Acetobacter xylinum*	Cotton linters	Flax	Pulp	Kapok	Ramie
*DPw(10^3)	8–9	25–27	8–15	2–6	1–5	7–8	2.1	9.5	9–11

DPw, DP weight average determined by viscometric techniques

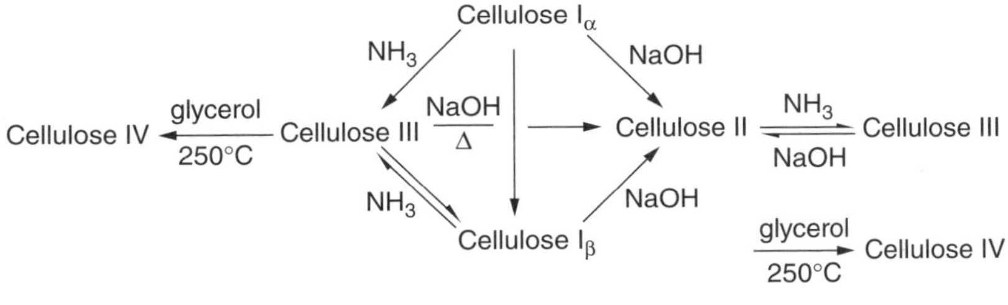

Fig. 2.2. Transformation of cellulose into its various polymorphs. (Reproduced, with permission, from Kamel *et al.*, 2008.)

Cellulose Derivatives

The cellulosics are formed after typical modifications of cellulose such as esterification and etherification at the hydroxyl group. Chemical modification generates cellulose derivatives with improved properties and processing abilities to form tailor-made polymers for drug delivery strategies. Cellulosics are biocompatible, reproducible and recyclable for various pharmaceutical applications (Argyropoulos, 2001). Typical cellulosics include methyl cellulose, ethyl cellulose, hydroxy ethyl cellulose, hydroxyethylmethyl cellulose, hydroxylpropyl cellulose and carboxymethyl cellulose ethers (Schuerch, 1973). Figures 2.3 and 2.4 represent the structure of cellulose ethers and esters, respectively. Cellulose ethers are formed via hydroxyl etherification using the specific alkyl halide using the cellulose obtained from wood pulp (Xiong and Ye, 2004). The average number of alkyl groups R represents the degree of substitution. The maximum degree of substitution (DS) is three on each glucan unit. DS can be varied for obtaining various different derivatives to be used in different pharmaceutical applications.

Cellulose derivatives in pharmaceutical applications are used as excipients to deliver drugs orally, topically and parenterally (Bastedo, 1939; Felton, 2006). In drug delivery, they are used to

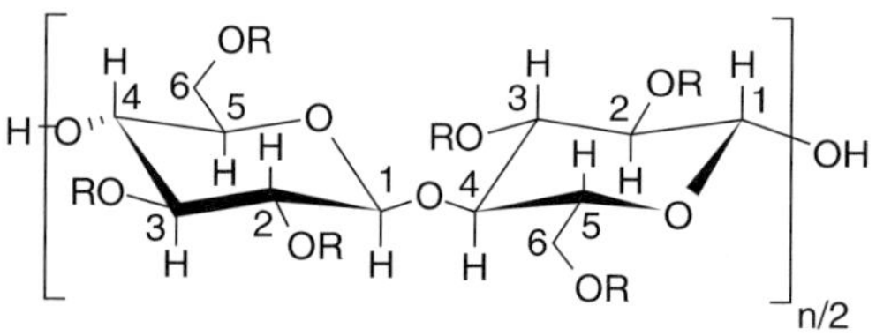

Cellulose ethers	R groups
Methylcellulose	H, CH_3
Ethylcellulose	H, CH_2CH_3
Hydroxyethylmethylcellulose	H, CH_3, $[CH_2CH_2O]_nH$
Hydroxypropylcellulose	H, $[CH_2CH(CH_3)O]_nH$
Carboxymethylcellulose	H, CH_2COONa

Fig. 2.3. Cellulose ether derivatives. (Reproduced, with permission, from Marques-Marinho and Vianna-Soares, 2013.)

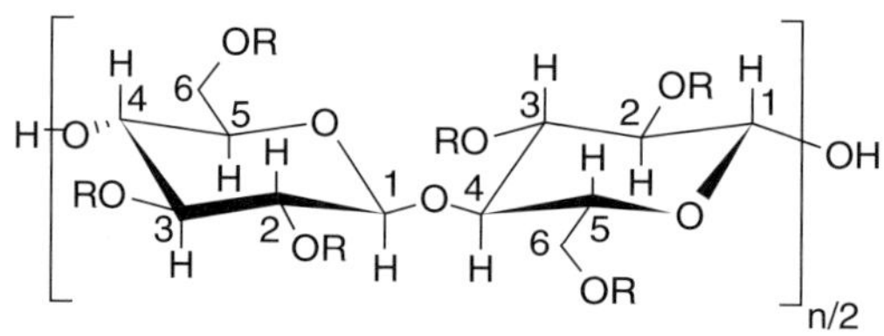

(Cellulose) esters	R groups
Acetate	H, I
Acetate trimellitate	H, I, II
Acetate phthalate	I, III
Hydroxypropylmethylphthalate	H, CH_3, $CH_2CH(OH)CH_3$, III, IV
Hydroxypropylmethylphthalate acetate succinate	H, CH_3, $CH_2CH(OH)CH_3$, I, V

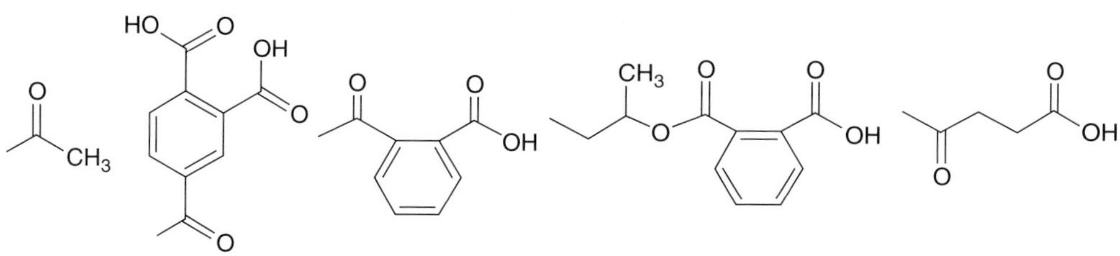

I II III IV V

Fig. 2.4. Cellulose ester derivatives. (Reproduced, with permission, from Marques-Marinho and Vianna-Soares, 2013.)

form matrix systems for solid and oral drug delivery forms. Cellulose derivatives swell in water, facilitating the release of drugs by diffusion mechanisms from these matrix systems. One example is the use of sodium carboxymethyl cellulose (NaCMC) as a powerful excipient in oral solid dosage forms. Cellulose esters such as cellulose acetate pathalate (CAP), cellulose acetate trimallitate and cellulose hydroxypropylmethyl phthalate (HPM) are also widely used in pharmaceutical applications. CAP is used to deliver drugs in pH-sensitive systems (Edgar, 2006). Cellulose ester derivatives found a use in the coatings of tablets and capsules to be delivered in the enteric environment of the stomach (Felton, 2006). The cellulose esters are used along with plasticizers such as diethyl phthalate, glycerine, triacetin, propylene glycol and dibutyl tartarate to produce the most effective coating films (Béchard *et al.*, 1995; Felton, 2006). Table 2.2 shows the byproducts released during formation of cellulose ethers.

Drug Delivery Applications

Cellulose

Cellulose is obtained from plant components and is used in various pharmaceutical applications. Microcrystalline cellulose (MCC) and oxycellulose are formed from cellulose pulp which is free of lignin and hemicelluloses. Crystalline cellulose is obtained by using 2–2.5 N solutions of mineral acids via controlled hydrolysis, followed by filtration and spray drying of the remaining aqueous solution (Felton, 2006). Powdered cellulose can be obtained by mechanical size-reduction techniques. Powdered cellulose has been used in compounded medicines as a suspending agent, adsorbent, diluents, thickening agent, etc. Both powdered cellulose and MCC are used as bulking agents to enhance the mass of the active ingredient used in the oral solid dosage formulations (Bastedo, 1939).

MCC is made of β-1,4 linked polymers of D-glucose residues, with a degree of polymerization of 150–250. MCC can support various different needs with different grades available in the market. It offers enhanced physical properties, being a good bulking agent, binder, disintegrant, lubricant and glidant, and is also a suspending agent and stability enhancer. It can be used in immediate release liquid and tablet dosage forms, as well as in sustained release matrix formulations. It offers applications in the formulation of solid dosage forms, but some of its properties, such as loss of compactibility during wet granulation, low bulk density, sensitivity to lubricants and moderate flowability, are potent limitations. The formation of silicified MCC by silicification enhances its particle size, density, flowability, moisture content, compressibility and compactibility. Co-drying a suspension having MCC and colloidal silicon dioxide to obtain a concentration of 2% dried silicon dioxide generates silicified MCC (Jaiyeoba *et al.*, 2006). The use of silicon dioxide in the formation of silicified MCC is facilitated by the adherence on the MCC

Table 2.2. Etherifying agents, co-products, and by-products from the production of cellulose ethers. (Reproduced, with permission, from Kamel *et al.*, 2008.)

			By-product	
Cellulose ether	Etherifying agent	Co-product	Name	Formula
Methyl (MC)	Methyl chloride	NaCl	Methanol dimethyl ether	CH_3OH CH_3OCH_3
Ethyl (EC)	Ethyl chloride	NaCl	Ethanol diethyl ether	C_2H_5OH $C_2H_5OC_2H_5$
Hydroxyethyl (HEC)	Ethylene oxide	none	Ethylene glycol and polymers thereof	CH_2OHCH_2OH
Hydroxypropyl (HPC)	Propylene oxide	none	Propylene glycol and polymers thereof	$CH_3CH_2OHCH_2OH$
Carboxymethyl (CMC)	Chloroacetic acid	NaCl	Glycolic acid	$HO–CH_2–COOH$

surface (Moreton, 2009). The silicified MCC therefore offers higher bulk density and appropriate tensile strength for pharmaceutical applications (Luukkonen *et al.*, 1999; Edge *et al.*, 2000).

Tablet formulations prepared using silicified MCC (SMCC) show reduced lubricant sensitivity and enhanced compactibility compared to MCC. It was found during the comparison of SMCC 90 and MCC Avicel PH102 that silicified derivative in the absence of the drug was 10–40% more compactible (Aljaberi *et al.*, 2009). During a blended time study SMCC 90 was proved to have 2–3 times enhanced compactibility, with lower lubricant sensitivity. SMCC performance was also compared with that of direct fill formulations of hard gelatin capsules using MCC, pregelatinized starch and anhydrous lactose. It was also demonstrated that silicified MCC was superior to PGS and lactose in terms of compactibility. SMCC had a higher dissolution rate compared to PGS and lactose with 30% and 50% acetaminophen and 5% piroxicam. It was concluded that SMCC was a superior agent in providing compactibility and faster dissolution rates, especially in direct fill formulations of hard shell capsules (Guo and Augsburger, 2003). It was also demonstrated, in a different study, that SMCC was 20% more compactible as compared to the regular grade MCC, and decreased the size and weight of the tablets, resulting in improved patient compliance (Zografi *et al.*, 1984). It also decreases the hygroscopicity of the active agent ingredient by decreasing the size and increasing the compactibility, showing good flow properties, which makes manufacture significantly cheaper (Haware *et al.*, 2010). In comparison to the SMCC generally in use, high-density SMCC improved flow properties and the sensitivity of the tablets. The tensile strength of SMCC 90 and Proslov HD 90 high-density SMCC was evaluated using sodium stearyl fumarate and magnesium stearate in 0.5% concentration using ascorbic acid and acetylsalicylic acid 50% as an active agent ingredient (Mužíková and Nováková, 2007). It was concluded that high-density SMCC was more sensitive to lubricant addition and sodium stearyl fumarate decreased the strength further. The disintegration time reported in compaction of high-density SMCC with or without lubricants was less than that of SMCC and increased as the compression force rose (Mužíková and Nováková, 2007).

Another form of cellulose used is oxycellulose. This contains some carboxyl groups instead of the terminal alcohol groups, making synthetic polyanhydrocellobiuronide as the product, which is brittle and very readily soluble. Products with lower carboxyl contents are more useful. The oxidized form of cellulose fabric cotton and gauze is insoluble in aqueous solutions and acids and soluble in dilute alkalis. It becomes translucent and gelatinous and swells in weakly alkaline solutions. When it is clustered using blood, it forms a dark brown gelatinous mass, swells and becomes sticky. Its use is, therefore, more recommended in surgical interventions by direct application to the oozing surface, but not recommended for open wound surface dressing for homeostasis. Its use is not reported or recommended for dressing of open wounds (Malhotra *et al.*, 1999). The oxidized form of cellulose forms thixotropic dispersions in water which can be used as film-forming materials in pharmaceutical applications. Controlled and sustained release formulations can be made from oxidized cellulose utilizing solid, liquid, neutral, acidic, or volatile compounds which are loaded into such systems. Topical formulations made using oxidized cellulose are prepared for human skin and hair application (Banker and Kumar, 1995). Anti-fungal creams, anti-acne lotions and sunscreen sprays are important applications (Banker and Kumar, 1995).

Cellulose Derivatives

Both cellulose ethers and esters have a wide range of applications in polymeric drug delivery. The cellulose ethers are used most importantly for the development of polymer matrix tablets. These swell and the hydrogel layer surrounds the dry core of the tablet. Water molecules encounter a hydrogel barrier which they penetrate to reach the drug polymer matrix, thereby releasing the drug (Ryzhov *et al.*, 1964; Colombo *et al.*, 1999; Saša *et al.*, 2006; Swarbrick, 2007; Kazlauskė and Liesienė, 2011).

Sodium carboxymethyl cellulose

The sodium salt of carboxymethyl cellulose (NaCMC) is a commercially available polyanionic derivative of cellulose used as an emulsifying agent in both the cosmetic and the pharmaceutical industries (Sharpe, 2001) (Table 2.3). It is one of the best exploited derivatives of cellulose used as a thickener, stabilizer, film-former, binder and suspension agent in a wide variety of applications (Lohani *et al.*, 2016).

The use of NaCMC has been exploited in biomedicines for prevention of epidural scar adhesions and development of postsurgical soft tissue complications (Sannino *et al.*, 2004). It is also used in treatment of oedemas as hydroxyethyl cellulose-based adsorbants of water (Jackson and Maillard, 2001). It is used to deliver controlled release of superoxide dismutase enzyme for therapeutic applications in the form of hydrogels (Jackson and Maillard, 2001). In a study involving the use of NaCMC and hydroxyl ethyl cellulose for incorporation of three model drugs (ciprofloxacin, metformin and esomeprazole) for studying the floating and swelling characteristics of a gastroretentive drug delivery system, it was found that the degree of swelling decreased at every 6 h in each formulation that possessed NaCMC as compared to those in deionized water. Furthermore, the hydration of NaCMC tablets was found to be retarded due to the concentration of NaCl in medium that resulted in swelling sizes and smaller gel layers. Metformin, a water-soluble drug, was released by diffusion mechanisms; ciprofloxacin and esomeprazole were released by anomalous diffusion. It was also concluded that the release of model drugs was facilitated by various concentrations of NaCl, affecting drug solubility and their mechanism of release (Chen *et al.*, 2015).

Table 2.3. Applications for NaCMC. (Adapted, with permission, from Kamel *et al.*, 2008.)

Applications	Properties
Toothpaste	Binder, thickener, suspending aid, flavour stabilizer
Sustained release	Diffusion barrier, thickener
Jellies	Film-former, thickener, protective colloid, gelling agent
Tablet coating	Film-former
Denture adhesives	Long-lasting adhesion
Syrups, suspensions	Suspending aid, thickener
Tablet binding and granulation	High-strength binder
Ointments and lotions	Film-former, stabilizer, emulsion, thickener

Superoxide dismutase is an enzyme that is sensitive to product inhibition and deactivation by hydrogen peroxide and clears very quickly from the bloodstream. When this enzyme was adsorbed into hydrogels using the superoxide dismutase enzyme (SOD) CMC conjugates and hydrogels, it was found that 50% of the enzyme was released from the SOD CMC hydrogel after 72 h, facilitating controlled release (Domínguez *et al.*, 2004). Table 2.4 presents an overview of the stability of commercially available drugs made from sodium carboxymethyl cellulose, a common cellulosic derivative used in oral suspensions.

Sodium carboxymethyl cellulose, being an excellent pharmaceutical excipient for drug delivery procedures, has enhanced mucoadhesivity, good filmability and binding properties. Its use has also been demonstrated for the treatment of colonic diseases. In one such study a novel polymer consisting of NaCMC with glycine was used to screen colonic *in vitro* drug release. NaCMC was used to develop synthesizing its derivative using glycine with azo linkage. Such an azo polymer system was then evaluated for its various characteristics, including Rf value, solubility, colour, melting point, infrared spectroscopy (IR) and proton nuclear magnetic resonance (^{1}HNMR) spectroscopy spectral analysis. Special studies were carried out analysing rat faeces. It was concluded that NaCMC showed promising specificity in treating colonic diseases for a period of 120 min for controlled drug release (Ojha *et al.*, 2012). NaCMC was also used in the treatment of patients suffering from Sjögren's syndrome, in which CMC and a glycerine mouthwash were used as a control. The CMC substitute relieved the oral discomfort at night (Little *et al.*, 1981). In comparison, the lubricating properties of saliva substitutes containing mucin and CMC showed similar effects, with changed friction values of about 15 min – more than twice as long as for water. Both the substitutes and water were successful in relieving the symptoms of dry mouth but did not have a lasting effect (Olsson and Axéll, 1991). In another intervention an analgesic and anti-inflammatory agent known as Ketorolac tromethamine was successfully encapsulated into microspheres, in which the diffusion coefficient decreased by increasing the crosslinking and adding more NaCMC into the matrix (Rokhade *et al.*, 2006).

Table 2.4. Overview of the stability of commercially available oral suspensions made from sodium carboxymethyl cellulose. (Adapted, with permission, from Marques-Marinho and Vianna-Soares, 2013.)

Drug name	Suspension (mg/ml)	Ora-plus/ CMC added	Stability @ 25°C (days)	Dosage form (mg)	Reference
Hydralazine HCl	4[a]	Ora-Sweet SF Ora-Sweet	Unstable	Tablets:10, 25, 50, 100	Allen and Erickson, 1998a
Metolazone	1[a]	Ora-Sweet SF Ora-Sweet	60[g]	Tablets: 2.5, 5, 10	Allen and Erickson, 1996a
Moxifloxacin	20	Ora-Sweet Ora-Sweet SF	90[e]	Tablets: 400	Hutchinson *et al.*, 2009
Norfloxacin	20[b]	Strawberry syrup	56[d]	Tablets: 400	Johnson *et al.*, 2001
Rifampin	25[a]	Ora-Sweet Ora-Sweet SF	28[g]	Capsules: 150, 300	Allen and Erickson, 1998b
Tetracycline HCl	25[a]	Ora-Sweet Ora-Sweet SF	28[g] & 7[g]	Capsules: 250, 500	Allen and Erickson, 1998b
Sunitinib malate	10	Ora-Sweet	60[c]	Capsules: 50	Navid *et al.*, 2008
Levodopa and carbidopa	5 & 2.5	Ora-Sweet	28[e]	Tablets: 100+25	Nahata *et al.*, 2000
Spironolactone	1	Simple syrup	91[c]	Tablets: 25, 50, 100	Allen and Erickson, 1996a
Tiagabine HCl	1	Ora-Sweet	70[e]	Tablets: 2, 4, 6, 8, 10, 12, 16	Nahata and Morosco, 2003
Theophylline	5	Ora-Sweet Ora-Sweet SF	90[e]	Capsules: 125, 200, 300	Johnson, 2005
Levofloxacin	50	Strawberry syrup	57[e]	Tablets: 200, 500, 750	VandenBussche *et al.*, 1999
Ketoconazole	20[a]	Ora-Sweet Ora-Sweet SF	60[g]	Tablets: 200	Allen and Erickson, 1996a
Nifedipine	4	Ora-Sweet	91[d]	Capsules: 10, 20	Morosco, 2002
Acetazolamide	25[a]	Ora-Sweet SF Ora-Sweet	60[f]	Tablets: 125, 250	Allen and Erickson, 1996b
Gabapentin	100	Ora-Sweet	56[e]	Capsules: 100, 300, 400	Morosco, 2002
Dipyridamole	10a	Ora-Sweet Ora-Sweet SF	60[g]	Tablets: 25, 50, 75	Allen and Erickson, 1996c
Alprazolam	1[a]	Ora-Sweet SF Ora-Sweet	60[g]	Tablets: 0.25, 0.5, 1, 2	Allen and Erickson, 1998a
Chloroquine phosphate	15[a]	Ora-Sweet SF Ora-Sweet	60[g]	Tablets: 250, 500	Allen and Erickson, 1998a
Flucytosine	10[a]	Ora-Sweet Ora-Sweet SF	60[f]	Tablets: 250, 500	Allen and Erickson, 1996b
Azathioprine	50[a]	Ora-Sweet Ora-Sweet SF	60[f]	Tablets: 50	Allen and Erickson, 1996b

[a]: storage in dark; [b]: in fluorescent light; [c]: amber glass; [d]: plastic; [e]: amber plastic; [f]: polyethylene terephthalate; [g]: amber PET
Ora-Sweet Sugar-free (SF) Formulation: glycerine, sodium saccharin, citric acid, sorbitol, xanthan gum, flavouring, methylparaben, water, propylparaben, sodium citrate, potassium sorbate in pH4.2.
Ora-Sweet Formulation: potassium sorbate, citric acid, water, flavouring, sodium phosphate, glycerin, methylparaben, sorbitol, sucrose in pH4.2.
Ora-Plus Formulation: NaCMC, flavouring, microcrystalline glucose, citric acid, methylparaben, water, potassium sorbate, xanthan gum, simethicone, sodium phosphate in pH4.2.
CMC, carboxymethyl cellulose

Indomethacin, a non-steroidal anti-inflammatory agent having a short half-life of 2.6–11.2 h was delivered using CMC (Flower *et al.*, 1990; Waree and Garnpimol, 2003). The prescribed dosage for adults is either 25 or 50 mg orally, two or three times a day. It was found that excellent controlled release was achieved using CMC, increasing patient compliance; this facilitated good plasma concentrations with no side effects. One such preparation resulted from incorporating indometacin into a mixture of CMC and chitosan crosslinked using gluteraldehyde (Flower *et al.*, 1990; Waree and Garnpimol, 2003).

Other interesting work involves the recent development of lyophilized wafers of NaCMC, and CMC, for mucosal wound healing. These wafers create a moist environment for wound healing in modern dressing procedures. In such a study a CMC-containing wafer was developed having potential antimicrobial agents for treating wound infections and facilitating healing. Intitally the pre-formulation was made using methylcellulose (MC), NaCMC, xanthan gum and sodium alginate. Physical examination of the wafer revealed uniform distribution of the drug, good swelling and hydration characteristics, solvent loss and efficacy of antimicrobials. Three drugs having good antimicrobial characteristics: silver nitrate, sulfacetamide sodium and neomycin trisulfate salt hydrate were used in the wafers. Neomycin trisulfate salt hydrate containing a NaCMC wafer proved to be the most desirable wound healing wafer of the three, having improved sponginess, flexibility, drug distribution uniformity and wafer texture against both Gram-positive and Gram-negative bacterial strains (Ng and Jumaat, 2014).

Hydroxypropylmethyl cellulose

Hydroxy propyl methyl cellulose (HPMC) is a propylene glycol ether of methyl cellulose whose properties are affected by hydroxypropoxy and methoxy group contents. HPMC, a cellulosic derivative, is used for developing controlled release tablet formulations for pharmaceutical applications. It is a widely used hydrophilic carrier as it possesses high 'swellability' and has profound effects on the release kinetics of the encapsulated drug. Some of its properties, such as being a nonionic polymer, maintain consistency of drug release in pH-dependent systems. For most soluble drugs HPMC has been used for burst release, increasing bioavailability. It has been demonstrated in a study that the use of HPMC in endoplasmic reticulum matrix tablet formulations for delivery of insoluble drugs has shown enhanced bioavailabilities exhibiting food effects (Davis *et al.*, 2009).

When water distribution was studied by Differential Scanning Calorimetry (DSC) techniques in HPMC-based gels, it was demonstrated that at low temperatures water was found to be loosely associated with the polymer in DSC scans (Ford and Mitchell, 1995). In some formulations water is also used to hydrate the polymers, which usually disengage from the matrix. The release of drugs from the HPMC matrices is due to simple drug diffusion and erosion of the swollen matrix layer (Ford *et al.*, 1987; Rao *et al.*, 1990; Skoug *et al.*, 1993; Tahara *et al.*, 1995). In such controlled release preparations the percentage of erosion and diffusion are important (Lee, 1980; Lee and Peppas, 1987). Drug release via erosion was the function of the average weight of the HPMC polymer; diffusion of HPMC via the diffusion layer of water was found to be the rate limiting step (Reynolds *et al.*, 1998). The surface area to volume ratio facilitates the release of drug from HPMC matrices. The drug release profile of tablets having different sizes, dosages and shapes can be formulated using such matrices. Tablet formulations showing similar surface area profiles but different surface area to volume ratios did not release the drug similarly, indicating tablet formulations with large surface areas showed faster release rates (Reynolds *et al.*, 2002).

In order to study the effect of HPMC-based drug release systems, the HPMC/lactose ratio and the HPMC viscosity grade were studied for drug release mechanisms from matrix-type systems. The compound used as a model drug was adinazolam mesylate. It was demonstrated that drug release followed the Higuchi expression and the most preferred mode was diffusion of the drug from the hydrated gel layer (Sung *et al.*, 1996). Viscosity of the HPMC residues can be used in controlled drug release and this was demonstrated in naproxen and naproxen sodium (Katzhendler *et al.*, 2000). In spite of the differing molecular weights, similar microvisocity was exhibited in HPMC. The pH measurement values explain the successful incorporation of naproxen into HPMC matrix. A lower pH was reported for the hydrated tablet in comparison with naproxen sodium. These properties of naproxen HPMC matrices released the drug through diffusion (Katzhendler *et al.*, 2000).

Anticipation of Cellulose in Drug Delivery

Since the cellulosics are available at a lower cost, and are recyclable, biocompatible, strong, non-toxic and reproducible, they are used widely in drug delivery applications. They can easily be modified and tailored for target-specific drug release by enhancing their film-forming properties, rheology, thickening and tablet-binding properties. The intervention of nanotechnology in generating nanocrystalline cellulose (NCC) has resulted in an important renewable nanomaterial showing applications in drug delivery, the cosmetics and chemical industries, personal care, etc. Modification of NCC allows its excellent physicochemical properties to be used in site-specific drug delivery. Negative charges stabilize the nanoparticles in aqueous suspensions during acid hydrolysis. NCC formulation changes the isotropic to the anisotropic chiral nematic crystalline liquid phase. Another cellulose nanomaterial, nanofibrillar cellulose (NFC), has attracted considerable research attention as a matrix-forming material for sustained drug release in therapeutic applications. It has proved to be an effective film-forming matrix material for resilient drug delivery, even for as long as 3 months. Characterizations of NFC films in drug delivery have shown an efficacy of 90%, mainly for the release of heat-sensitive drugs. These systems have proved to be significant in maintaining the physical state of the incorporated drug, with sustained drug release capabilities and improved mechanical properties (Kolakovic *et al.*, 2012). NCC is a promising biomaterial showing enhanced performance as a nanocomposite. Nanotechniques will not only facilitate drug release without side effects but can also be used to obtain the polymers needed for enhancing drug delivery to inaccessible sites.

Acknowledgements

We thank Dr Ashok K. Chauhan, Founder President, Ritanand Balved Educational Foundation, for support. The guidance from Shri. Atul Chauhan, Chancellor, Amity University Uttar Pradesh and Prof. Dr (Mrs) Balvinder Shukla, Vice Chancellor, Amity University Uttar Pradesh is greatly appreciated. Insights from the articles cited in this chapter have been of considerable assistance, and we are grateful to their authors.

References

Aljaberi, A., Chatterji, A., Shah, N. and Sandhu, H. (2009) Functional performance of silicified microcrystalline cellulose versus microcrystalline cellulose: a case study. *Drug Development and Industrial Pharmacy* 35(9), 1066–1071.

Allen, L.V. Jr and Erickson M.A. 3rd (1996a) Stability of ketoconazole, metolazone, metronidazole, procainamide hydrochloride and spironolactone in extemporaneously compounded oral liquids. *American Journal of Health-System Pharmacy* 53(17), 2073–2078.

Allen, L.V. Jr and Erickson M.A. 3rd (1996b) Stability of acetazolamide, allopurinol, azathioprine, clonazepam, and flucytosine in extemporaneously compounded oral liquids. *American Journal of Health-System Pharmacy* 53, 1944–1949.

Allen, L.V. Jr and Erickson M.A. 3rd (1996c) Stability of baclofen, captopril, diltiazem hydrochloride, dipyridamole, and flecainide acetate in extemporaneously compounded oral liquids. *American Journal of Health-System Pharmacy* 53(18), 2179–2184.

Allen, L.V. Jr and Erickson M.A. 3rd (1998a) Stability of alprazolam, chloroquine phosphate, cisapride, enalapril maleate, and hydralazine hydrochloride in extemporaneously compounded oral liquids. *American Journal of Health-System Pharmacy* 55, 1915–1920.

Allen, L.V. Jr and Erickson M.A. 3rd (1998b) Stability of bethanechol chloride, pyrazinamide, quinidine sulfate, rifampin, and tetracycline hydrochloride in extemporaneously compounded oral liquids. *American Journal of Health-System Pharmacy* 55(17), 1804–1809.

Argyropoulos, D. (2001) Wood and cellulosic chemistry, second edition, edited by David N.-S. Hon and Nobuo Shiraishi reviewed in *Journal of the American Chemical Society* 123(36), 8880–8881.

Banker, G.S. and Kumar, V. (1995) Microfibrillated oxycellulose. US patent 5405953.

Bastedo, W. (1939) The United States Pharmacopeial Convention, Inc., decennial period, 1930–1940 committee of revision of the United States Pharmacopeia reviewed in *The Journal of the American Medical Association* 113(2), 164.

Béchard, S., Levy, L. and Clas, S. (1995) Thermal, mechanical and functional properties of cellulose acetate phthalate (CAP) coatings obtained from neutralized aqueous solutions. *International Journal of Pharmaceutics* 114(2), 205–213.

Bochek, A. (2003) Effect of hydrogen bonding on cellulose solubility in aqueous and nonaqueous solvents. *Russian Journal of Applied Chemistry* 76(11), 1711–1719.

Buleon, A. and Chanzy, H. (1980) Single crystals of cellulose IVII: preparation and properties. *Journal of Polymer Science: Polymer Physics Edition* 18(6), 1209–1217.

Buurma, H., de Smet, P., van den Hoff, O., Sysling, H., Storimans, M. and Egberts, A. (2003) Frequency, nature and determinants of pharmacy compounded medicines in Dutch community pharmacies. *Pharmacy World & Science* 25(6), 280–287.

Chanzy, H., Henrissat, B., Vincendon, M., Tanner, S. and Belton, P. (1987) Solid-state 13C-N.M.R. and electron microscopy study on the reversible cellulose I → cellulose IIII transformation in Valonia. *Carbohydrate Research* 160, 1–11.

Chen, Y., Ho, H., Liu, D., Siow, W. and Sheu, M. (2015) Swelling/floating capability and drug release characterizations of gastroretentive drug delivery system based on a combination of hydroxyethyl cellulose and sodium carboxymethyl cellulose. PLOS ONE 10(1), e0116914.

Colombo, P., Bettini, R. and Peppas, N. (1999) Observation of swelling process and diffusion front position during swelling in hydroxypropyl methyl cellulose (HPMC) matrices containing a soluble drug. *Journal of Controlled Release* 61(1–2), 83–91.

Davis, J., Burton, J., Connor, A., Macrae, R. and Wilding, I. (2009) Scintigraphic study to investigate the effect of food on a HPMC modified release formulation of UK-294,315. *Journal of Pharmaceutical Sciences* 98(4), 1568–1576.

Domínguez, A., Valdivia, A., Hernández, J. and Villalonga, R. (2004) Biocompatibilid in vitro de Superóxido dismutase interactuando con polímero e hidrogeles de carboximetilcelulosa ensayado con fibroblastos humanos. *Revista de Biotecnología Aplicada* 21, 218–2381.

Edgar, K. (2006) Cellulose esters in drug delivery. *Cellulose* 14(1), 49–64.

Edge, S., Steele, D., Chen, A., Tobyn, M. and Staniforth, J. (2000) The mechanical properties of compacts of microcrystalline cellulose and silicified microcrystalline cellulose. *International Journal of Pharmaceutics* 200(1), 67–72.

Felton, L. (2006) A review of Handbook of Pharmaceutical Excipients, 5th edition. *Drug Development and Industrial Pharmacy* 32(8), 1003–1003.

Flower, R.J., Moncada, S. and Vane, J.R. (1990) Indomethacin. In: Hardman, J.G. and Limbird, L.E. (eds) *Goodman and Gilman's the Pharmacological Basis of Therapeutics*. Macmillan Publishing, New York, pp. 695–697.

Ford, J. and Mitchell, K. (1995) Thermal analysis of gels and matrix tablets containing cellulose ethers. *Thermochimica Acta* 248, 329–345.

Ford, J., Rubinstein, M., Mccaul, F., Hogan, J. and Edgar, P. (1987) Importance of drug type, tablet shape and added diluents on drug release kinetics from hydroxypropylmethylcellulose matrix tablets. *International Journal of Pharmaceutics* 40(3), 223–234.

Guo, M. and Augsburger, L. (2003) Potential application of silicified microcrystalline cellulose in direct-fill formulations for automatic capsule-filling machines. *Pharmaceutical Development and Technology* 8(1), 47–59.

Haware, R., Bauer-Brandl, A. and Tho, I. (2010) Comparative evaluation of the powder and compression properties of various grades and brands of microcrystalline cellulose by multivariate methods. *Pharmaceutical Development and Technology* 15(4), 394–404.

Hinterstoisser, B. and Salmén, L. (2000) Application of dynamic 2D FTIR to cellulose. *Vibrational Spectroscopy* 22(1–2), 111–118.

Hutchinson, D., Johnson, C. and Klein, K. (2009) Stability of extemporaneously prepared moxifloxacin oral suspensions. *American Journal of Health-System Pharmacy* 66(7), 665–667.

Jackson, I. and Maillard, G. (2001) Carboxy-methyl cellulose hydrogels used to fill breast implants: a 15-year experience. *European Journal of Plastic Surgery* 24(4), 176–176.

Jaiyeoba, K., Alfa, J. and Odeniyi, M. (2006) Direct compression properties of microcrystalline cellulose and its silicified product. *East and Central African Journal of Pharmaceutical Sciences* 7(3), 56–59.

Johnson, C. (2005) Stability of anhydrous theophylline in extemporaneously prepared alcohol-free oral suspensions. *American Journal of Health-System Pharmacy* 62(23), 2518–2520.

Johnson, C.E., Price, J. and Hession, J.M. (2001) Stability of norfloxacin in an extemporaneously prepared oral liquid. *American Journal of Health-System Pharmacy* 58, 577–579.

Kamel, S., Ali, N., Jahangir, K., Shah, S.M. and El-Gendy, A.A. (2008) Pharmaceutical significance of cellulose: A review. *eXPRESS Polymer Letters* 2(11), 758–778. Available at: www.expresspolymlett.com/articles/EPL-0000662_article.pdf (accessed 19 October 2016). DOI: 10.3144/expresspolymlett.2008.90.

Katzhendler, I., Mäder, K. and Friedman, M. (2000) Structure and hydration properties of hydroxypropyl methylcellulose matrices containing naproxen and naproxen sodium. *International Journal of Pharmaceutics* 200(2), 161–179.

Kazlauskė, J. and Liesienė, J. (2011) Application of cellulose derivatives for oral protein-based drug delivery. Chemical Technology 57(1–2), 36–42.

Kennedy, J. and Knill, C. (2001) Physical chemistry of non-aqueous solutions of cellulose and its derivatives. *Carbohydrate Polymers* 46(2), 199.

Klemm, D., Heublein, B., Fink, H. and Bohn, A. (2005) Cellulose: fascinating biopolymer and sustainable raw material. *Angewandte Chemie International Edition* 44(22), 3358–3393.

Kolpak, F. and Blackwell, J. (1975) The structure of regenerated cellulose. *Macromolecules* 8(4), 563–564.

Kolakovic, R., Peltonen, L., Laukkanen, A., Hirvonen, J. and Laaksonen, T. (2012) Nanofibrillar cellulose films for controlled drug delivery. *European Journal of Pharmaceutics and Biopharmaceutics* 82(2), 308–315.

Kuga, S. and Malcolm Brown, R. (1988) Silver labeling of the reducing ends of bacterial cellulose. *Carbohydrate Research* 180(2), 345–350.

Langan, P., Nishiyama, Y. and Chanzy, H. (1999) A revised structure and hydrogen-bonding system in cellulose II from a neutron fiber diffraction analysis. *Journal of the American Chemical Society* 121(43), 9940–9946.

Lee, P. and Peppas, N. (1987) Prediction of polymer dissolution in swellable controlled-release systems. *Journal of Controlled Release* 6(1), 207–215.

Lee, P.I. (1980) Diffusional release of a solute from a polymeric matrix — approximate analytical solutions. *Journal of Membrane Science* 7(3), 255–275.

Little, J., Klestov, A., Webb, J., Latt, D., Schiller, G., McNamara, K., Young, D., Hobbes, J. and Fetherston, J. (1981) Treatment of xerostomia: a double-blind trial in 108 patients with Sjögren's syndrome. *Oral Surgery, Oral Medicine, Oral Pathology* 51(6), 594–599.

Lohani, A., Singh, G., Bhattacharya, S., Rama Hegde, R. and Verma, A. (2016) Tailored-interpenetrating polymer network beads of κ-carrageenan and sodium carboxymethyl cellulose for controlled drug delivery. *Journal of Drug Delivery Science and Technology* 31, 53–64.

Luukkonen, P., Schæfer, T., Hellén, L., Juppo, A. and Yliruusi, J. (1999) Rheological characterization of microcrystalline cellulose and silicified microcrystalline cellulose wet masses using a mixer torque rheometer. *International Journal of Pharmaceutics*, 188(2), 181–192.

Malhotra, B., Kharkwal, H. and Yadav, M.P. (1999) Textbook of Organic Medicinal and Pharmaceutical Chemistry, tenth edition, edited by Jaime N. Delgado and William A. Remers, reviewed in *Journal of Medicinal Chemistry* 42(13), 2491–2491.

Marchessault, R. (1994) *Wood Chemistry: Fundamentals and Applications*. 2nd edn, by Ero Sjöström. Academic Press, NY, 1993.

Marques-Marinho, F.D. and Vianna-Soares, C.D. (2013) Cellulose and its derivatives use in the pharmaceutical compounding practice. In: van de Ven, T. and Godbout, L. (eds) *Cellulose - Medical, Pharmaceutical and Electronic Applications*. InTech, DOI: 10.5772/56637. Available from: www.intechopen.com/books/cellulose-medical-pharmaceutical-and-electronic-applications/cellulose-and-its-derivatives-use-in-the-pharmaceutical-compounding-practice.

Moreton, R.C. (2009) Cellulose, silicified microcrystalline. In: Rowe, R.C., Sheskey, P.J. and Quinn, M.E. (eds) *Handbook of Pharmaceutical Excipients,* sixth edition. Pharmaceutical Press, London.

Morosco, R. (2002) Stability of nifedipine in two oral suspensions stored at two temperatures. *Journal of the American Pharmacists Association* 42(6), 865.

Mužíková, J. and Nováková, P. (2007) A study of the properties of compacts from silicified microcrystalline celluloses. *Drug Development and Industrial Pharmacy* 33(7), 775–781.

Nahata, M.C. and Morosco, R.S. (2003) Stability of tiagabine in two oral liquid vehicles. *American Journal of Health-System Pharmacy* 60(1), 75–77.

Nahata, M.C., Morosco, R.S. and Leguire, L.E. (2000) Development of two stable oral suspensions of levodopa-carbidopa for children with amblyopia. *Journal of Pediatric Ophthalmology and Strabismus* 37(6), 333–337.

Navid, F., Christensen, R., Minkin, P., Stewart, C., Furman, W. and Baker, S. (2008) Stability of sunitinib in oral suspension. *Annals of Pharmacotherapy* 42(7), 962–966.

Ng, S. and Jumaat, N. (2014) Carboxymethyl cellulose wafers containing antimicrobials: A modern drug delivery system for wound infections. *European Journal of Pharmaceutical Sciences* 51, 173–179.

Ojha, M., Madhav, N. and Singh, A. (2012) Synthesis and evaluation of sodium carboxymethyl cellulose azo polymer for colon specificity. *International Current Pharmaceutical Journal* 1(8), 209–179.
Olsson, H. and Axéll, T. (1991) Objective and subjective efficacy of saliva substitutes containing mucin and carboxymethylcellulose. *European Journal of Oral Sciences*, 99(4), 316–319.
Rao, K., Devi, K. and Buri, P. (1990) Influence of molecular size and water solubility of the solute on its release from swelling and erosion controlled polymeric matrices. *Journal of Controlled Release* 12(2), 133–141.
Reynolds, T., Gehrke, S., Hussain, A.S. and Shenouda, L. (1998) Polymer erosion and drug release characterization of hydroxypropyl methylcellulose matrices. *Journal of Pharmaceutical Sciences* 87(9), 1115–1123.
Reynolds, T., Mitchell, S. and Balwinski, K. (2002) Investigation of the effect of tablet surface area/volume on drug release from hydroxypropylmethylcellulose controlled-release matrix tablets. *Drug Development and Industrial Pharmacy* 28(4), 457–466.
Rokhade, A., Agnihotri, S., Patil, S., Mallikarjuna, N., Kulkarni, P. and Aminabhavi, T. (2006) Semi-interpenetrating polymer network microspheres of gelatin and sodium carboxymethyl cellulose for controlled release of ketorolac tromethamine. *Carbohydrate Polymers* 65(3), 243–252.
Ryzhov, V., Burkhanova, N. and Kozlov, P. (1964) Crystallization tendency of some simple cellulose ethers. *Polymer Science U.S.S.R.* 6(8), 1630–1639.
Sannino, A., Madaghiele, M., Conversano, F., Mele, G., Maffezzoli, A., Netti, P., Ambrosio, L. and Nicolais, L. (2004) Cellulose derivative–hyaluronic acid-based microporous hydrogels cross-linked through divinyl sulfone (DVS) to modulate equilibrium sorption capacity and network stability. *Biomacromolecules* 5(1), 92–96.
Sarko, A., Southwick, J. and Hayashi, J. (1976) Packing analysis of carbohydrates and polysaccharides. 7. Crystal structure of cellulose IIII and its relationship to other cellulose polymorphs. *Macromolecules* 9(5), 857–863.
Saša, B., Odon, P., Stane, S. and Julijana, K. (2006) Analysis of surface properties of cellulose ethers and drug release from their matrix tablets. *European Journal of Pharmaceutical Sciences* 27(4), 375–383.
Saxena, I. (2005) Cellulose biosynthesis: current views and evolving concepts. *Annals of Botany* 96(1), 9–21.
Schuerch, C. (1973) Cellulose and cellulose derivatives. *Carbohydrate Research* 28(1), 183–184.
Sharpe, D. (2001) Comment on: “Carboxymethyl cellulose (CMC) hydrogels used to fill breast implants: a 15-year experience” by H. Arion. *European Journal of Plastic Surgery* 24(4), 179–180.
Skoug, J., Mikelsons, M., Vigneron, C. and Stemm, N. (1993) Qualitative evaluation of the mechanism of release of matrix sustained release dosage forms by measurement of polymer release. *Journal of Controlled Release* 27(3), 227–245.
Sung, K., Nixon, P., Skoug, J., Ju, T., Gao, P., Topp, E. and Patel, M. (1996) Effect of formulation variables on drug and polymer release from HPMC-based matrix tablets. *International Journal of Pharmaceutics* 142(1), 53–60.
Swarbrick, J. (2007) *Encyclopedia of Pharmaceutical Technology*. Vols 1–6. Informa Healthcare, Hoboken, New Jersey, USA.
Tahara, K., Yamamoto, K. and Nishihata, T. (1995) Overall mechanism behind matrix sustained release (SR) tablets prepared with hydroxypropyl methylcellulose 2910. *Journal of Controlled Release* 35(1), 59–66.
Thompson, M. (2006) Making Medicines – A brief history of pharmacy and pharmaceuticals edited by Stuart Anderson reviewed in *Drug Development and Industrial Pharmacy* 32(7), p. 909.
VandenBussche, H.L., Johnson, C.E., Fontana, E.M. and Meram, J.M. (1999) Stability of levofloxacin in an extemporaneously compounded oral liquid. *American Journal of Health-System Pharmacy* 56(22), 2316–2318.
Wada, M., Heux, L., Isogai, A., Nishiyama, Y., Chanzy, H. and Sugiyama, J. (2001) Improved structural data of cellulose III I prepared in supercritical ammonia. *Macromolecules* 34(5), 1237–1243.
Waree, T. and Garnpimol, C.R. (2003) Development of indomethacin sustained release microcapsules using chitosan-carboxymethyl cellulose complex coacervation. *Songklanakarin Journal of Science Technology* 25, 245–254.
Wu, C. (1995) *Handbook of Size Exclusion Chromatography*. Dekker, New York, USA.
Xiong, J. and Ye, J. (2004) Synthesis and characterization of cellulose ethers/Eu(III). *Journal of Applied Polymer Science* 95(3), 743–747.
Zografi, G., Kontny, M., Yang, A. and Brenner, G. (1984) Surface area and water vapor sorption of macrocrystalline cellulose. *International Journal of Pharmaceutics* 18(1–2), 99–116.

3 Hydrocolloid-based Hydrogels in Drug Delivery

Neerupma Dhiman*
Amity Institute of Pharmacy, Amity University, Noida, India

Abstract
The application of hydrocolloids in pharmaceutical formulations includes their use in the manufacture of implants, films, beads, microparticles, nanoparticles, and inhalable and injectable systems, as well as viscous liquid formulations. The biomedical and pharmaceutical applications of hydrocolloid-based hydrogels and their importance are the highlights of this chapter.

Introduction

The design and development of new drug molecules is an expensive and time-consuming procedure. Later, they have to be transported in the human and/or animal body and in this regard the drug delivery is an important process. It is the method of administering the active pharmaceutical ingredient (API) to achieve the desired therapeutic effect. The controlled delivery systems or controlled release technology (CRT) provide release at a predetermined, predictable and controlled rate to achieve high therapeutic efficiency with minimal toxicity (Pandey *et al.*, 2012). Hence, the development of novel drug delivery vehicles is an essential step towards controlled and site-specific administration of therapeutics. The desirable characteristics are that these should be introduced into the body through minimally invasive means and that these vehicles should be capable of releasing APIs to the intended location at a controlled rate.

Hydrocolloid-based hydrogels are of particular interest for delivery applications as they can fulfill the above-mentioned characteristics in addition to having good biocompatibility and a tunable network structure to control the drug diffusion. Some examples of CRTs are transdermal and transmucosal controlled release delivery systems, nasal and buccal aerosol sprays, drug-impregnated lozenges, encapsulated cells, oral soft gels and iontophoretic devices to administer drugs through skin, and a variety of programmable, implanted delivery devices (Tiwari *et al.*, 2012). Hydrocolloids are widely used to perform various functions in the pharmaceutical and biomedical sectors. Their specific applications include implants, films, beads, microparticles, nanoparticles, and inhalable and injectable systems as well as viscous liquid formulations (Pandey and Khuller, 2004).

*E-mail: ndhiman@amity.edu

Uses of Hydrocolloids

A hydrocolloid is defined as a colloidal system in which the particles are dispersed in an aqueous phase. The colloidal particles are spread throughout the aqueous phase and, depending on the quantity and on the degree of hydration, the system could take up the form of either a gel or a solution. These are hydrophilic polymers obtained from vegetable, animal, microbial and/or synthetic origins, contain many hydroxyl groups and could be polyelectrolytic in nature. At sufficiently high solution concentrations, they become entangled and form networks that influence the solution properties (Cassiday, 2012; Kapoor *et al.*, 2013). They have the ability to alter the solution viscosity of many times their own weight of water due to interactions with the water molecules through hydrogen-bonding (Torres *et al.*, 2012), and can perform various functions in formulations, namely: (i) thickening and gelling aqueous solutions; (ii) stabilizing foams, emulsions and dispersions; (iii) inhibiting ice and sugar crystal formation; (iv) controlling the release of flavours; (v) flocculation; and (vi) film formation (Fig. 3.1, Torres *et al.*, 2012).

Classification of Hydrocolloids

Hydrocolloids can be classified as natural, semisynthetic and synthetic (Table 3.1). The naturally occurring ones are those obtained from plants, animals and microbes. The semisynthetic are those synthesized via modification of naturally occurring hydrocolloids, e.g. methyl cellulose and carboxymethyl guar. Synthetics are synthesized on the industrial scale, starting with petroleum-based materials such as acrylate copolymers (Cassiday, 2012). Naturally occuring polymers possess several advantages as well as disadvantages. They are inexpensive, easy to handle and non-toxic but, at the same time, large quantities are needed to be effective as emulsifiers and they are also susceptible to microbial growth, whereas the semi-synthetic hydrocolloids are stronger emulsifiers, non-toxic and less likely to undergo microbial growth. The synthetic hydrocolloids are the strongest emulsifiers, are non-toxic and do not support microbial growth, but are expensive. They are mainly limited to use as oil-in-water emulsifiers (Kapoor *et al.*, 2013).

What are Hydrogels?

Hydrogels are three-dimensional and cross-linked networks of polymers. They can be formulated in a variety of physical forms, including slabs, microparticles, nanoparticles, coatings and films (Bindu Sri *et al.*, 2012). They are commonly used in clinical practice and experimental medicine for a wide range of applications including tissue engineering and regenerative medicine, diagnostics, cellular immobilization, separation of biomolecules or cells and barrier materials to regulate biological adhesions (Qiu and Park, 2012).

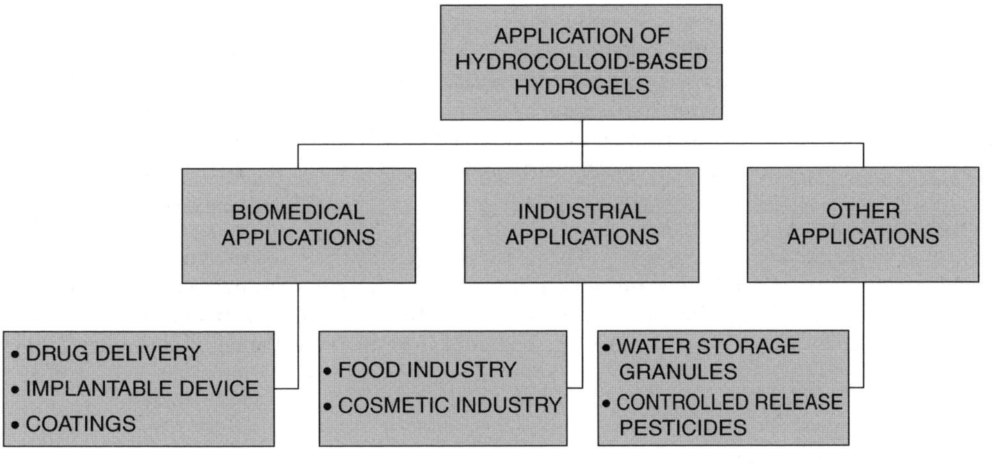

Fig. 3.1. Application of hydrocolloid-based hydrogels.

Table 3.1. Classification of natural hydrocolloids.

	Classification	Hydrocolloids
1	Plants	Cellulose, gums (arabic, karaya, guar, locustbean, tara, ghatti and tragacanth), starch, pectin, konjac mannan
2	Algae	Agar and carrageenan from red algae and alginates from brown algae
3	Microbes	Xanthan gum, dextran, gellan gum
4	Animals	Gelatin and chitosan

Properties of Hydrogels

Hydrogels are polymeric materials that do not dissolve in water under physiological conditions. However, they swell considerably in an aqueous medium (Simões *et al.*, 2012) and demonstrate extraordinary capacity (>20%) for imbibing water into their network structure (Peppas *et al.*, 2000). Those exhibiting a phase transition in response to change in external conditions such as pH, ionic strength, temperature and electric current are known as 'stimuli-responsive' or 'smart' gels (Fig 3.2). Being insoluble, these three-dimensional hydrophilic networks contribute good blood compatibility and maintain a certain degree of structural integrity and elasticity (Li *et al.*, 2006). This phenomenon is due to the presence of hydrophilic functional groups, such as -OH, -COOH, $-CONH_2$ and $-SO_3H$, which are capable of absorbing water without undergoing dissolution.

The unique physical properties of hydrogels are of particular interest for drug delivery applications. Their highly porous network structure can easily be tuned by controlling the density of cross-links in the gel matrix and the affinity of the hydrogels for the aqueous environment in which they are swollen. Their porosity also permits loading of drugs into the gel matrix and subsequent release at a rate dependent on the diffusion coefficient of the entrapped molecules (Hoare and Kohane, 2008).

Hydrogels are, as discussed, water-swollen polymer matrices with a tendency to imbibe water when placed in aqueous environments. This ability to swell, under biological conditions, makes them ideal as materials for drugs and also for the immobilization of proteins, peptides and other biological compounds (Bindu Sri *et al.*, 2012). Their networks have a three-dimensional structure, crosslinked together either physically (entanglements, crystallites), or chemically (tie-points, junctions) (Bindu Sri *et al.*, 2012). This insoluble network structure allows effective immobilization of active agents and biomolecules, as well as allowing their release in a well-defined, specific manner. Their biocompatibility and crosslinked structure are thus responsible for varied applications (Yang and Pierstorff, 2012).

The two main properties of the hydrogels that make them as efficient drug delivery systems are porosity and swelling ratio of the polymer.

Porosity

The highly porous nature of hydrogels makes the drug release rate depend significantly on the diffusion coefficient through the gel network. The porosity can easily be tuned by controlling the degree of crosslinking, which will affect the release rate of the entrapped drug particles (Hoare and Kohane, 2008). Hydrogels can be prepared by two types of crosslinking: physical and chemical crosslinking. Physical crosslinking can be triggered by changes in pH, temperature, ionic strength and a variety of physicochemical interactions (e.g. hydrophobic interaction, charge condensation, hydrogen bonding, stereocomplexation and supramolecular chemistry). Chemical crosslinking involves adding additional linking entities covalently to the hydrocolloids. Small molecule and polymer–polymer crosslinking are the two major strategies in this category (Hoare and Kohane, 2008).

Swelling ratio

The ability of hydrogels to rapidly swell with water can lead to fast release of the loaded drug and rapid degradation of the polymer, but the swelling ratio of hydrogels is affected by the crosslinking ratio. This is one of the most important factors that affects the swelling of hydrogels and is defined as the ratio of moles of crosslinking agent to the moles of polymer repeating units. The higher the crosslinking ratio,

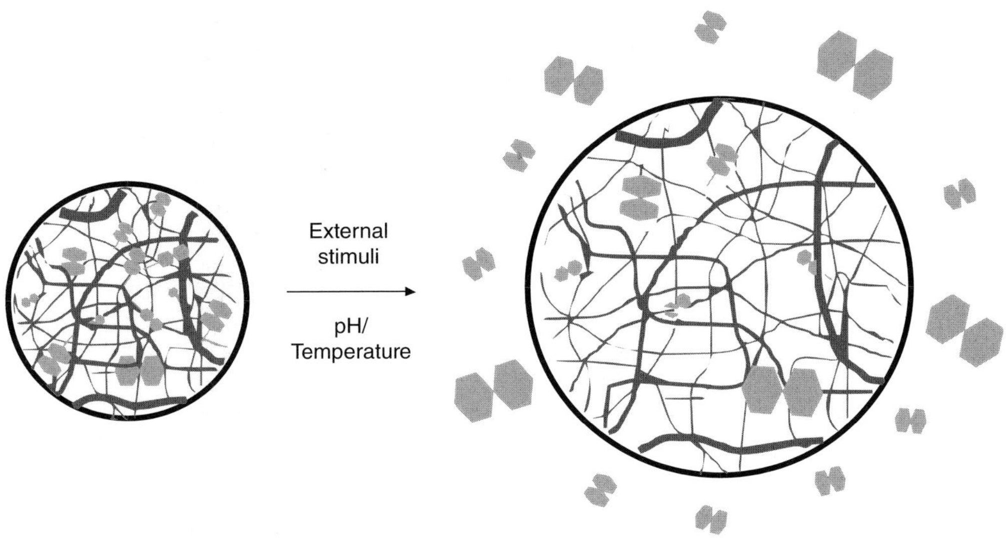

Fig. 3.2. Release of the drug from the crosslinked network of the polymer when there is change in the physical environment (pH, temperature, etc.).

the more crosslinking agent is incorporated in the hydrogel structure. Highly crosslinked hydrogels have a tighter structure and will swell less compared to the same hydrogels with lower crosslinking ratios (Bindu Sri *et al.*, 2012). Crosslinking hinders the mobility of the polymer chain, hence lowering the swelling ratio (Bindu Sri *et al.*, 2012). By changing the degree of crosslinking, the desired mechanical property of the hydrogel can be achieved. Increasing the degree of crosslinking will result in a stronger gel, generating a more brittle structure. Hence, there is an optimum degree of crosslinking to achieve a relatively strong and yet elastic hydrogel. Copolymerization has also been used to achieve the desired mechanical properties of hydrogels.

Chemical structure

The chemical structure of the polymer may also affect the swelling ratio of the hydrogels. Hydrogels containing hydrophilic groups swell to a higher degree compared to those containing hydrophobic groups (Bindu Sri *et al.*, 2012). Hydrophobic groups collapse in the presence of water, thus minimizing their exposure to the water molecule, resulting in the hydrogels swelling much less than those containing hydrophilic groups (Bindu Sri *et al.*, 2012). Swelling of environmentally sensitive products can be affected by specific stimuli. Swelling of temperature-sensitive hydrogels can be affected by changes in the temperature of the swelling media. Ionic strength and pH affect the swelling of ionic strength and pH-sensitive hydrogels, respectively. There are many other specific stimuli that can affect the swelling of other environmentally responsive hydrogels (Bindu Sri *et al.*, 2012).

Advantages of Hydrogels

The hydrogels provide a great advantage in drug delivery systems: the drugs can elute slowly, maintaining a high-level local concentration in the surrounding tissues over an extended period. They are highly biocompatible and possess physiochemical properties similar to the native extracellular matrix, both compositionally (particularly in the case of carbohydrate-based hydrogels) and mechanically (Bindu Sri *et al.*, 2012). They can thus be utilized *in vivo* in the peritoneum and other sites. The biodegradability or dissolution of the hydrogels can be designed via enzymatic, hydrolytic and/or environmental (e.g. pH, temperature or electric field) pathways (Bindu Sri *et al.*, 2012). They are relatively deformable and can conform to the shape of the surface to which they are adsorbed. For example, the muco- or bio-adhesive properties of some hydrogels could be advantageous in immobilizing them at the site of application or in applying onto surfaces that are not horizontal (Hoare and Kohane, 2008).

Disadvantages of Hydrogels

Hydrogels also possess some limitations, such as their low tensile strengths which limit their use in load-bearing applications, and could also result in premature dissolution or flowing away from the targeted site (Bindu Sri *et al.*, 2012). The quantity and homogeneity of drug loading could be limited, particularly in the case of hydrophobic drugs. The high water content and large pore sizes often result in relatively rapid drug release, over a few hours to days, and ease of application could be problematic (Bindu Sri *et al.*, 2012). Although some of them are sufficiently deformable and could be injectable, many are not, and need surgical implantation (Hoare and Kohane, 2008).

Polymers Used in the Preparation of Hydrogels

Different polymers are used for the development of hydrogel systems including cellulose derivatives, natural gums, polycrylates and gelatin (Bindu Sri *et al.*, 2012). Natural polymers are used widely as these are important to achieve predetermined rates of drug delivery. Their physicochemical characteristics, and ready availability, provide a platform for various controlled and targeted delivery systems (Fig. 3.3; Sharma *et al.*, 2011). They remain attractive primarily because they are economical, readily available, capable of a multitude of chemical modifications, potentially degradable and biocompatible (Sharma *et al.*, 2011). They exhibit a large diversity of structures and physiological functions, and offer a variety of potential applications in the field of tissue engineering due to properties such as their pseudoplastic behaviour, gelling ability, water-binding capacity and biodegradability. Hydrogels obtained from natural polymers resemble natural living tissues more than any other class of synthetic biomaterials due to their high water content and soft consistency similar to natural tissue (Simões *et al.*, 2012). Further, their high water content contributes to biocompatibility and hence they could be used as contact lenses, linings for artificial hearts, materials for artificial skin and membranes for biosensors (Xinming *et al.*, 2008; Simões *et al.*, 2012).

Classification of Hydrogels

The hydrogels can be classified on the basis of their method of preparation, ionic charges and structure, and on the mechanism of controlling drug release (Table 3.2).

Hydrogel Beads in Drug Delivery

Hydrogel beads are prepared from hydrophilic polymers and their sizes range from 0.2 to 3 mm;

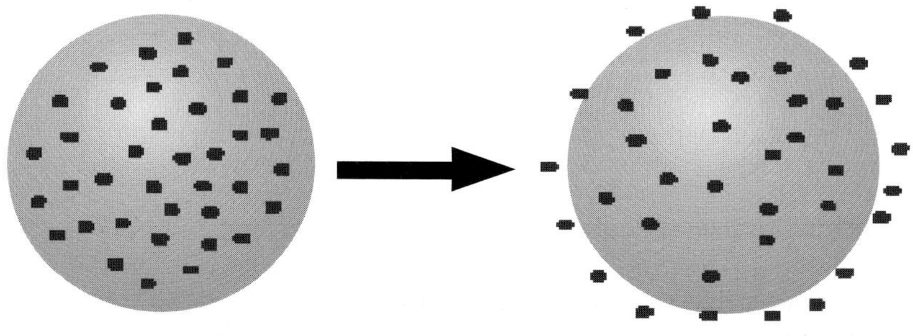

Fig. 3.3. The drug is released in a controlled manner when incorporated in the hydrogels.

Table 3.2. Classification of hydrogels.

S. No.	Basis of classification	Examples	Reference
1	Method of preparation	Homopolymer hydrogels, copolymer hydrogels, multipolymer hydrogels	Ahmed, 2015
2	Ionic charges	Neutral hydrogels, anionic hydrogels, cationic hydrogels, ampholytic hydrogels	Ahmed, 2015
3	Structure	Amorphous hydrogels, semi-crystalline hydrogels, hydrogen-bonded hydrogels	Ahmed, 2015
4	Mechanism of controlling drug release	Diffusion controlled release systems, swelling controlled release systems, chemically controlled release systems, environment-responsive systems	Schuetz *et al.*, 2008; Ahmed, 2015

they are basically spherical in shape (Nussinovich and Gal, 2010). The shape and size of the hydrogel beads are the essential parameters as these will specify the rate of release of the drug from the beads. The drug or the required ingredient for the cosmetic will be integrated into the microbeads by one of two processes: (i) by incorporating the drug or required material during the process of preparation of the microbeads; or (ii) by adsorption of a solution or suspension of the active ingredients on the prepared microbeads (Jaiswal *et al.*, 2014).

The polymers used for the preparation of beads are sodium alginate, pectin, chitosan, gellan gum, guar gum and karaya gum. The main advantage of the alginate beads is that these are non-toxic when taken orally; have high biocompatiblity; and do not re-swell in an acidic environment but easily re-swell in an alkaline environment. The latter quality makes them the perfect delivery system for acid-sensitive drugs in the stomach (Jaiswal *et al.*, 2014). Drugs such as theophylline (Soni *et al.*, 2010) and gatilfloxacin (Kesavan *et al.*, 2010) have been included in sodium alginate beads. Malviya *et al.* prepared matrix tablets of diclofenac sodium using pectin polymer in different concentrations and studied its release profile. It was found that a 1:1.5 drug:polymer ratio was the best formulation for oral sustained release of the tablet (Malviya *et al.*, 2010).

Hydrogel Microparticles

When the hydrogels are in the form of microscopic links or confined to smaller dimensions, they are termed microgels or hydrogel microparticles (Diana *et al.*, 2014). The majority of polymers used in fabrication of polymeric networks are beta-cyclodextrins, chitosan, carboxymethylcellulose, hydroxymethylcellulose and carrageenan. Microparticles are widely studied for a variety of medical applications including imaging, biodetection, tissue engineering and drug delivery. Specially, microparticles are of interest because they provide a way to overcome some of the major challenges in drug delivery. Microgels have a central network for the inclusion of drug molecules, allowing physical loading of bioactive molecules such as drugs, proteins, carbohydrates and DNA in the polymeric network, and hence have become targeted drug delivery carriers for biomedical applications. They not only incorporate organic molecules but also inorganic nanocrystals such as quantum dots and magnetic nanoparticles for optical and magnetic imaging of living cells, and gold nanorods for photodynamic therapy (Das *et al.*, 2007). There are many advantages to microparticles, such as protection of the drug from unfavourable environmental conditions, and control of the drug release profile for a specified period of time (Ying *et al.*, 2014). Controlled release for a specified time is an important criterion and microparticles show this behaviour very well, making them an appropriate delivery system. Hydrogel microparticles consist of a hydrophilic mixture and show the characteristics of both solids and liquids. The hydrogel structure consists of networks that are formed by random cross-linking of macromolecules with three phases: polymeric-network matrix solid phase, interstitial fluid phase and ionic phase. The solid phase consists of a network of crosslinked polymeric chains.

Polymeric chains produce a three-dimensional matrix with interstitial gaps occupied by water and frequently by biological fluids. The fluid phase imbibes into polymeric network pores and gives wet and elastic properties to the hydrogel microparticles. As the microparticles have wet and elastic properties, the structure of the microparticle hydrogels resembles a living tissue. In hydrogel microparticles, drugs can be incorporated by the same methods as used in the preparation of hydrogel beads.

Release of drug from the matrix of hydrogel microparticles is based upon composition of the formulation, type of polymer, drug, monomer, initiator, cross-linker, etc., size and shape of the particles, method of preparation and environmental conditions. The release occurs by different phenomena: exterior diffusion, interior diffusion, desorption and chemical reaction (Roman *et al.*, 2010).

There are various methods for the preparation of hydrogel microparticles: ionic gelation, spray drying, dispersion photopolymerization (Lee and Kim, 2011), ionotropic gelation method (Patil *et al.*, 2011), suspension cross linking, free radical precipitation polymerization (Thomas *et al.*, 2006), membrane emulsification, inverse emulsion polymerization, Michael addition reaction (Oh *et al.*, 2008) and free radical polymerization (Mundargi *et al.*, 2011).

Hydrogel microparticles are employed in enhancement of hydrophobic and poorly water-soluble drugs such as acyclovir, whose solubility has been enhanced by the formation of chitosan hydrogel microparticles (Parize *et al.*, 2012). They exhibit controlled and prolonged release of drugs by modifying the swelling properties of the microparticles. Hydrogel microparticles are defending carriers for acid-labile drugs, proteins and enzymes, as encapsulation prevents degradation of such moieties in acidic conditions. For example, insulin could be protected from the gastric environment and enzymes (Yamagata *et al.*, 2006).

Hydrogel Nanoparticles in Drug Delivery

Hydrogel nanoparticles (NPs) are also known as nanogels and are formed when the size of the microgels is reduced to the submicrometer range. As the name nanogels indicates, these have the features and characteristics of both hydrogels and NPs at the same time. These have gained considerable attention in pharmaceutical and biomedical fields in recent years and have become one of the most promising nanoparticulate drug delivery systems. Their intrinsic properties (including hydrophilicity, flexibility, versatility, high water absorptivity and biocompatibility), along with a long life span in circulation and the possibility of being actively or passively targeted to the desired biophase (e.g. tumour sites), are advantageous in pharmaceutical and biomedical applications. There are different methods for their preparation, such as covalent crosslinking, water-in-oil (w/o) emulsion, ionic crosslinking, desolvation, emulsion-droplet coalescence, reverse micellar and self-assembly via chemical modification (Table 3.3; Patel and Patel, 2010). Several polymeric hydrogel nanoparticulate systems have been prepared by using chitosan, alginic acid, etc., and have been characterized very well. Nanogels are often used as they have several advantages such as particle size, and their surface characteristics can easily be manipulated to accomplish passive and active drug targeting after parenteral administration. These provide controlled and sustained release of drugs during transportation and at the site of action, with increased therapeutic efficacy and reduced side effects. Controlled release and particle degradation characteristics could be modulated by choosing appropriate matrix constituents. Drug loading is relatively high and drugs could be incorporated into the systems without any chemical reaction, thus their activity could be preserved for longer durations. Site-specific delivery could be achieved by attaching targeting ligands on the particle surface. Nanogels also offer the possibility of versatile administration including oral, nasal, parenteral and intra-ocular.

Chitosan-based hydrogel nanoparticles

Chitosan, $\alpha(1\rightarrow4)$-2-amino-2-deoxy β-D-glucan, is a deacetylated form of chitin and is an abundant polysaccharide present in crustacean shells. This is a promising biopolymer for drug delivery as it is biocompatible with low toxicity, and is specifically used to deliver macromolecules (Patel *et al.*, 2007). It is a hydrophilic polymer and is positively charged; therefore it can bind to negatively charged molecules such as polyanions. This interaction results in the sol–gel transition stage, which is important for the nanoencapsulation process (Shutava and Lvov, 2006). The special property of chitosan is that it adheres to the mucosal surfaces of the body, hence it is the choice for mucosal drug delivery. This potential of chitosan enables it to open tight junctions between epithelial cells though well-organized epithelia. There are different techniques for the preparation of chitosan-based hydrogel nanoparticles, as discussed in Table 3.3.

Alginate-based hydrogel nanoparticles

Alginic acid is an anionic biopolymer consisting of linear chains of α-L-glucuronic acid and β-D-mannuronic acid. It is a hydrophilic polymer (i.e. with a high degree of aqueous solubility), has a property of gelation to form high porosity gels and, above all, it is biocompatible and nontoxic (Tonnesen and Karlsen, 2002). Sequential crosslinking and formation of polymeric networks by alginate results in hydrogel structured drug delivery carriers such as micro- and nanoparticles. It was seen that the addition of counter-ions to any possible cationic species could initiate the reaction sequence of this biopolymer, but calcium chloride is most utilized as the gelation with calcium chloride is very fast and the gelation speed affects gel uniformity and strength directly (Hamidi *et al.*, 2008). The method of preparation of calcium alginate nanoparticles is very important as it determines the control of the gelation phenomenon, which leads to desired size ranges of the nanoparticles depending on various factors including alginate concentration/viscosity, counter-ion concentration and the speed of adding counter-ion solution onto the alginate solution. Insulin-loaded NPs were prepared by Sarmento *et al.* (2007) with alginate ionotropic pre-gelation followed by chitosan polyelectrolyte complexation. This process generates particles in the nanometer size range under optimized conditions with an insulin encapsulation efficiency of 14.3%. In another study dextran polysaccharide was used as the complexing agent and insulin was loaded in the alginate–dextran nanospheres via nanoemulsion dispersion followed by triggered *in situ* gelation. This process forms particles ranging from 267 nm to 2.76 μm with the insulin encapsulation efficiency of 82.5% (Reis *et al.*, 2007). Reis *et al.* (2007) showed that the bioavailability of drugs encapsulated in alginate NPs is significantly higher than those of free drugs.

The antifungal drugs clotrimazole and econazole and the antituberculosis drugs rifampicin, ethambutol, isoniazid, and pyrazinamide are encapsulated within alginate (ALG) nanoparticles (Ahmed, 2015) by means of a modified cation-induced controlled gelification (Arora *et al.*, 2011).

Inhalable Hydrogels

Currently, pharmaceutical technology relies mainly on systemic delivery of drugs. However, this approach is not capable of providing sufficient drug quantities. To increase the efficacy, a major goal is the development of delivery vehicles that can protect the drug from degradation, deliver to a specific site and then release in a controlled fashion. The use of particles for pulmonary delivery is of particular interest for both local and systemic treatments. In case of local treatment of disease such as lung cancer, tuberculosis, asthma or chronic obstructive pulmonary diseases (COPD), inhaling the drug is advantageous as it allows the active molecule to go directly to the site of action as compared to systemic administration. Additionally, this localized delivery to the lungs for pulmonary diseases would certainly be advantageous for enhancing bioavailability as there is rapid absorption and no first pass metabolism; side effects would also decrease (Guntur and Dhand, 2007).

Pulmonary delivery has also been explored as a method for the systemic circulation because of the lung's natural permeability to small molecules, peptides and proteins (Tuncer and Nevin, 2007; Siekmeier and Scheuch, 2008). It facilitates direct, targeted application of bioactive materials to the lungs in a controlled manner and could as well serve as an effective treatment for lung cancer patients (Kumar *et al.*, 2013). Certain medical conditions such as asthma implement targeted pulmonary delivery as the first choice and there is a growing interest in inhalable lung

Table 3.3. Chitosan-based hydrogel nanoparticles prepared by different techniques.

S.no.	Method	Preparation technique	Application	References
1	Covalent crosslinking	Reacting tetramethoxysilan with hydroxyl groups on the chitosan monomers	Drug loaded nanoparticles	Patel and Patel, 2010
2	Water-in-oil (w/o) emulsion method	Glutaraldehyde crosslinking of the chitosan amino groups	Nanospheres loaded by 5-fluorouracil (5-FU), an anti-cancer drug	Patel and Patel, 2010
3	Ionic crosslinking	Ionotropic gelation occurs by the addition of an alkaline phase (pH=7–9) containing tripolyphosphate (TPP) into an acidic phase (pH=4–6) containing chitosan	Insulin-loaded chitosan NPs	Gan and Wang, 2005
4	Desolvation method	Dropwise addition of sodium sulfate into a solution of chitosan and polysorbate 80	The chitosan–DNA NPs could partially protect the encapsulated plasmid DNA from nuclease degradation	Patel and Patel, 2010
5	Emulsion-droplet coalescence method	A stable emulsion containing aqueous solution of chitosan along with the drug to be loaded is produced in liquid paraffin. At the same time, another stable emulsion of chitosan aqueous solution containing NaOH is produced in the same manner. When both emulsions are mixed under high speed stirring, droplets of each collide at random and coalesce, thereby precipitating chitosan droplets to give small solid particles	Drug loaded nanoparticles are obtained	Patel and Patel, 2010
6	Reverse micellar method	The surfactant is dissolved in an organic solvent to prepare reverse micelles. To this, aqueous solutions of chitosan and drug are added gradually with constant vortexing to avoid any turbidity. The organic solvent is then evaporated to obtain the micellar transparent drug mass. The remaining material is dispersed in water and then, by adding a suitable salt, the surfactant precipitates out. The mixture is then subjected to centrifugation. The supernatant solution containing the drug-loaded NPs is decanted	Tumour targeted delivery of encapsulated dextran–doxorubicin conjugate using chitosan nanoparticles as carrier	Mitra and Gaur, 2001
7	Self-assembly via chemical modification	Fractional conjugation of polyethylene glycol, PEG, via an amide linkage to soluble chitosan was shown to yield self-aggregation at basic pH. These aggregates could trap insulin following incubation in phosphate buffer saline (PBS), probably due to the electrostatic interaction between the unconjugated chitosan monomers and the anionic residues of the protein.	Insulin-loaded chitosan NPs	Yu and Hu, 2006

cancer treatment modalities. Pulmonary administration of anti-cancer agents offers several advantages with the most important being higher local concentrations along with reduced systemic side effects. Traditionally, inhaled therapies consist of small-molecule drugs and excipients; however, targeted pulmonary delivery provides a platform for localizing novel nanoparticles to the lungs through direct, topical application (Azarmi *et al.*, 2006).

Injectable Hydrogels

In past decades, injectable hydrogels have emerged as promising biomaterials because of their biocompatibility, excellent permeability, minimal invasion and easy integration into surgical procedures. These systems provide an effective and convenient way to administer a wide variety of bioactive agents such as proteins, genes and even living cells. Additionally, they could be designed to be degradable and eventually cleared from the body. Given their unique characteristics, injectable biodegradable hydrogels have been actively explored as drug reservoir systems (Nguyen and Lee, 2010).

Injectable biodegradable hydrogels have been widely used in biomedical applications such as drug/cell delivery and tissue engineering because of their highly hydrophilic characteristics. These are of particular interest because drugs, proteins and cells can be easily incorporated into polymer solutions prior to administration. Importantly, no surgical procedures are required for their insertion into the body and they are administered through simple injection (Ruel-Gariepy and Leroux, 2004; He *et al.*, 2008; Yu and Ding, 2008).

Insulin (Ins) spontaneously and easily complexes with the hydrogel nanoparticle of hydrophobized cholesterol-bearing pullulan (CHP) in water. The complexed nanoparticles (diameter 20–30 nm) thus obtained form a very stable colloid. The thermal denaturation and subsequent aggregation of Ins were effectively suppressed upon complexation, and the complexed Ins was significantly protected from enzymatic degradation. Spontaneous dissociation of Ins from the complex was barely observed, except in the presence of bovine serum albumin. The original physiological activity of complexed Ins was preserved *in vivo* after IV injection (Ogaji *et al.*, 2011).

Recent Advances

Tissue engineering

Over the past few decades, the field of tissue engineering (TE) has met the need for generation of new organs and tissues. The term TE is associated with the replacement of various fractions of tissues or the whole tissue by using a combination of cells and engineering methods. The advances in TE have recognized the role of hydrogels as a scaffold material. Hydrogels structurally resemble the extracellular matrix (ECM) of the human body, and their three-dimensional structures organize cells and offer stimuli to direct the creation of new tissues. The hydrogels derived from both natural and synthetic polymers are frequently used in TE applications (Vashist *et al.*, 2014).

During the fabrication of implantable scaffolds, particular attention should be paid to body compatibility, mechanical properties, scaffold morphology and porosity, as well as to healing and tissue replacement capacity (Dutta *et al.*, 2011). The main requirements for the elaboration of TE scaffolds also include that the scaffold: should not induce an acute or chronic response; should be biodegradable so that the cured tissue will be able to replace the biomaterial; should possess surface properties that will promote cell attachment, differentiation and proliferation; should have suitable mechanical properties for handling, mimicking the damaged tissue; and should be capable of manufacture into a variety of shapes (Olteanu and Enescu, 2007). Chitosan appears a relevant candidate for the preparation of such biomaterials, which could substitute for missing or damaged tissue and organs (Dutta *et al.*, 2011) and allow cell attachment and proliferation provided that three-dimensional scaffolds could be produced. Such development thus relies on robust processing methods for chitosan, which must ensure that the scaffold properties are adjusted to meet the requirements of the diseased tissue. Therefore, methodologies have been developed to finely shape chitosan hydrogels and foams (or sponges) as three-dimensional scaffolds applicable for TE (Croisier and Jérôme, 2013).

Cartilage repair

The need for TE cartilage is immense and of great clinical significance. Traumatic and degenerative lesions of articular cartilage are leading causes of disability. TE methods to improve cartilage repair and regeneration will therefore have a high clinical impact. The advantage of injectable therapies for cartilage repair is not only that the implant is maintained within the defect, but also that it allows immediate weight-bearing due to the stiffness and strength that is achieved almost instantly (Sun and Tan, 2013). The physical properties of the alginate hydrogel can be designed to easily match those of articular cartilage, in addition to matching the mechanical properties of the scaffold with the native tissue. Alginate-based injectable hydrogels, solid- and gel-microspheres have been used in cartilage regeneration. Many studies have been carried out into the combination of alginate-based microspheres and hydrogels for controlled growth factor delivery in TE (Henrionnet *et al.*, 2012). One example is a study demonstrating the positive effect of immobilizing arginine-glycine-aspartic acid (RGD) to a macro-porous alginate scaffold in promoting transforming growth factor-beta (TGF-β) induced human mesenchymal stem cell (MSC) differentiation (Reem *et al.*, 2010). The cell-matrix interactions facilitated by the immobilized RGD peptide were shown to be an essential feature of the cell microenvironment, allowing better cell accessibility to the chondrogenic-inducing molecule TGF-β.

Bian and colleagues investigated the co-encapsulation of TGF-β-containing alginate microspheres with human MSCs in hyaluronic acid (HA) hydrogels with regard to the development of implantable constructs for cartilage repair (Bian *et al.*, 2011). TGF-β-loaded alginate microspheres combined with hydrogels form a composite carrier, which may help to retain TGF-β bioactivity in the scaffold and promote chondrogenesis of MSCs when implanted.

Bone regeneration

Bone regeneration is a significant challenge in reconstructive surgery. There are several reasons for lack of bone tissue, such as trauma and tumour removal. A desirable strategy to repair bone tissue is to induce osteogenesis *in situ*. One method to accomplish this is to use stem cells that can differentiate to form bone tissue, and seed those cells into an injectable scaffold, resulting in bone tissue formation (Jin *et al.*, 2012; Rubert *et al.*, 2012). There have been numerous studies involving the use of injectable alginate-based scaffolds for bone regeneration (Brun *et al.*, 2011; Florczyk *et al.*, 2012; Nguyen and Lee, 2012). Adequate bone tissue formation was observed using MSCs and alginate as the scaffold (Tang *et al.*, 2012; Xia *et al.*, 2012). Alginate, therefore, is applicable for generating tissue in gels, displaying osteogenic as well as angiogenic properties. Many researchers reported bone regeneration using injectable scaffolds combining alginate-based hydrogels or microspheres, which were mixed with undifferentiated MSCs or ASCs (Chen *et al.*, 2012; Rubert *et al.*, 2012). These studies demonstrated the potential of bone morphogenetic protein (BMP) and TGF-β delivery to induce osteogenic differentiation to mature osteocytes from MSCs and ASCs (Florczyk *et al.*, 2012; Xia *et al.*, 2012). Kolambkar *et al.* (2011) introduced a hybrid growth factor delivery system that consists of an electrospun nanofibre mesh tube for guiding bone regeneration combined with a peptide-modified alginate hydrogel injected inside the tube for sustained recombinant BMP-2 (rhBMP-2) release. The results indicated that sustained delivery of rhBMP-2 via alginate hydrogel was required for substantial regeneration to occur. This hybrid technique may be clinically useful for bone regeneration in the case of fracture of non-unions and large bone defects. Present findings showed that the co-immobilization of osteogenic and endothelial cells within RGD-alginate microspheres is a promising new injectable strategy for bone TE (Grellier *et al.*, 2013).

Breast implants

Numerous health risks are raised by the use of silicone gel, including siliconoma, a condition arising from the leaking of the silicone gel into the pectoral muscle and the breast parenchyma. There is, therefore, an unmet need for the development of a different filler with minimum risk and which is cosmetically beneficial (Arion, 2001). Carboxymethylcellulose (CMC) is a cellulose polysaccharide weighing approximately

10,000 Da. It is soluble in water or serum and remains stable at temperatures ranging from 10°C to 130°C. It is a biodegradable, non-toxic, nonmutagenic, viscoelastic hydrogel, which has been tested in a variety of animal studies. Implants based on carboxymethyl-cellulose (CMC) hydrogel display higher radio translucency than silicone gel and the integrity of the device is easy to prove by clinical examination (Arion, 2001).

Conclusions

Hydrocolloid-based hydrogel delivery is an efficient system with minimal limitations and having an easy route of administration. These hydrogels have wide applications in all branches of science and play an important role in biotechnology and pharmaceuticals. Scientists are now developing formulations that can deliver the drug at a specific site, and hydrocolloid-based hydrogels certainly match the need for targeted drug delivery. Resources for hydrocolloids are large and renewable, and even their derivatives provide great opportunities in the field of organ-specific drug delivery. The development of efficient drug delivery systems, in conjunction with the development of newly found drugs, can benefit both economically as well as reduce the time it takes to develop new drug formulations. In addition, these hydrogels serve as the scaffolds in bone regeneration and cartilage regeneration, and also play an important role in TE.

Acknowledgements

We thank Dr Ashok K. Chauhan, Founder President, Ritanand Balved Educational Foundation for support. The guidance from Shri. Atul Chauhan, Chancellor, Amity University Uttar Pradesh and Prof. Dr (Mrs) Balvinder Shukla, Vice Chancellor, Amity University Uttar Pradesh is greatly appreciated. We gained several insights from the articles cited in this manuscript and these eminent scholars are acknowledged profoundly.

References

Ahmed, E.M. (2015) Hydrogel: preparation, characterization, and applications: a review. *Journal of Advanced Research* 6, 105–121.

Arion, H. (2001) Carboxy-methyl cellulose hydrogels used to fill breast implants: 15 years of experience. *European Journal of Plastic Surgery* 24, 172–175.

Arora, S., Gupta, S., Narang, R.K. and Budhiraja, R.D. (2011) Amoxicillin loaded chitosan-alginate polyelectrolyte complex nanoparticles as mucopenetrating delivery system for H. pylori. *Scientia Pharmaceutica* 79, 673–694.

Azarmi, S., Tao, X., Chena, H., Wang, Z., Finlay,W.H., Löbenberg, R. and Roa, W.H. (2006) Formulation and cytotoxicity of doxorubicin nanoparticles carried by dry powder aerosol particles. *International Journal of Pharmaceutics* 319, 155–161.

Bian, L., Zhai, D.Y., Tous, E., Rai, R., Mauck, R.L. and Burdick, J.A. (2011) Enhanced MSC chondrogenesis following delivery of TGF-β3 from alginate microspheres within hyaluronic acid hydrogels in vitro and in vitro. *Biomaterials* 32, 6425–6434.

Bindu Sri, M., Vadithya, A. and Chatterjee, A. (2012) As a review on hydrogels as drug delivery in the pharmaceutical field. *International Journal of Pharmaceutical and Chemical Sciences* 1, 642–661.

Brun, F., Turco, G., Accardo, A. and Paoletti, S. (2011) Automated quantitative characterization of alginate/hydroxyapatite bone tissue engineering scaffolds by means of micro-CT image analysis. *Journal of Materials Science: Materials in Medicine* 22, 2617–2629.

Cassiday, L. (2012) Personal care products: hydrocolloids get personal. *Inform* 23, 349–351.

Chen, W., Zhou, H., Weir, M.D., Bao, C. and Xu, H. (2012) Umbilical cord stem cells released from alginate-fibrin microbeads inside macroporous and biofunctionalized calcium phosphate cement for bone regeneration. *Acta Biomaterialia* 8, 2297–2306.

Croisier, F. and Jérôme, C. (2013) Chitosan-based biomaterials for tissue engineering. *European Polymer Journal* 49, 780–792.

Das, M., Sanson, N., Fava, D. and Kumacheva, E. (2007) Microgels loaded with gold nanorods: photothermally triggered volume transitions under physiological conditions. *Langmuir* 23, 196–201.

Diana, C., Valente, A.J.M., Graça, M.M. and João, Q. (2014) Plasmid DNA hydrogels for biomedical applications. *Advances in Colloid and Interface Science* 205, 257–264.

Dutta, P., Rinki, K. and Dutta, J. (2011) Chitosan: a promising biomaterial for tissue engineering scaffolds. Chitosan for biomaterials II. In: Jayakumar, R., Prabaharan, M. and Muzzarelli, R.A.A. (eds) *Advances in Polymer Science*. Springer Berlin, Heidelberg, Germany, pp. 45–79.

Florczyk, S.J., Leung, M., Jana, S., Li, Z., Bhattarai, N., Huang, J., Hopper, R.A. and Zhang, M. (2012) Enhanced bone tissue formation by alginate gel-assisted cell seeding in porous ceramic scaffolds and sustained release of growth factor. *Journal of Biomedical Materials Research Part A* 100, 3408–3415.

Gan, Q. and Wang, T. (2005) Modulation of surface charge, particles size and morphological properties of chitosan-TPP nanoparticles intended for gene delivery. *Colloids and Surfaces B: Biointerface* 44, 65–73.

Grellier, M., Granja, P.L., Fricain, J.C., Bidarra, S.J., Renard, M., Bareille, R., Bourget, C., Amédée, J. and Barbosa, M.A. (2013) The effect of the co-immobilization of human osteoprogenitors and endothelial cells within alginate microspheres on mineralization in a bone defect. *Biomaterials* 30, 3271–3278.

Guntur, V.P. and Dhand, R. (2007) Inhaled insulin: extending the horizons of inhalation therapy. *Respiratory Care* 52, 911–922.

Hamidi, M., Azadi, A. and Rafiei, P. (2008) Hydrogel nanoparticles in drug delivery. *Advanced Drug Delivery Reviews* 60, 1638–1649.

He, C., Kim, S.W. and Lee, D.S. (2008) In situ gelling stimuli-sensitive block copolymer hydrogels for drug delivery. *Journal of Controlled Release* 127, 189–207.

Henrionnet, C., Wang, Y., Roeder, E., Gambier, N., Galois, L., Mainard, D., Bensoussan, D., Gillet, P. and Pinzano, A. (2012) Effect of dynamic loading on MSCs chondrogenic differentiation in 3-D alginate culture. *Bio-Medical Materials and Engineering* 22, 209–218.

Hoare, T.R. and Kohane, D.S. (2008) Hydrogels in drug delivery: progress and challenges. *Polymer* 49, 1993–2007.

Jaiswal, A.R., Vyas, M.R., Behera, A.K. and Mesharam, D.B. (2014) Colon targeted multiparticulate system: hydrogel beads. *World Journal of Pharmacy and Pharmaceutical Science* 3, 507–525.

Jin, H.H., Kim, D.H., Kim, T.W., Shin, K.K., Jung, J.S., Park, H.C. and Yoon, S.Y. (2012) In vitro evaluation of porous hydroxyapatite/chitosan-alginate composite scaffolds for bone tissue engineering. *International Journal of Biological Macromolecules* 51, 1079–1085.

Kapoor, M., Khandal, D., Seshadri, G., Aggarwal, S. and Khandal, R.K. (2013) Novel hydrocolloids: preparation and applications – a review. *International Journal of Research and Reviews in Applied Sciences* 16, 432–434, 82.

Kesavan, K., Nath, G. and Pandit, J.K. (2010) Sodium alginate based mucoadhesive system for Gatifloxacin and its in vitro antibacterial activity. *Scientia Pharmaceutica* 78, 941–957.

Kolambkar, Y.M., Dupont, K.M., Boerckel, J.D., Huebsch, N., Mooney, D.J., Hutmacher, D.W. and Guldberg, R.E. (2011) An alginate-based hybrid system for growth factor delivery in the functional repair of large bone defects. *Biomaterials* 32, 65–74.

Kumar, A., Chen, F., Mozhi, A., Zhang, X., Zhao, Y., Xue, X., Hao, Y., Zhang, X., Wang P.C. and Liang, X. (2013) Innovative pharmaceutical development based on unique properties of nanoscale delivery formulation. *Nanoscale* 5, 8307–8325.

Lee, E. and Kim, B. (2011) Preparation and characterization of pH-sensitive hydrogel microparticles as a biological on–off switch. *Polymer Bulletin* 67, 67–76.

Li, H., Luo, R. and Lam, K.Y. (2006) Modeling of environmentally sensitive hydrogels for drug delivery: an overview and recent developments. *Front Drug Discovery* 2, 295–331.

Malviya, R., Srivastava, P., Bansal, M. and Sharma, P.K. (2010) Formulation and optimization of sustained release matrix tablets of diclofenac sodium using pectin as release modifier. *International Journal of Drug Development and Research* 2, 330–335.

Mitra, S. and Gaur, U. (2001) Tumor targeted delivery of encapsulated detxran-doxorubicin conjugate using chitosan nanoparticles as carrier. *Journal of Controlled Release* 74, 317–323.

Mundargi, R.C., Rangaswamy, V. and Aminabhavi, T.M. (2011) Poly(N-vinylcaprolactam-co-methacrylic acid) hydrogel microparticles for oral insulin delivery. *Journal of Encapsulation* 28, 384–394.

Nguyen, M.K. and Lee, D.S. (2010) Injectable biodegradable hydrogels. *Macromolecular Bioscience* 10, 563–579.

Nguyen, T.P. and Lee, B.T. (2012) Fabrication of oxidized alginate-gelatin-BCP hydrogels and evaluation of the microstructure, material properties and biocompatibility for bone tissue regeneration. *Journal of Biomaterials Applications* 27, 311–321.

Nussinovich, A. and Gal, A. (2010) Hydrocolloid carrier beads with inert filler material. Patent number: US 20100015192.

Ogaji, I.J., Nep, E.I. and Audu-Peter, J.D. (2011) Advances in natural polymers as pharmaceutical excipients. *Pharmaceutica Analytica Acta* 3, 1–16.

Oh, J.K., Drumright, R., Siegwart, D.J. and Matyjaszewski, K. (2008) The development of microgels/nanogels for drug delivery applications. *Progress in Polymer Science* 33, 448–477.

Olteanu, C.E. and Enescu, D. (2007) Chitosan involved tissue engineering and regenerative medicine. *Romanian Biotechnological Letters* 12, 3217–3233.

Pandey, R. and Khuller, G.K. (2004) Polymer based drug delivery systems for mycobacterial infections. *Current Drug Delivery* 1, 195–201.

Pandey, A., Kumar, G., Kothiyal, P. and Barshiliya, Y. (2012) A review on current approaches in gastro retentive drug delivery system. *Asian Journal of Pharmacy and Medical Science* 2, 60–77.

Parize, A.L., Stulzer, H.K., Laranjeira, M.C.M., Brighente, I.M. da C. and Souza, T.C.R. (2012) Evaluation of chitosan microparticles containing curcumin and crosslinked with sodium tripolyphosphate produced by spray drying. *Química Nova* 35, 1127–1132.

Patel, H.A. and Patel, J.K. (2010) Nanogel as a controlled drug delivery system. *International Journal of Pharmaceutical Sciences Review and Research* 4, 37–41.

Patel, M., Shah, T. and Amin, A. (2007) Therapeutic opportunities in colon-specific drug delivery systems. *Critical Reviews in Therapeutic Drug Carrier Systems* 24, 147–202.

Patil, J.S., Kamalapur, M.V., Marapur, S.C. and Kadam, D.V. (2011) Ionotropic gelation and polyelectrolyte complexation: The novel techniques to design hydrogel particulate sustained, modulated drug delivery system: a review. *Digest Journal of Nanomaterials and Biostructures* 5, 241–248.

Peppas, N.A., Huang, Y., Torres-Lugo, M., Ward, J.H. and Zhang, J. (2000) Physicochemical, foundations and structural design of hydrogels in medicine and biology. *Annual Review of Biomedical Engineering* 2, 9–29.

Qiu, Y. and Park, K. (2012) Environment-sensitive hydrogels for drug delivery. *Advanced Drug Delivery Reviews* 64, 49–60.

Reem, T., Tsur-Gang, O. and Cohen, S. (2010) The effect of immobilized RGD peptide in macroporous alginate scaffolds on TGFβ1-induced chondrogenesis of human mesenchymal stem cells. *Biomaterials* 31, 6746–6755.

Reis, C.P., Ribeiro, A.J., Houng, S., Veiga, F. and Neufeld R.J. (2007) Nanoparticulate delivery system for insulin: design, characterization and in vitro/in vivo bioactivity. *European Journal of Pharmaceutics and Biopharmaceutics* 30, 392–397.

Roman, Z., Zofia, M. and Katarzyna, N. (2010) Drug release from hydrogel matrices. *Ecological Chemistry and Engineering S* 17, 117–136.

Rubert, M., Monjo, M., Lyngstadaas, S.P. and Ramis, J.M. (2012) Effect of alginate hydrogel containing polyproline-rich peptides on osteoblast differentiation. *Biomedical Materials* 7(5), 055003.

Ruel-Gariepy, E. and Leroux, J. (2004) In situ-forming hydrogels – review of temperature-sensitive systems. *European Journal of Pharmaceutics and Biopharmaceutics* 58, 409–426.

Sarmento, B., Ribeiro, A., Veiga, F., Sampaio, P., Neufeld, R. and Ferreira, D. (2007) Alginate/chitosan nanoparticles are effective for oral insulin delivery. *Pharmaceutical Research* 24, 2198–2206.

Schuetz, Y.B., Gurny, R. and Jordan, O. (2008) A novel thermoresponsive hydrogel of chitosan. *European Journal of Pharmaceutics and Biopharmaceutics* 68, 19–25.

Sharma, K., Singh, V. and Arora, A. (2011) Natural biodegradable polymers as matrices in transdermal drug delivery. *International Journal of Drug Development & Research* 3, 85–103.

Shutava, T.G. and Lvov, Y.M. (2006) Nano-engineered microcapsules of tannic acid and chitosan for protein encapsulation. *Journal of Nanoscience and Nanotechnology* 6, 1655–1661.

Siekmeier, R. and Scheuch, G. (2008) Systemic treatment by inhalation of macromolecules – principles, problems and examples. *Journal of Physiology and Pharmacology* 59, 53–79.

Simões, S., Figueiras, A. and Veiga, F. (2012) Modular hydrogels for drug delivery. *Journal of Biomaterials and Nanobiotechnology* 3, 185–199.

Soni, M.L., Kumar, M. and Namdeo K.P. (2010) Sodium alginate microspheres for extending drug release: formulation and in vitro evaluation. *International Journal of Drug Delivery* 2, 64–66.

Sun, J. and Tan, H. (2013) Alginate-based biomaterials for regenerative medicine applications. *Materials* 6, 1285–1309.

Tang, M., Chen, W., Weir, M.D., Thein-Han, W. and Xu, H. (2012) Human embryonic stem cell encapsulation in alginate microbeads in macroporous calcium phosphate cement for bone tissue engineering. *Acta Biomaterialia* 8, 3436–3445.

Thomas, J.B., Creecy, C.M., McGinity, J.W. and Peppas, N.A. (2006) Synthesis and properties of lightly crosslinked poly((meth)acrylic acid) microparticles prepared by free radical precipitation polymerization. *Polymer Bulletin* 57, 11–20.

Tiwari, G., Tiwari, R., Sriwastawa, B., Bhati, L., Pandey, S., Pandey, P. and Bannerjee, S.K. (2012) Drug delivery systems: an updated review. *International Journal of Pharmaceutical Investigation* 2, 2–11.

Tonnesen, H.H. and Karlsen, J. (2002) Alginate in drug delivery systems. *Drug Development and Industrial Pharmacy* 28, 621–630

Torres, M.D., Moreira, R., Chenlo, F. and Vazquez, M.J. (2012) Water absorption isotherms of CMC, guar, locust bean, tragacanth and xanthan gums. *Carbohydrate Polymers* 89, 592–598.

Tuncer, D.I. and Nevin, C. (2007) Controlled delivery of peptides and proteins. *Current Pharmaceutical Design* 13, 99–117.

Vashist, A., Vashist A., Gupta, Y.K. and Ahmad, S. (2014) Recent advances in hydrogel based drug delivery systems for the human body. *Journal of Materials Chemistry B* 2, 147–166.

Xia, Y., Mei, F., Duan, Y., Gao, Y., Xiong, Z., Zhang, T. and Zhang, H. (2012) Bone tissue engineering using bone marrow stromal cells and an injectable sodium alginate/gelatin scaffold. *Journal of Biomedical Materials Research Part A* 100, 1044–1050.

Xinming, L., Yingde, C., Lloyd, A.W., Mikhalovsky, S.V., Sandeman, S.R., Howel, C.A. and Liewen, L. (2008) Polymeric hydrogels for novel contact lens-based ophthalmic drug delivery systems: a review. *Contact Lens & Anterior Eye* 31, 57–64.

Yamagata, T., Morishita, M., Kavimandan, N.J., Nakamura, K., Fukuoka, Y., Takayama, K. and Peppas, N.A. (2006) Characterization of insulin protection properties of complexation hydrogels in gastric and intestinal enzyme fluids. *Journal of Controlled Release* 112, 343–349.

Yang, W.-W. and Pierstorff, E. (2012) Reservoir-based polymer drug delivery systems. *Journal of Laboratory Automation* 17, 50–58.

Ying, L., Michael, S. and Kinam, P. (2014) Microparticles produced by the hydrogel template method for sustained drug delivery. *International Journal of Pharmaceutics* 258, 461–269.

Yu, L. and Ding, J. (2008) Injectable hydrogels as unique biomedical materials. *Chemical Society Review* 37, 1473–1481.

Yu, S.J. and Hu, X.P. (2006) Stable and pH sensitive nanogels prepared by self-assembly of chitosan and ovalbumin. *Langmuir* 22, 2754–2759.

4 Water-soluble Biodegradable Polymers for Drug Delivery

Bhanu Malhotra[1]**, Harsha Kharkwal**[2,*] **and Anuradha Srivastava**[3]

[1]*Amity Institute of Biotechnology and Amity Center for Carbohydrate Research, Amity University, Noida, India;* [2]*Amity Center for Carbohydrate Research and Amity Institute of Phytomedicine and Phytochemistry, Amity University Uttar Pradesh, Noida, India;* [3]*Biological Sciences and Geology, Queensborough Community College, Bayside, New York, USA*

Abstract

At the heart of polymer chemistry and biomedical applications lie water-soluble polymer drug conjugates for novel drug delivery systems. Designing multifunctional water-soluble polymer drug conjugates via copolymerization of bioactive compounds, and incorporating hydrophilic groups, makes them extremely water soluble and with improved biocompatibilities. Hydrophobic charged groups can be introduced into the polymers, which enable them to carry out specialized interactions and responses. Water-soluble polymer drug conjugates have the ability to store prodrugs (inactive drugs), facilitating the transfer of drugs passively or actively to the target site then activating them through cellular signalling cascades and bringing about the desired response. This chapter throws light on the advances made in natural and synthetic water-soluble polymer drug conjugates for various different biomedical applications.

Introduction

Advancing polymer science has led to the creation of novel drug delivery systems. The use of water-soluble polymers to hold and deliver the drugs at the target site, overcoming different barriers to site-specific drug delivery, has been challenging. In the beginning the polymers were used as solubilizers, mechanical supports and stabilizers for sustained drug release. Now, with deepening research in the field, the functionalities of polymers have been changed (Torchilin, 2006). Water-soluble polymers find a good range of applications in textiles, construction, paper, coatings and pharmaceuticals. Various natural and synthetic polymers which are biodegradable and water soluble have profound drug delivery impacts. Water-soluble polymers have various advantages in drug delivery:

- They improve the water solubility of various hydrophobic low molecular weight drugs.
- There is a concomitant improvement in drug bioavailability.
- They have long-lasting circulation times in the bloodstream.
- The can protect unstable drugs from degradation and deterioration.
- They can actively accumulate a drug at a tumour site by passively accumulating it by the enhanced permeability and retention (EPR) effect (exclusive phenomena of solid tumours owing to their pathophysiological and anatomical

*Corresponding author. E-mail: hkharkwal@amity.edu

differences with respect to normal tissues — discussed in detail in chapter 8).

- There is lower nonspecific toxicity of conjugated drugs.

Natural Water-soluble Polymers

Xanthan gum

Xanthan gum consists of repeating pentasaccharide units consisting of two D-glucopyranosyl molecules, two D-mannopyranosyl molecules and one molecule of D-glucopyranosyluronic unit (Sharma *et al.*, 2006). At low concentrations xanthan is a flowing, powder-like viscous solution in hot and cold water. It has considerable industrial importance in controlling the rheology of water-soluble polymeric drug delivery systems (Cairns *et al.*, 1986). In comparison to other polysaccharides, xanthan gives highly viscous solutions even at low concentrations, making it a stabilizer and a thickener (Lachke, 2004). This polymer also shows pseudo-plastic behaviour, enhancing the effect of mouth feel and flavour release (Badwaik *et al.*, 2013; Jackson *et al.*, 2015). Xanthan gum drug delivery solutions are not much affected by pH and are stable in both alkaline and acidic environments in the range of 1–13. Xanthan can be used with carboxymethyl cellulose, starch and sodium alginate (Shalviri *et al.*, 2010). Figure 4.1 shows the structure of xanthan gum.

Xanthan gum is used in cosmetic preparations and toothpaste formulation (Katzbauer, 1998). It shows good flow properties and can be extruded from the tube easily. The shear thinning properties enables it to improve the dispersion and flow characteristics of rinsing removal from the teeth. The gum typically feels very gentle and is used as a stabilizer in eye gels. It is also used to separate barium sulfate, an insoluble ingredient in X-ray contrast media, basically in emulsions and suspensions.

In a comparative study to assess xanthan gum performance with hydroxy propyl methyl cellulose (HPMC) as a hydrophilic matrix agent for *in vitro* drug release, it was found that xanthan gum showed enhanced overall compaction and flow behaviour for drug release as compared to HPMC, because of their difference in hydrophilicity and improved hydration properties (Talukdar *et al.*, 1996). In another study later with xanthan gum and HPMC tablets it was illustrated that they release the drug differently inside the body (Talukdar and Kinget, 1997). The diffusion characteristics of three drugs – indomethacin sodium, indomethacin and caffeine – were studied in the hydrated gels of xanthan gum and HPMC for comparison. The diffusivity of the drugs (that is, release of the drugs from the polymer matrix) was monitored using UV absorbance under identical experimental conditions. It was concluded that the drug released faster from the HPMC matrix as compared to the xanthan gum, showing that something from the xanthan gum hindered drug transport.

Highlighting the salt effects, it was found that diffusion of the drug from the hydrated xanthan gum matrices was due to the backing up of water release characteristics of all the water-soluble drugs from the xanthan gum tablet preparation. The main mechanism of drug release

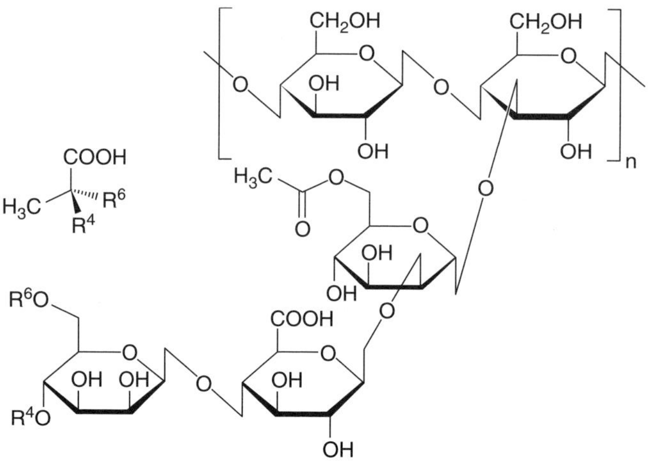

Fig. 4.1. Structure of xanthan gum.

from the xanthan gum tablets was diffusion of drugs from the hydrated gel of the xanthan gum matrix for sodium salt of indomethacin and caffeine. Therefore xanthan gum facilitated targeted controlled release of drugs over a stipulated period of time (Talukdar and Kinget, 1997).

In another controlled release system, starch and xanthan gum were cross-linked and gelatinized to explore their potentials. Sodium trimetaphosphate and xanthan gum were used to form cross-linked polymers with starch and were examined using Fourier transform infrared spectroscopy (FTIR) and nuclear magnetic resonance spectroscopy (NMR) analysis. Swelling properties and other rheological characteristics were examined using SEM. The swelling and mesh size increased from 2.84 to 6.74, and the polymeric film permeability of negatively charged drugs was lower due to electrostatic repulsion between the polymer matrix and the drug. Improved characteristics and results confirmed the use of cross-linked starch and xanthan gums for controlled drug delivery procedures (Shalviri *et al.*, 2010).

Pectins

The source of pectin is apple pomades and citrus peels obtained as a byproduct of juice manufacturing processes. Pectin is a mixture of polysaccharides mainly made up of D-galacturonic acid units linked via α-(1-4) glycosidic linkage. Only 10–15% pectin is released from apple pomade and about 20–30% from citrus peel (Levene and Kreider, 1937; Hirst and Jones, 1939; Liu *et al.*, 2005; Wu and Mort, 2014). Figure 4.2 represents: (a) a repeating segment of pectin with functional groups; (b) a carboxyl group; (c) an ester group; and (d) an amide group.

Carboxyl groups are present in uronic acids as methyl esters; when treated with ammonia they produce carboxamide groups.

Pectins are hydrophilic water-soluble polymers in nature. Monovalent cations of pectic acids and pectinic are water soluble while di- and tri-valent cations show partially weak soluble characteristics. In its dry form, pectin powder forms clumps when mixed with water. This can be prevented by using improved pectin dispersibility (Whistler and BeMiller, 1993). In pectin, the factors increasing the gel strength also increase the gelling ability and viscosity, and decrease the solubility, and vice versa. The properties of pectins are attributed to their structure, which is of a linear polycarboxylate polyanion. Monovalent salts of pectins are ionisable in solutions and charge distribution along the molecule in extended form because of coulombic repulsion (Fishman and Jen, 1986). Pectin, being

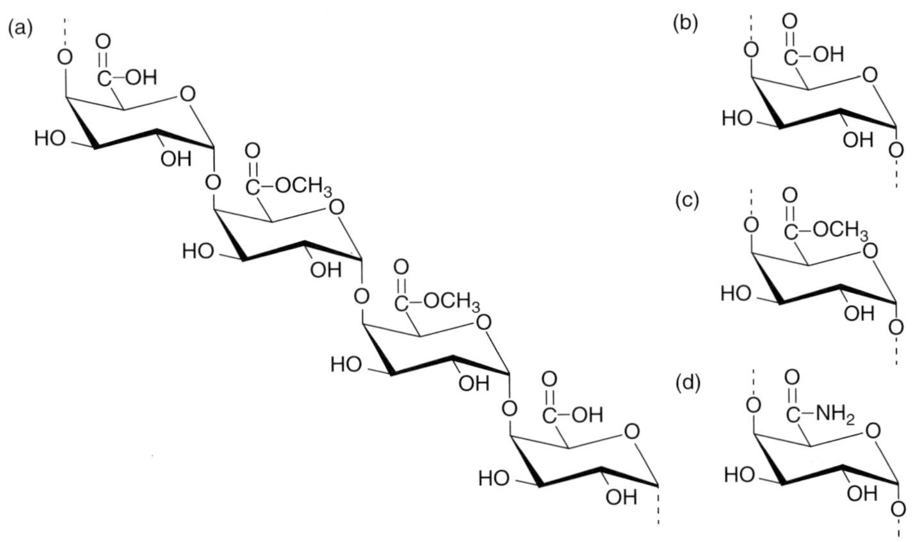

Fig. 4.2. (a) Repeating segment of pectin with functional groups; (b) carboxyl group; (c) ester group; (d) amide group.

a water-soluble polymer, finds a good range of applications in the pharmaceutical industries (Fishman and Jen, 1986). Standardized pectin is used as a binding agent in tablets. With HPMC, methoxy pectin is utilized in the production of compressible tablets in monolithic bioerodible systems. In calcium pectinate gel beads, methoxy pectin can be formulated to develop sustained drug release tablets by an ionotropic gelation technique. Microspheres of pectin have been prepared by emulsification (Bhatia, 2008).

Specialized coated film-forming tablets can be made by using either hydroxyl methyl or Low Methoxy (LM) pectin along with chitosan, or by using High Methoxy (HM) pectin and ethyl cellulose in aqueous dispersion. Pectin can deliver enormous varieties of drugs, cells and proteins as a matrix system by entrapping them. Cholesterol levels can also be reduced by 6 g/day pectin consumption (Ginter, 1979; Calce *et al.*, 2014). It can be used to thicken and stabilize beverages and food products. It is used to remove mercury and lead from the gastrointestinal and respiratory tract (Kohn, 1982). If pectin is administered intravenously it reduces the coagulation time of the blood, thereby controlling localized bleeding and haemorrhage. Hydrogels of pectins find their use as binders in formulating tablets (Slany, 1981a; Slany, 1981b).

It was found that pectin and its derivatives containing amine groups primarily were used as mucoadhesive in mucosal and nasal drug delivery. The derivatives of pectin sustain the release of the incorporated fragrances over prolonged durations. Pectin is also used in aromatherapy formulations as its less esterified derivatives penetrate deeper into the skin than normal pectin and less esterified derivatives. Corn zein protein and pectin are used to form hydrogel beads. Pectin shields the zein from proteases, while zein retains the bead porosity. This type of complex is used in the treatment of colonic diseases (Liu *et al.*, 2006).

Chitosan derivatives

Chitin is a ubiquitous, white, abundant (second after cellulose) natural polysaccharide made up of β-1, 4 linked N-acetylglucosamine residues. In contrast to cellulose it possesses acetamide groups and is found in the internal crusts of invertebrates and exoskeletons. Chitosan comprises α-1, 4 linked 2-amino, 2-deoxy β-D- glucopyranose residues obtained from the deacetylation of chitin to variable degrees, consequently becoming a copolymer of N-acetylglucosamine and glucosamine. Figures 4.3 and 4.4 illustrate their structures (Sakurai, 1990; Jayakumar, 2013; Younes and Rinaudo, 2015; Zargar *et al.*, 2015).

Chitin and chitosan possess enormous applications in medicine, pharmaceutical science, waste management and food processing. Because of its biocompatibility and biodegradability, chitosan is generally used as an excipient because it is non-toxic for drug delivery. It can be used as a mucoadhesive agent in oral absorption enhancers in protein and gene delivery procedures (Köping-Höggård *et al.*, 2004; Foster *et al.*, 2015). Researchers have synthesized soluble derivatives of chitosan, which have improved properties through chemical modifications to deliver hydrophilic chitin; this shows enhanced dissolving power in organic solvents and in aqueous solutions (Sugimoto *et al.*, 1998). Derivatization of chitosan forms N- carboxymethyl chitosan (N-CMC) which can easily be solubilized in wide ranges of pH (An *et al.*, 2009). The N-CMC derivative of

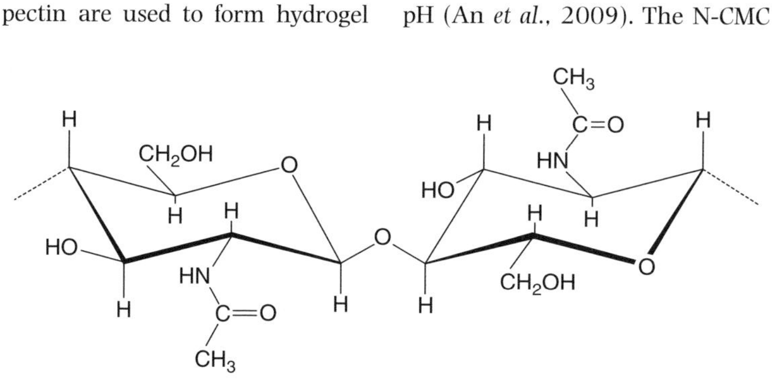

Fig. 4.3. Structure of chitin.

chitosan is used for controlled drug release in connective tissues and orthopaedic devices (Chen and Park, 2003; Abreu and Campana-Filho, 2005; Ge and Luo, 2005; Sun and Wang, 2006; Sun *et al.*, 2006). The detailed outline of chitosan and their derivatives in therapeutic drug delivery is enlightened in Chapter 11 of this book.

Hyaluronic acid

Hyaluronic acid (HA) is a heteropolysaccharide, occurring naturally in all living organisms. It is usually made up of several thousand carbohydrates, and in the unbound state it is water soluble making it viscous (rather like jelly). It is a linear polyanion having the structure [(1→3)-β-DGlcNAc-(1→4)-β-D-GlcA-] and is a component of the pericellular and extracellular matrix, also occurring intracellularly (Fig. 4.5). HA serves the important biological function of maintaining elasticity and viscosity of the liquid connective tissues, mainly in the synovial joints, and also in the vitreous eye fluid. It is also involved in the assembling of proteoglycans, and in tissue hydration. It has profound effects on receptor-mediated cell detachment and on processes of mitosis, migration and tumorigenesis, having effects on metastasis and inflammation in the body (Balazs *et al.*, 1986; Toole *et al.*, 2002; Turley *et al.*, 2002; Hascall *et al.*, 2004).

HA binds to water and lubricates the movable parts of joints and muscles. It is an excellent moisturizer and is used extensively in skin care products (Shokri *et al.*, 2014). It is one of the most hydrophilic polymers used as a natural moisturizer (Worcester, 2008). HA also has good biocompatibilities and is completely non-immunogenic in a variety of clinical applications due to its viscoelastic nature. It has a profound role in the treatment of arthritis by supplementing the joint fluid (Neo *et al.*, 1997; Barbucci *et al.*, 2002; Uthman *et al.*, 2003; Medina *et al.*, 2006). It is a powerful aid in eye surgery, facilitating the healing and regeneration of surgical wounds inside the eye. HA has been investigated for therapeutic drug administration routes, including pulmonary, ophthalmic, parenteral, nasal, and topical applications (Brown and Jones, 2005).

It is approved by the Food and Drug Administration (FDA) regulations for use in therapeutic drug delivery. The overexpression of CD-44 receptor ligand for HA makes it a pioneer for drug targeting (Kong *et al.*, 2003). The pharmacokinetics of HA are very efficient for circulation, with 2–6 min of half-life and a turnover of about 10–100 mg per day in an adult. Uptake of HA from the blood occurs in the liver endothelial cells. Renal elimination of HA is limited by alterations in blood flow in the eliminating organ of HA, and it competes with proteoglycans, chondroitin sulfates and other macromolecules

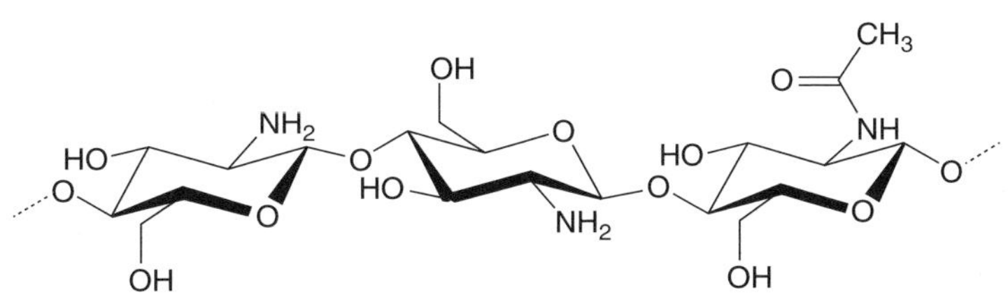

Fig. 4.4. Structure of chitosan.

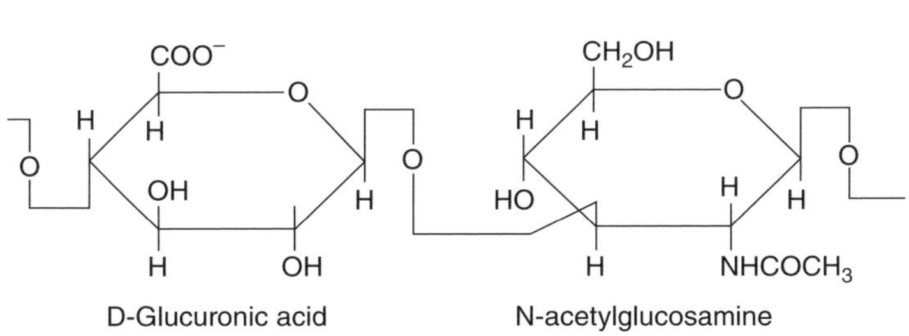

Fig. 4.5. Hyaluronic acid structure.

for their excretion. These factors are strongly influenced during diseased conditions in the body and play a very efficient role in maintaining normal and pathological levels of HA in the body.

The attributes of HA make it physicochemically and pharmacologically appealing. CD 44, a popular cell surface marker, is present on the surface of chondrocytes. This is basically a cell surface receptor for HA, involved in various biochemical interactions (Akmal *et al.*, 2005). HA has a profound role in cancer dynamics because its chondroprotective effects stimulate the production of tissue inhibitors of matrix metalloproteineses, specifically TIMP-1. It not only causes inhibition of neutrophil-mediated cartilage degradation but also provides subsequent attenuation of IL-1 induced matrix degeneration *in vitro* (Gerwin *et al.*, 2006).

Cultured articular chondrocytes with HA have shown greater rates of DNA proliferation and extracellular matrix production in comparison to control chondrocytes cultured without HA (Akmal *et al.*, 2005).

HA has an important biological function in regeneration of tendons, maintenance of synovial fluids and development of the cartilages (Toole, 1997, 2001). It is found in enormous concentrations, especially in the adult joint tissues, including in the outer layer of cartilage and in the sinovial fluid (Leach and Schmidt, 2004). It is a shock absorber and a mucoadhesive lubricant due to its viscoelasticity and ability to form highly hydrated matrices. Clinically, viscosupplementation first came into Japan and Italy in 1987, followed by Canada in 1992, Europe in 1995 and the USA in 1997. Synthetic hylan G-F 20- Synvisc and naturally found hyaluronan- Hyalgan are supplied in the USA. Hylans are elastoviscous, providing longer persistence in the joint space when compared to HA because of their low molecular weights (Wen, 2000). HA is found in high concentrations in the skin and soft connective tissues naturally; it is utilized as a matrix to support material for dermal regeneration in wound healing applications. HA has also been used as an important agent in cosmetic applications of soft tissue augmentation (Leach and Schmidt, 2004; Dechert *et al.*, 2006).

Conclusion

Since they are biodegradable, biocompatibile, non-immunogenic, non-toxic and easily chemically modified, natural water-soluble polymers are a boon in drug delivery at various locations in the body. A variety of commercially available natural polymers and their derivatives with tailor-made properties are used to deliver non-toxic insoluble drugs at different locations in the body, utilizing their flexibilities. These polymers can deliver site-specific, controlled and sustained drug deliveries, and form films, coatings, matrices, fibres, hydrogels and microspheres with multidisciplinary applications in the pharmaceutical industry.

Acknowledgements

We thank Dr Ashok K Chauhan, Founder President, Ritanand Balved Educational Foundation for support. Guidance from Shri. Atul Chauhan, Chancellor, Amity University Uttar Pradesh and Prof Dr (Mrs) Balvinder Shukla, Vice Chancellor, Amity University Uttar Pradesh is greatly appreciated. Insights from articles cited in this chapter have been of considerable assistance, and we are grateful to their authors.

References

Abreu, F. and Campana-Filho, S. (2005) Preparation and characterization of carboxymethylchitosan. *Polímeros* 15(2), 79–83.

Akmal, M., Singh, A., Anand, A., Kesani, A., Aslam, N., Goodship, A. and Bentley, G. (2005) The effects of hyaluronic acid on articular chondrocytes. *Journal of Bone and Joint Surgery* 8, 1143–1149.

An, N., Thien, D., Dong, N. and Dung, P. (2009) Water-soluble N-carboxymethylchitosan derivatives: preparation, characteristics and its application. *Carbohydrate Polymers* 75(3), 489–497.

Badwaik, H., Giri, T., Nakhate, K., Kashyap, P. and Tripathi, D. (2013) Xanthan gum and its derivatives as a potential bio-polymeric carrier for drug delivery system. *Current Drug Delivery* 10(5), 587–600.

Balazs, E.A., Laurent, T.C. and Jeanloz, R.W. (1986) Nomenclature of hyaluronic acid. *Biochemical Journal* 235, 903.

Barbucci, R., Lamponi, S., Borzacchiello, A., Ambrosio, L., Fini, M., Torricelli, P. and Giardino, R. (2002) Hyaluronic acid hydrogel in the treatment of osteoarthritis. *Biomaterials* 23, 4503–4513.

Bhatia, M. (2008) Chemical modification of pectins, characterization and evaluation for drug delivery. *Scientia Pharmaceutica* 76(4), 775–784.

Brown, M.B. and Jones, S.A. (2005) Hyaluronic acid: a unique topical vehicle for the localized delivery of drugs to the skin. *Journal of European Academy of Dermatology and Venereology* 19, 308–318.

Cairns, P., Miles, M. and Morris, V. (1986) Intermolecular binding of xanthan gum and carob gum. *Nature* 322(6074), 89–90.

Calce, E., Mignogna, E., Bugatti, V., Galdiero, M., Vittoria, V. and De Luca, S. (2014) Pectin functionalized with natural fatty acids as antimicrobial agent. *International Journal of Biological Macromolecules* 68, 28–32.

Chen, X. and Park, H. (2003) Chemical characteristics of O-carboxymethyl chitosans related to the preparation conditions. *Carbohydrate Polymers* 53(4), 355–359.

Dechert, T.A., Ducale, A.E., Ward, S.I. and Yager, D.R. (2006) Hyaluronan in human acute and chronic dermal wounds. *Wound Repair and Regeneration* 14, 252–258.

Fishman, M.L. and Jen, J.J. (1986) *Chemistry and Function of Pectins*. American Chemical Society, Washington, D.C.

Foster, L., Ho, S., Hook, J., Basuki, M. and Marçal, H. (2015) Chitosan as a biomaterial: influence of degree of deacetylation on its physiochemical, material and biological properties. PLOS One 10(8), e0135153.

Ge, H. and Luo, D. (2005) Preparation of carboxymethyl chitosan in aqueous solution under microwave irradiation. *Carbohydrate Research* 340(7), 1351–1356.

Gerwin, N., Hops, C. and Lucke, A. (2006) Intraarticular drug delivery in osteoarthritis. *Advanced Drug Delivery Reviews* 58, 226–242.

Ginter, E. (1979) Natural hypocholesterolemic agent: pectin plus ascorbic acid. *International Journal of Viticulture and Natural Resource* 49, 406–408.

Hascall, V.C., Majors, A.K., de la Motte, C.A., Evanko, S.P., Wang, A., Drazba, J.A., Strong, S.A. and Wight, T.N. (2004) Intracellular hyaluronan: a new frontier for inflammation? *Biochimica and Biophysica Acta* 1673, 3–12.

Hirst, E. and Jones, J. (1939) 103 Pectic substances. Part III. Composition of apple pectin and the molecular structure of the araban component of apple pectin. *Journal of the Chemical Society* (Resumed), 454–460.

Jackson, C., Emeje, M. and Ofoefule, S. (2015) Use of xanthan gum in the formulation of diethylcarbamazine for targeted drug delivery. *British Journal of Pharmaceutical Research* 5(4), 280–285.

Jayakumar, R. (2013) Welcome to the Journal of Chitin and Chitosan Science. *Journal of Chitin and Chitosan Science* 1(1), 1–1.

Katzbauer, B. (1998) Properties and applications of xanthan gum. *Polymer Degradation and Stability* 59(1–3), 81–84.

Kohn, R. (1982) Binding of toxic cations to pectin, its oligomeric fragments and plant tissues. *Carbohydrate Polymers* 2(4), 273–275.

Kong, Q., Liu, J., Chen, X., Wang, X., Sun, Y. and Li, H. (2003) Differential expression patterns of hyaluronan receptors CD44 and RHAMM in transitional cell carcinomas of urinary bladder. *Oncology Reports* 10(1), 51–55.

Köping-Höggård, M., Vårum, K., Issa, M., Danielsen, S., Christensen, B., Stokke, B. and Artursson, P. (2004) Improved chitosan-mediated gene delivery based on easily dissociated chitosan polyplexes of highly defined chitosan oligomers. *Gene Therapy* 11(19), 1441–1452.

Lachke, A. (2004) Xanthan – a versatile gum. *Resonance* 9(10), 25–33.

Leach, J.B. and Schmidt, C.E. (2004) Hyaluronan. *Encyclopedia of Biomaterials and Biomedical Engineering*. Marcel Dekker, New York, pp. 779–789.

Levene, P. and Kreider, L. (1937) On the structure of pectin polygalacturonic acid. *Science* 85(2217), 610–610.

Liu, L., Chen, G., Fishman, M. and Hicks, K. (2005) Pectin gel vehicles for controlled fragrance delivery. *Drug Delivery* 12(3), 149–157.

Liu, L., Fishman, M. and Hicks, K. (2006) Pectin in controlled drug delivery – a review. *Cellulose* 14(1), 15–24.

Medina, J.M., Thomas, A. and Denegar, C.R. (2006) Knee osteoarthritis: should your patient opt for hyaluronic acid injection? *Journal of Family Practice* 8, 667–675.

Neo, H., Ishimaru, J.I., Kurita, K. and Goss, A.N. (1997) The effect of hyaluronic acid on experimental temporomandibular joint osteoarthrosis in the sheep. *Journal of Oral Maxillofacial Surgery* 55, 1114–1119.

Sakurai, K. (1990) Structure of chitin and chitosan. *Sen'i Gakkaishi* 46(12), P553–P557.

Shalviri, A., Liu, Q., Abdekhodaie, M. and Wu, X. (2010) Novel modified starch–xanthan gum hydrogels for controlled drug delivery: synthesis and characterization. *Carbohydrate Polymers* 79(4), 898–907.

Sharma, B.R., Naresh, L., Dhuldhoya, N.C., Merchant, S.U. and Merchant, U.C. (2006) Xanthan gum – a boon to food industry. *Food Promotion Chronicle*. 1, 27–30.

Shokri, N., Akbari Javar, H. and Ghadermazi, R. (2014) Effects of skin penetration enhancers in topical antiaging products containing α-hydroxy acids and hyaluronic acid. *Avicenna Journal of Medical Biochemistry* 2(2), 1–7.

Slany, J. (1981a) Evaluation of tablets with pectin as a binding agent. *Farmaceuticky Obzor* 50, 491–498.

Slany, J. (1981b) Study of functional action of citrus pectins in tablets. *Ceska a Slovenska Farmacie* 30, 195–200.

Sugimoto, M., Morimoto, M., Sashiwa, H., Saimoto, H. and Shigemasa, Y. (1998) Preparation and characterization of water-soluble chitin and chitosan derivatives. *Carbohydrate Polymers* 36(1), 49–59.

Sun, S. and Wang, A. (2006) Adsorption properties and mechanism of cross-linked carboxymethyl-chitosan resin with Zn(II) as template ion. *Reactive and Functional Polymers* 66(8), 819–826.

Sun, S., Wang, L. and Wang, A. (2006) Adsorption properties of crosslinked carboxymethyl-chitosan resin with Pb(II) as template ions. *Journal of Hazardous Materials* 136(3), 930–937.

Talukdar, M. and Kinget, R. (1997) Comparative study on xanthan gum and hydroxypropylmethyl cellulose as matrices for controlled-release drug delivery. II. Drug diffusion in hydrated matrices. *International Journal of Pharmaceutics* 151(1), 99–107.

Talukdar, M., Michoel, A., Rombaut, P. and Kinget, R. (1996) Comparative study on xanthan gum and hydroxypropylmethyl cellulose as matrices for controlled-release drug delivery I. Compaction and in vitro drug release behaviour. *International Journal of Pharmaceutics* 129(1–2), 233–241.

Toole, B.P. (1997) Hyaluronan in morphogenesis. *Journal of Internal Medicine* 242, 35–40.

Toole, B.P. (2001) Hyaluronan in morphogenesis. *Cell and Developmental Biology* 12, 79–87.

Toole, B.P., Wight, T.N. and Tammi, M.I. (2002) Hyaluronancell interactions in cancer and vascular disease. *Journal of Biological Chemistry* 277, 4593–4596.

Torchilin, V. (2006) Recent approaches to intracellular delivery of drugs and DNA and organelle targeting. *Annual Review of Biomedical Engineering* 8(1), 343–375.

Turley, E.A., Noble, P.W. and Bourguignon L.Y. (2002) Signaling properties of hyaluronan receptors. *Journal of Biological Chemistry* 277, 4589–4592.

Uthman, I., Raynauld, J.P. and Haraoui, B. (2003) Intra-articular therapy in osteoarthritis. *Postgraduate Medicine Journal* 79, 449–453.

Wen, D.Y. (2000) Intra-articular hyaluronic acid injections for knee osteoarthrosis. *American Family Physician* 62, 565–570.

Whistler, R.L. and BeMiller, J.N. (1993) *Industrial Gums: Polysaccharides and Their Derivatives*, Third Edition. Academic Press, Inc, San Diego, pp. 145–181.

Worcester, S. (2008) Daily application of azelaic acid plus moisturizer soothes Rosacea. *Family Practice News* 38(16), 21.

Wu, X. and Mort, A. (2014) Structure of a rhamnogalacturonan fragment from apple pectin: implications for pectin architecture. *International Journal of Carbohydrate Chemistry* 2014, 1–6.

Younes, I. and Rinaudo, M. (2015) Chitin and chitosan preparation from marine sources. Structure, properties and applications. *Marine Drugs* 13(3), 1133–1174.

Zargar, V., Asghari, M. and Dashti, A. (2015). A review on chitin and chitosan polymers: structure, chemistry, solubility, derivatives, and applications. *ChemBioEng Reviews* 2(3), 204–226.

5 Polysaccharide-based Drug Carriers

Srinivas Janaswamy*
Department of Dairy and Food Science, South Dakota State University, South Dakota, USA

Abstract

Many challenges arise during the development of new drug carrier systems and paramount among them are safety, solubility and controlled release requirements. Although synthetic polymers are effective, the possibility of side effects imposes restrictions on their acceptable use and dose limits. Thus, there is a clear need for a new drug carrier system that is safe to handle and free from side effects, and in this regard food-grade polysaccharides stand tall as worthy alternatives. Organized polysaccharide networks in particular and the available water pockets are effective in encapsulating and protecting the drug molecules as well as releasing them in a sustained manner. Overall, human compatible carbohydrate polymers possessing stable architectures will indeed cause a paradigm shift in the design of effective drug delivery systems.

Introduction

Drug discovery and development involve highly challenging, laborious and expensive protocols. There are numerous new drug molecules arising from these high throughput-screening processes. However, the majority fail to become potential drugs due to poor pharmacokinetics. A formulation scientist continuously aims to optimize the delivery with defined dose levels, chosen rate, selected time intervals and targeted sites. In this regard, drug carriers are critical for developing effective pharmaceutical products, since their main functionality is to overcome the natural barriers presented by the human body to the assimilation of active compounds. The carrier must meet certain important requirements such as safety, solubility and stability, and – most vital – the delivery profile (Langer, 1990). Synthetic and biodegradable polymers serve this purpose to a great extent but the possibility of side effects, toxicity and associated higher costs impose restrictions on their acceptable dosage formulations (Langer and Tirrell, 2004). There is thus a great need for a new carrier system that is safe to handle, less expensive and with virtually no side effects. There are several reports with biomacromolecules as possible carriers (Takakura and Hashida, 1996; Goldberg and Gomez-Orellana, 2003; Juliano, 2007) but many of them do not possess a stable molecular structure and organized network. Carriers with stable architecture generally have advantages over those with non-rigid structures for retaining the active structural state of the encapsulated molecules and delivering effectively at the intended site. Exploitation of other biopolymer systems, preferably GRAS (Generally Recognized As Safe) materials

*E-mail: janaswam@purdue.edu

that are abundant, low-in-cost and compatible with human digestibility (as well being able to maintain the ordered molecular and packing structures) would be the best scientific and industrial approach to circumvent this quandary (Fig. 5.1); and polysaccharides stand out as favourable choices (Smelcerovic *et al.*, 2008). Polysaccharides are ubiquitous biopolymers that are used extensively in food and pharmaceutical applications as thickeners, viscosifiers and gelling agents (Stephen, 1995; Rinaudo *et al.*, 2004).

Strategy

The proposed research is based on our hypothesis that drug molecules (DMs) could be effectively embedded into the ordered networks of crystalline polysaccharide matrices, protected from external stressors (e.g. heat, moisture and pH, to name a few) and released in a controlled manner. This hypothesis is based on the following three observations: (i) although polysaccharides are mostly amorphous, ordered structures could be accomplished by preparing crystalline and well-oriented fibres under suitable experimental conditions (Chandrasekaran, 1997); (ii) in the crystalline state, polysaccharides adopt well-organized helical structures coupled with well-orchestrated networks stabilized by intra- and inter-helical hydrogen bonds. These interactions are further stabilized via ordered water molecules and cations, in some cases (Rees, 1981); and (iii) in the polysaccharide networks, especially in anionic systems, there are 15–20 Å-wide voids often filled with water molecules (Janaswamy and Chandrasekaran, 2005). Such water pockets, along with well-ordered arrangements of polysaccharide helices, are amenable for embedding DMs. Thus, the main thrust of this research is to utilize ordered polysaccharide fibre matrices for encapsulating and delivering DMs (Fig. 5.2). Natural systems such as cellulose, chitin and starch possess prearranged and steady networks during biosynthesis. On the other hand, several GRAS polysaccharides that are utilized in food and pharmaceutical applications do not possess organized networks; however, under suitable experimental conditions extended networks composed of sturdy molecular and packing structures could be prepared. For example, iota-carrageenan, gellan and locust bean

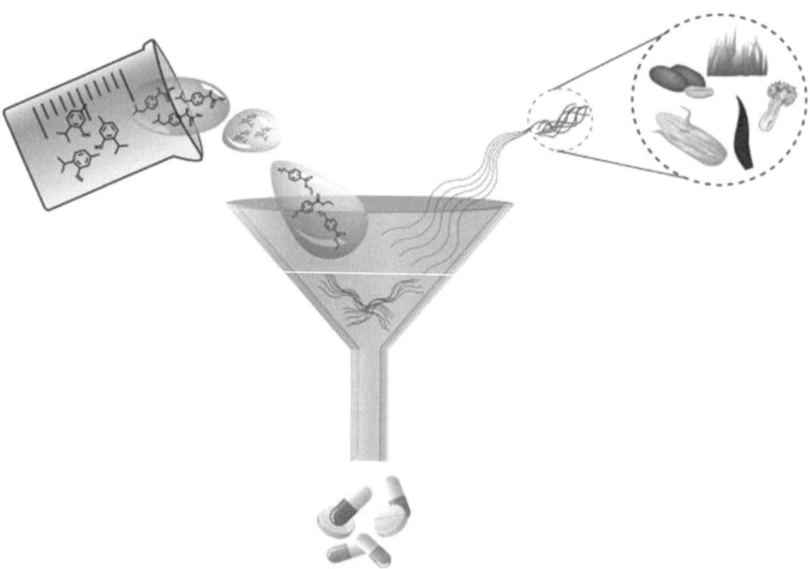

Fig. 5.1. Cartoon highlighting the research idea of encapsulating drug molecules in the oriented polysaccharide fibres that possess network structures. The outcome provides an elegant and cost-effective approach for developing drug carriers based on inexpensive, non-toxic and GRAS (Generally Recognized As Safe) polysaccharide excipients.

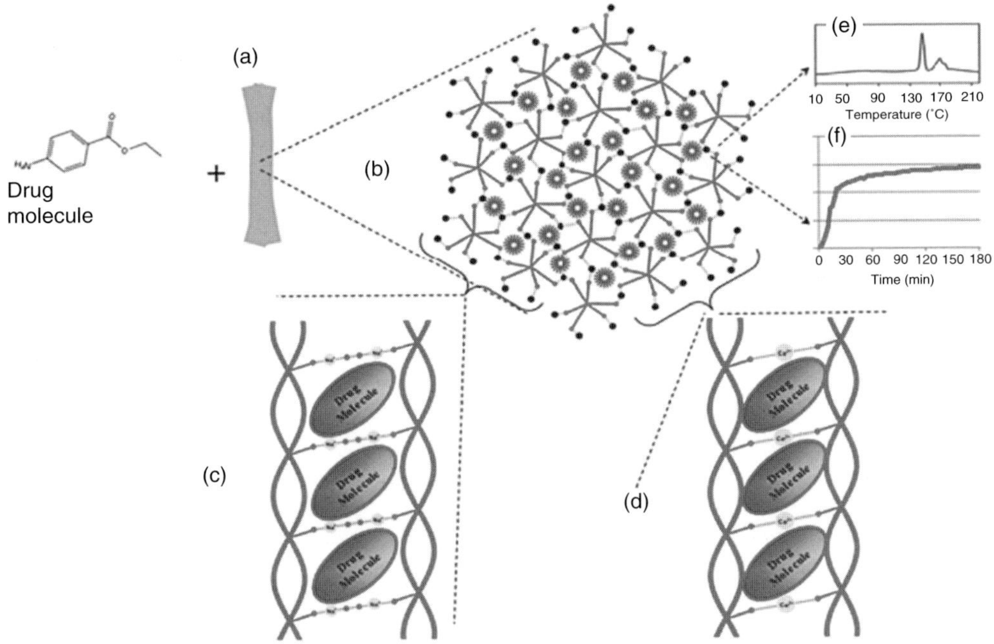

Fig. 5.2. Schematic encapsulation of drugs in the polysaccharide fibre: (a) The drug molecule is encapsulated in a well-oriented and crystalline polysaccharide fibre. (b) Cross-section of complex assembly containing ordered drugs (small star) in the polysaccharide, e.g. iota-carrageenan network, viewed down the helix-axis. The larger stars represent carrageenan double helix and small black circles correspond to cations. (c) Drugs are securely trapped between a pair of helices, viewed normal to the helix-axis, and gain the required protection. (d) Compared to sodium ions, adjacent helices are drawn closer in the presence of calcium ions, resulting in much stronger interactions. The encapsulated drugs (e) are thermally protected and (f) released in a controlled manner by/from the network.

gum fibres, to name a few, could yield highly ordered and robust molecular and packing structures (Chandrasekaran, 1997; Janaswamy and Chandrasekaran, 2001, 2002, 2006, 2008). Overall, the significant outcome of this research lies in providing a scientifically elegant and nature-based approach for the design and development of innovative drug delivery systems.

Iota-carrageenan–Ibuprofen/ Griseofulvin Complexes

Iota-carrageenan (IC) is an important member of the carrageenan family of 15 sulfated polysaccharides extracted from marine algae (Stortz and Cerezo, 2000). Carrageenans have long been used in food and pharmaceutical applications as thickeners, viscosifiers, gelling agents and stabilizers. The chemical structure of IC is composed of a disaccharide repeat of →3)-β-D-Gal*p*-4-SO_3^-(1→4)-3,6-anhydro-α-D-Gal*p*-2-SO_3^-(1→ where Gal*p* is galactopyranose. X-ray structural analysis of sodium/calcium salt forms of IC fibres suggests that it has a rigid core structure with flexible peripheral sulfate groups (Janaswamy and Chandrasekaran, 2001, 2002, 2006, 2008). The IC network is quite stable, with well-organized cation-mediated inter-helical interactions. Interestingly, there are pockets of free space of about 15–20 Å between the double helices that could be exploited for entrapping DMs (Fig. 5.2) through van der Waals, ionic and hydrogen bonding interactions.

Complex confirmation

The first example is about the encapsulation of ibuprofen and griseofulvin in the sodium IC fibres (Janaswamy *et al.*, 2013). The encapsulation experiments were performed at room temperature by dissolving the DMs (~0.1%) in isopropyl alcohol. X-ray fibre diffraction is the only available methodology to structurally characterize the polysaccharide fibres and hence the complexes have been subjected to this analysis. Figure 5.3 compares the diffraction patterns of IC and its complexes. In general, in any fibre diffraction pattern the first reflection on the meridian – an imaginary line perpendicular to the reflection layers that passes through the centre in a north–south direction – suggests the helix fold. In the case of IC the meridional reflection is seen on the 3rd layer line, suggesting a threefold helical arrangement. In the complexes the meridional reflection is also on the 3rd layer line, signifying an intact IC helical structure. This observation is indicative of the fact that DMs enter the voids in the carrageenan lattice and are held through van der Waals and hydrogen bonding interactions with little to no influence on the carrageenan molecular structure. However, there are significant variations in the intensity distribution and positions of the reflections on the individual layer lines, portraying alterations in the packing arrangement upon encapsulation. The calculated unit cell dimensions indicate that the *c* values (that is, the layer line spacing) are relatively unchanged, meaning that the molecular structure of iota-carrageenan (IC) is rigid but the variable basal net dimensions of 53.25, 24.57 and 21.32 Å for IC, IC:ibuprofen and IC:griseofulvin, respectively, imply the packing alterations. Overall, the results highlight the remarkable combination of rigidity and flexibility of the IC network towards encapsulating the DMs. While the rigidity of the IC molecular structure confers a highly structured configuration, its flexibility (due to peripheral sulfate groups) illustrates the ability of the IC network to adjust the cavity size to the dimensions of DMs, leading to complex formation.

Thermal stability upon encapsulation

The exothermic peak (180°C) in the IC fibres is due to crystallization of the amorphous phase and the endothermic event (187°C) corresponds to the melting of junction zones (Fig. 5.4). The ibuprofen melts at around 75°C; however, upon encapsulation, melting of neither ibuprofen nor IC is observed. Instead, crystallization at 170°C is noticed, around 10°C less than the IC. Thus, the ibuprofen molecules present in the complex fibre are responsible for the crystallization of the amorphous phase at a lower temperature. Absence of ibuprofen melting in the complex is mainly due to lack of inter-molecule interactions between the ibuprofen molecules, as the neighbours in the newly formed lattice are separated by the IC helices. The griseofulvin, which melts at around 220°C, displays the glass transition temperature change by about 8°C after encapsulating in IC. The crystallization and melting temperatures of the complex are 180°C and 187°C respectively. Overall, in the case of low melting-point DMs the IC network protects them from heat, and most probably from other stresses too, and in the case of higher melting-point DMs the solubility increases in addition to the required protection.

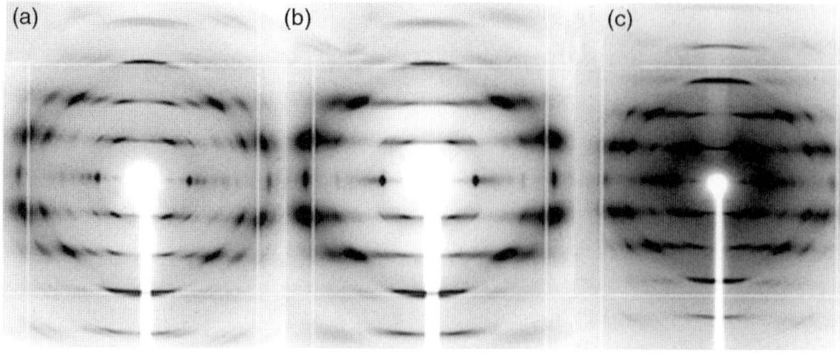

Fig. 5.3. X-ray fibre diffraction patterns of (a) iota-carrageenan and its complexes with (b) ibuprofen and (c) griseofulvin.

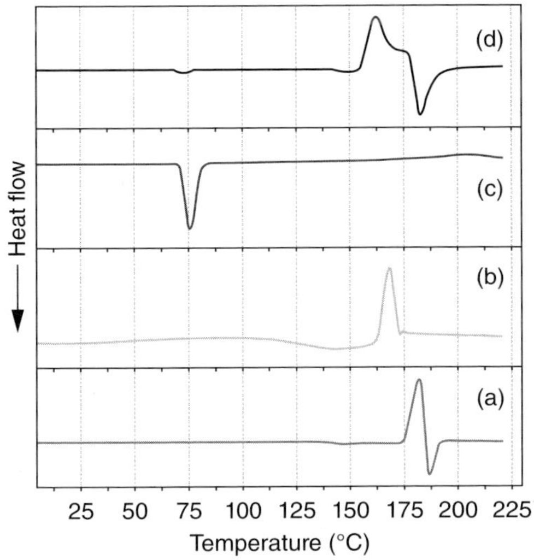

Fig. 5.4. Melting profiles of (a) iota-carrageenan fibre; (b) iota-carrageenan:ibuprofen complex; (c) ibuprofen powder; and (d) a physical mixture of iota-carrageenan fibres and ibuprofen powder.

Control release of encapsulated molecules

The griseofulvin release amount versus time is shown in Fig. 5.5. The complex displays faster release within the first 30 min and after 2 h a plateau is reached. It appears that around 1.8 μg of griseofulvin is released per 1 mg of IC. It is believed that different DMs will have variable loading and release profiles.

Iota-carrageenan–Curcumin Complexes

Complex formation and curcumin release

Experiments have been undertaken at 4°C, 25°C, 35°C and 45°C (Janaswamy and Youngren, 2012). The IC and 4°C and 25°C complexes yield a trigonal unit cell with dimensions $a = b =$ 24.1(2), c = 13.1(1) Å; $a = b$ = 24.8(1), c = 13.0(1) Å and $a = b$ = 24.5(1), c = 13.2(1) Å, respectively. On the other hand, 35°C and 45°C samples correspond to an orthorhombic net with dimensions a = 21.2(1), b = 26.8(1), c =13.1(1) Å; and a = 21.1(1), b = 27.2(2),

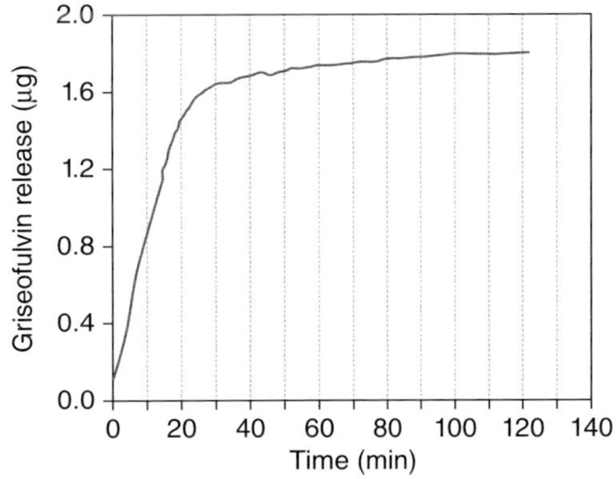

Fig. 5.5. The nature of release of griseofulvin from the iota-carrageenan:griseofulvin complex.

c = 13.1(1) Å, respectively. The changes in the basal net dimensions unequivocally suggest a complex formation. The encapsulated curcumin (Fig. 5.6) releases very fast, within the first 20 min, and slowly later to reach saturation at around 3 h. The complexation temperature has an effect on the total loading amounts. The 25°C complex has a maximum of 0.15 μg followed by 0.9, 0.12 and 0.9 μg/mg for 4°C, 35°C and 45°C, respectively. At 25°C the trigonal unit cell accommodates three IC helices laterally spaced at 14.1 Å apart. A more compact arrangement of four double helices separated by 12.6 Å occurs upon the orthorhombic lattice transformation at 45°C. Thus, the reduction in the inter-helical space leads to less encapsulation of curcumin. Although the inter-helical spacing in the 4°C complex is the highest at 14.3 Å, slower kinetics or insufficient experiment time could have been the cause of its inability to hold sufficient curcumin.

Thermal protection of curcumin from the carrageenan network

Curcumin melts around 175°C (Fig. 5.7) while IC fibres display crystallization at 125°C followed by melting at 140°C and 160°C. On the other hand, thermal behaviour of the 4°C complex is very interesting: upon encapsulation melting from neither curcumin nor IC is noticed, instead crystallization around 145°C and 170°C has been observed. As seen from the diffraction patterns,

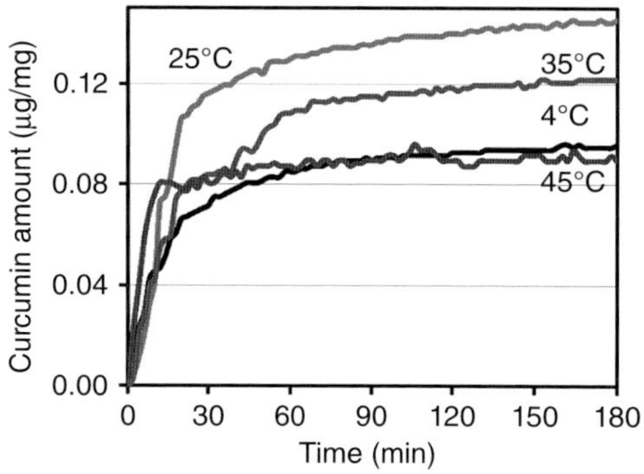

Fig. 5.6. The amount of curcumin released from the iota-carrageenan:curcumin complex.

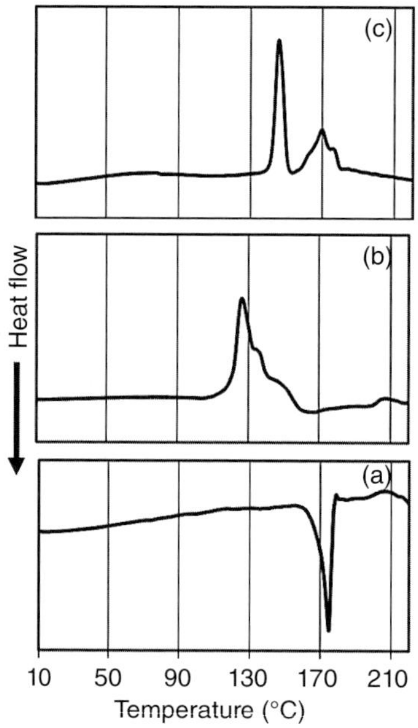

Fig. 5.7. DSC profiles (exothermic up) of (a) curcumin powder, (b) iota-carrageenan fibers, and (c) iota-carrageenan:curcumin complex prepared at 4°C.

the complex prepared at 4°C is comparatively less crystalline, suggesting the presence of a more amorphous network. It appears that most of the amorphous carrageenan chains crystallize at higher temperatures, but the important point to note is that the presence of curcumin molecules pushes the crystallization of carrageenan helices to higher temperatures. The absence of curcumin melting suggests that the IC network is able to provide the thermal protection. Thus, it is believed that the active structural state of curcumin will be preserved for longer durations and the bioavailability will be enhanced in actual biological applications.

Iota-carrageenan–Eugenol Complexes

The research strategy was to encapsulate eugenol molecules in the organized networks of monovalent (Na^+) and divalent (Ca^{2+}) salt forms of IC to understand the influence of cations on the loading amounts and release profiles (Polowsky and Janaswamy, 2015). The diffraction patterns and the melting profiles (not shown) clearly indicate the encapsulation and thermal protection of eugenol in and from the IC network. Figure 5.8 depicts the time-dependent release profiles of eugenol from Na and Ca IC fibre complexes prepared at 25°C. The results clearly showcase the intrinsic ability of cation type in modulating the encapsulated amounts and the associated release patterns. In the case of Na IC, major release occurs during the initial 20 min that later slows down and saturates. Around 3.2 μg of eugenol per mg of IC has been loaded and measured at 2.5 h. On the other hand, Ca IC fibres do not release any eugenol for about 18 min but later a gradual release is noticed that slowly peaks at around 36 min with subsequent saturation. Surprisingly, the load is only 1.6 μg/mg, 50% less than the Na IC.

The reason behind distinct encapsulated amounts and release rates between the Na IC and Ca IC complexes is altered inter-helical interactions of IC in the presence of Na and Ca ions. The three-dimensional structure analysis of Na IC (Janaswamy and Chandrasekaran, 2001) reveals that a pair of helices is connected by hydrogen bonding (···) interactions such as 4-S···*W*/Na···4-S, 2-S···*W*···*W*···*W*···4-S, 2-S···Na···2-S, 2-S···Na···4-S, 2-S···Na···*W*···2-S, 4-S···*W*···*W*···4-S, 4-S···Na···*W*···*W*···*W*···4-S, 2-S···*W*···Na···4-S, 4-S···Na···*W*···Na···4-S and 2-S···*W*···*W*···*W*···4-S; wherein 4-S, 2-S and *W* represent 4-sulfate group on β-D-Gal, 2-sulfate group on α-D-Gal and water molecule, respectively. In the case of Ca IC, direct interactions such as 2-S···Ca···2-S, 2-S···Ca···4-S and 4-S···Ca···4-S are made possible by the divalent Ca ions (Janaswamy and

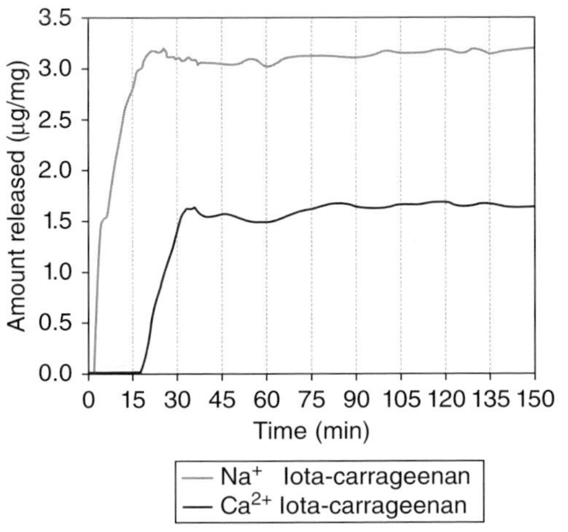

Fig. 5.8. The release rate of eugenol from the iota-carrageenan:eugenol complex prepared at 25°C.

Chandrasekaran, 2002). Thus, each Ca ion not only balances the charge on two sulfate groups of adjacent helices but also draws them closer and tighter; whereas a Na ion merely binds to a sulfate moiety on one helix and hence the inter-helical interactions are comparatively weaker and further mediated by space fillers such as water molecules. Thus, the resulting Na IC network is more flexible than that of Ca IC. These network variations reflect significantly on the encapsulated amounts and subsequent release profiles.

The IC:eugenol release behaviour compares well with those of IC:curcumin (Janaswamy and Youngren, 2012) and IC:griseofulvin (Janaswamy *et al.*, 2013). While the former accommodates around 3.2 μg/mg the latter two encompass the far smaller amounts of 0.15 and 1.8 μg/mg, respectively. Thus, it appears that the guest molecule's chemical structure, size and available functional groups, along with its associative kinetics with the polysaccharide helices, appear to dictate the embedding amounts as well as the release characteristics.

Conclusions

Polysaccharides exhibit a wide variety of unique chemical structures and physiological functions. They are capable of significantly altering the texture, gelation and viscosity of aqueous-based solutions, and a wide range of products could be developed using polysaccharides as functional ingredients. They form the bulk of many foods consumed by humans and play a central role in human health as well; for example, their regular consumption is accompanied by a reduction in chronic diseases such as diabetes, cardiovascular disease and cancer.

The present research is about utilizing the structural organization of polysaccharide helices in oriented fibres, and the innate water pockets present in the polysaccharide network, to encapsulate active molecules and protect them from external influences such as temperature, to maintain their functional state until delivered at the target site. The outcome will be equally applicable to protect and deliver vitamins, antioxidants and flavour compounds, for example, using several routinely used polysaccharides such as kappa-carrageenan, lambda-carrageenan and xanthan. In general, polysaccharides are inexpensive and many of their uses in food and non-food applications have already been approved by the US Food and Drug Administration. Obtaining further authorization for large-scale production of polysaccharide-based delivery systems is, therefore, quite feasible and would certainly aid in capitalizing heavily used and low-cost biomaterials as value-added products.

Acknowledgements

The artwork in Fig. 5.1 was created by Patrick J. Polowsky. Research was supported by the Whistler Center for Carbohydrate Research, the Dane O. Kildsig Center for Pharmaceutical and Processing Research and the Discovery Park Undergraduate Research Internship programme. The help of Drs Irina Kosheleva and Robert Henning with synchrotron X-ray intensity data was greatly appreciated. Use of the Advanced Photon Source was supported by the US Department of Energy, Basic Energy Sciences, Office of Science, under Contract No. DE-AC02-06CH11357. Use of the BioCARS Sector 14 was supported by the National Institutes of Health, National Center for Research Resources, under grant number RR007707.

References

Chandrasekaran, R. (1997) Molecular architecture of polysaccharide helices in oriented fibers. *Advances in Carbohydrate Chemistry and Biochemistry* 52, 311–439.

Goldberg, M. and Gomez-Orellana, I. (2003) Challenges for the oral delivery of macromolecules. *Nature Reviews: Drug Discovery* 2, 289–295.

Janaswamy, S. and Chandrasekaran, R. (2001) Three-dimensional structure of the sodium salt of iota-carrageenan. *Carbohydrate Research* 335, 181–194.

Janaswamy, S. and Chandrasekaran, R. (2002) Effect of calcium ions on the organization of iota-carrageenan helices: an X-ray investigation. *Carbohydrate Research* 337, 523–535.

Janaswamy, S. and Chandrasekaran, R. (2005) Cation-induced polymorphism in iota-carrageenan. *Carbohydrate Polymers* 60, 499–505.

Janaswamy, S. and Chandrasekaran, R. (2006) Sodium iota-carrageenan: a paradigm of polymorphism and pseudopolymorphism. *Macromolecules* 39, 3345–3349.

Janaswamy, S. and Chandrasekaran, R. (2008) Heterogeneity in iota-carrageenan molecular structure: insights for polymorph II→III transition in the presence of calcium ions. *Carbohydrate Research* 343, 364–373.

Janaswamy, S., Gill, L., Campanella, O. and Pinal, R. (2013) Organized polysaccharide fibers as stable drug carriers. *Carbohydrate Polymers* 94, 209–215.

Janaswamy, S. and Youngren, R.S. (2012) Hydrocolloid-based nutraceutical delivery systems. *Food & Function* 3, 503–507.

Juliano, R. (2007) Cellular delivery of therapeutic macromolecules: challenges to macromolecular drug delivery. *Biochemical Society Transactions* 35, 41–43.

Langer, R. (1990) New methods of drug delivery. *Science* 249, 1527–1533.

Langer, R. and Tirrell, D.A. (2004) Designing materials for biology and medicine. *Nature* 428, 487–492.

Polowsky, P.J. and Janaswamy, S. (2015) Hydrocolloid-based nutraceutical delivery systems: Effect of counter-ions on the encapsulation and release. *Food Hydrocolloids* 43, 658–663.

Rees, D.A. (1981) Polysaccharide shapes and their interactions – some recent advances. *Pure and Applied Chemistry* 53, 1–14.

Rinaudo, M., Auzely, R. and Mazeau, K. (2004) Polysaccharides. In: *Encyclopedia of Polymer Science and Technology*. Wiley, New York, pp. 200–261.

Smelcerovic, A., Knezevic-Jugovic, Z. and Petronijevic, Z. (2008) Microbial polysaccharides and their derivatives as current and prospective pharmaceuticals. *Current Pharmaceutical Design* 14, 3168–3195.

Stephen, A.M. (ed.) (1995) *Food Polysaccharides and Their Applications*. Marcel Dekker Inc., New York.

Stortz, C.A. and Cerezo, A.S. (2000) Novel findings in carrageenans, agaroids and "hybrid" red seaweed galactans. *Current Topics in Phytochemistry* 4, 121–134.

Takakura, Y. and Hashida, M. (1996) Macromolecular carrier systems for targeted drug delivery: Pharmacokinetic considerations on biodistribution. *Pharmaceutical Research* 13, 820–831.

6 Polymer-based Nanoparticles for Drug Delivery Systems and Cancer Therapeutics

Ram Prasad[1,3,*], Rishikesh Pandey[2], Ajit Varma[3] and Ishan Barman[1,4]
[1]Department of Mechanical Engineering, Johns Hopkins University, Baltimore, Maryland, USA; [2]Department of Pediatrics, University of Connecticut Health, Farmington, Connecticut, USA; [3]Amity Institute of Microbial Technology, Amity University Uttar Pradesh, Noida, India; [4]Department of Oncology, Johns Hopkins University, Baltimore, Maryland, USA

Abstract

Polymer-based nanoparticle-sustained drug delivery systems offer several advantages over conventional delivery systems such as maintenance of optimum therapeutic concentration of drug in the blood or cell, elimination of frequent dosing and better patient compliance. Therefore, they are good candidates for more efficient drug release devices. Preparation and characterization of polymeric nanoparticles (formulated with biocompatible and biodegradable polymers) whose size and surface properties can be intelligently designed allows them not only to achieve long circulation times in the blood and site-specific drug delivery but also to exploit physiological or biochemical features of infectious diseases. The use of biodegradable polymeric nanoparticles for controlled drug delivery has shown significant therapeutic potential. Concurrently, targeted delivery technologies are gradually significant as a scientific area of investigation. They may contribute to the development of other useful polymeric nanoparticles to deliver a spectrum of chemotherapeutic, diagnostic, multi-model imaging agents and drug/gene delivery as part of the next generation of delivery systems. To date, therapeutics based on polymer assemblies have mainly been studied for tumour therapy. With continuous efforts by multidisciplinary teams, it is clear that nanotechnology will shed new light on diagnostics and therapeutics in cancer research.

Introduction

Polymers play an integral role in the advancement of pharmaceutical applications, especially in the field of drug delivery (Liechty *et al.*, 2010; Duro-Castano *et al.*, 2014). Pharmaceutical nanotechnology focuses on formulating therapeutically active agents in biocompatible nano forms such as nanoparticles, nanocapsules, micellar systems and conjugates (Hamidi *et al.*, 2008). These systems offer many advantages in drug delivery, mainly focusing on improved safety and efficacy of the drugs (e.g. providing targeted delivery, improving bioavailability, extending drug or gene effect in the target tissue and improving the stability of therapeutic agents against chemical/enzymatic degradation) (Moghimi *et al.*, 2001). Around 60 million patients benefit from advanced drug delivery systems at present, receiving safer and more effective doses of medicines that are needed to fight a variety of human ailments, including life-threatening diseases such as cancer (Shaik *et al.*, 2012).

The polymeric nanoparticles (PNPs) are prepared from biocompatible and biodegradable

*Corresponding author. E-mail: rprasad@amity.edu

 Natural Polymers for Drug Delivery (eds H. Kharkwal and S. Janaswamy)

polymers between 10 nm and 1000 nm in size where the drug is dissolved, entrapped, encapsulated or attached to a nanoparticle matrix (Nagavarma *et al.*, 2012). Depending upon the method of preparation, nanoparticles, nanospheres or nanocapsules can be obtained. Nanocapsules are systems in which the drug is confined to a cavity surrounded by a unique polymer membrane, while nanospheres are matrix systems in which the drug is physically and uniformly dispersed (Kumaresh *et al.*, 2001; Sahoo and Labhasetwar, 2003). Polymer-based NPs have great advantages over other nanomaterials. They are more stable than other volatile agents; easily fabricated in larger quantities; oral and intravenous administration can be achieved with significant efficiency and effectiveness; and drugs can be delivered to the target at higher concentrations. The ability to choose the polymer and to vary release from the NPs mean the system is ideal for delivery of targeted drugs and vaccines for cancer therapy. Several types of nanoparticulate systems have been trialled as potential delivery systems including biodegradable polymeric nanoparticles, polymeric micelles, solid nanoparticles and lipid-based nanoparticles (e.g. solid lipid nanoparticles, nanostructured lipid carriers and lipid drug conjugates, nanoliposomes, inorganic nanoparticles, dendrimers, magnetic nanoparticles, ferrofluids and quantum dots) (Patel and Jivani, 2009; Salouti and Ahangari, 2014). Metallic nanoparticles are found to be decent candidates for a drug delivery system (Dreaden *et al.*, 2012) but their actual clinical translations are seldom implemented because of biocompatibility and environmental concerns. Attempts are, however, being made to synthesize biogenic metallic nanoparticles that are non-toxic, biocompatible and eco-friendly (Prasad *et al.*, 2015; Aziz *et al.*, 2015).

PNPs are playing an important role in a range of areas including biotechnology, electronics, sensors, photonics, conducting materials, cancer therapy, delivery of vaccines, contraceptives, delivery of targeted antibiotics, pollution control and environmental technology (Schmid, 2004; Wang *et al.*, 2004; Kayser *et al.*, 2005; Geckeler and Rosenberg, 2006; Geckeler and Nishide, 2010). PNPs are promising vehicles for drug delivery by easy manipulation to prepare carriers with the objective of delivering the drugs to a specific target, which improves drug safety (Shokri *et al.*, 2011). They effectively carry drugs, proteins and DNA to target cells and organs. Their nanometer-size promotes effective permeation through cell membranes and stability in the bloodstream (Nagavarma *et al.*, 2012). Polymers are very convenient materials for the manufacture of countless and varied molecular designs that could be integrated into unique nanoparticle constructs with many potential medical applications (Peer *et al.*, 2007; Abhilash, 2010).

The major emphasis in designing nanoparticles as a delivery system is to control particle size and shape, surface properties and release of pharmacologically active agents so as to achieve the site-specific action of the drug at the therapeutically optimal rate and dose regimen (Soppimath *et al.*, 2001; Jahanshahi *et al.*, 2005). Although liposomes have been used as potential carriers – with unique advantages including protecting drugs from degradation, targeting to site of action and reduction in toxicity or side effects – their applications are limited due to inherent problems such as low encapsulation efficiency, rapid leakage of water-soluble drug in the presence of blood components and poor storage stability (Rastogi *et al.*, 2014). On the other hand, PNPs offer some specific advantages over liposomes, mainly in increasing the stability of drugs/proteins and in having useful controlled release properties (Vila *et al.*, 2002; Mu and Feng, 2003). Nanoscopic therapeutic systems that incorporate therapeutic agents, molecular targeting and diagnostic imaging capabilities are emerging as the next generation of multifunctional nanomedicine towards significantly enhancing the therapeutic outcome of drug therapy (Nasongkla *et al.*, 2006) (Fig. 6.1).

Three Groups for Drug Delivery Systems

The three groups (Zambaux *et al.*, 1998) are:

- First generation systems: including microcapsules and microspheres for control of chemo- mobilization and control release of proteins and peptides for drug delivery within the brain.
- Second generation systems: liposomes, nanocapsules and nanospheres (called passive

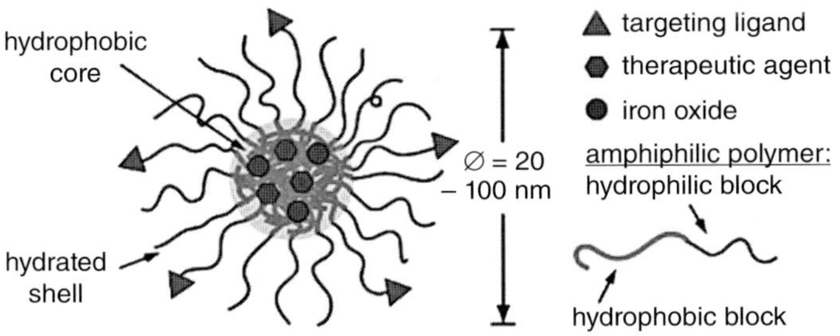

Fig. 6.1. Multifunctional no medicine platform for targeted drug delivery (reprinted with permission from Nasongkla *et al.*, 2006, © 2006, American Chemical Society).

colloidal carriers), and certain active carriers, which release their substances after a specific signal such as one given by temperature sensitive liposomes and magnetic nanospheres. They are less than 1 nm in diameter and are capable of releasing the active ingredient at the intended target after administration through the general route.

- Third generation systems: these are also true carriers based on monoclonal antibodies, which are characterized by a capability to specifically recognize a site.

Advantages of Using Nanoparticles as a Drug Delivery System

The advantages of nanoparticles as a drug delivery system (Li *et al.*, 2001; Kwon *et al.*, 2001; Ravi Kumar *et al.*, 2004) include:

- Easy manipulation of particle size and surface characteristics to achieve both passive and active drug targeting after parenteral administration.
- Controlled and sustained release of the drug during transportation and at the site of localization, altering organ distribution of the drug and subsequent clearance to achieve increase in drug therapeutic efficacy and reduction in side effects.
- Relatively high loading; drugs can be incorporated into the systems without any chemical reaction.
- Site-specific targeting achieved by attaching targeting ligands to surface of the particles, or by use of magnetic guidance.
- Use of the system for various routes of administration including oral, nasal, parenteral and intra-ocular.
- Avoidance of coalescence, leading to enhanced physical stability.
- Reduced mobility of incorporated drug molecules, leading to reduction in drug leakage.
- Possibility of surface modification of nanosuspensions for site-specific delivery.

Disadvantages of Using Nanoparticles as a Drug Delivery System

- Potential toxicity: while the small size of nanoparticles is what makes them useful in medicine, it is also a factor that might make them potentially dangerous to human health.
- Environmental concerns: nanoparticles derived from synthetic polymers would constitute a new class of non-biodegradable pollutants.
- There are no convenient methods by which exposure to nanoparticles in the workplace could be measured or assessed.

Biodegradable nanoparticles could be prepared from a variety of materials such as proteins, polysaccharides and synthetic biodegradable polymers. The selection of the base polymer is dependent on various designs and final application criteria. It further depends on factors such as: size of the desired nanoparticles, properties of the drug (aqueous solubility, stability, etc.) to be encapsulated, surface characteristics and functionality, degree of biodegradability and biocompatibility, and drug release profile of the final product. Subsequently, the methods could be classified as: dispersion of preformed polymers, polymerization of monomers and ionic gelation method for hydrophilic polymers. The general advantages and disadvantages of individual methods are summarized in Table 6.1.

Polymers Used in the Preparation of Nanoparticles

The polymers should be compatible with the body in the terms of adaptability (non-toxicity and non-antigenicity) and be biodegradable (Ghosh, 2000).

Natural polymers

The most common natural polymers used in preparation of polymeric nanoparticles are gums (e.g. *Acacia* and guar), lectin, viciline, gelatin, dextran, chitosan, sodium alginate and albumin (Aynie *et al.*, 1998; Farrugia and Grover, 1999; Fernandez-Urrusuno *et al.*, 1999; Luppi *et al.*, 2011).

Synthetic polymers

Any polymer used must be chemically inert, non-toxic and free of leachable impurities. A few examples that are in current use include polylactides (PLA), polyglycolides (PGA), poly(lactide co-glycolides) (PLGA), polyanhydrides, polyorthoesters, polycyanoacrylates, polycaprolactone, poly glutamic acid, poly malic acid, poly (N-vinyl pyrrolidone), poly(methyl methacrylate), poly(vinyl alcohol), poly(acrylic acid), poly acrylamide, poly(N-vinyl pyrrolidone), poly(ethylene glycol), poly(methacrylic acid), poly(methyl methacrylate), poly(2-hydroxy ethyl methacrylate) and poly(ethylene-co-vinyl acetate) (Annick, 2005; Mansour *et al.*, 2010; Luppi *et al.*, 2011; Babak *et al.*, 2011).

Table 6.1. Polymeric nanoparticles: general advantages and drawbacks of preparation methods (reproduced from Pinto Reis *et al.*, 2006, with permission from Elsevier).

Method	Simplicity of procedure	Need for purification	Facility scaling-up	EE (%)	Safety of compounds
Polymerization of monomers					
Emulsion polymerization					
Organic	Low	High	NR	Low	Low
Aqueous	High	High	High	High	Medium
Interfacial polymerization	Low	High	Medium	High	Low
Preformed polymers					
Synthetic					
Emulsification/solvent evaporation	High	Low	Low	Medium	Medium
Solvent displacement and interfacial deposition	High	NR	NR	High	Medium
Salting out	High	High	High	High	Low
Emulsion/solvent diffusion	Medium	Medium	High	High	Medium
Natural					
Albumin	NR	High	NR	Medium	Low
Gelatin	NR	High	NR	Medium	Low
Polysaccharides					
Alginate	High	Medium	High	High	High
Chitosan	High	Medium	High	High	High
Agarose	Medium	High	NR	NR	High
Desolvation	NR	High	NR	Low	Low

EE, Encapsulation Efficiency; NR, no reference available.

Polymeric nanoparticles provide significant flexibility in the design because polymers can be biodegradable or non-biodegradable, and can also be made synthetically or derived from natural sources. Some common polymers used for nanoparticle formation include poly(lactic acid), dextran and chitosan. Langer and Folkman (1976) were the first to demonstrate the controlled release of macromolecules using polymers, which allowed the development of anti-angiogenic drug delivery systems for cancer therapy and opened new avenues for the delivery of macromolecules. Biodegradable polymers are typically degraded into oligomers and individual monomers, which are metabolized and removed from the body via normal metabolic pathways (Duncan, 2006; Grund *et al.*, 2011). Degradation and drug release kinetics can be precisely controlled by the physicochemical properties of the polymer, such as molecular weight, dispersity index, hydrophobicity and crystallinity. In general, drugs can be released in a controlled manner with first-order kinetics due to drug diffusion through the polymeric matrix or triggered in response to the local environment. The nanoparticle surface is usually sterically stabilized by grafting, conjugating or adsorbing hydrophilic polymers such as polyethylene glycol (PEG) to its surface, which could also reduce hepatic uptake and improve circulation half-life (Peracchia *et al.*, 1999; Gref *et al.*, 2000; Dinarvand *et al.*, 2011).

Polymeric Drug Delivery Systems

Development of novel systems for delivery of anticancer drugs is a recent topic of research. Abraxane was the first polymeric nanoparticle product from American Pharmaceutical Partners, Inc. and American Bioscience, Inc., and was approved by the US Food and Drug Administration (FDA) in 2005. It consists of albumin-bound paclitaxel nanoparticles (Gradishar, 2006; Utreja *et al.*, 2010). It is free from toxic solvents such as cremophor-EL, known for life-threatening allergic reactions (Green *et al.*, 2006), routinely used to solubilize paclitaxel to administer intravenously.

The challenge of modern drug therapy is the optimization of the pharmacological action of the drugs coupled with the reduction of their toxic effects *in vivo*. The prime objectives are controlled delivery to the site of action at a therapeutically optimal rate and dosage to avoid toxicity as well as to improve the drug effectiveness and therapeutic index. The polymeric drug carriers deliver the drug at the tissue site by any one of the three general physicochemical mechanisms: (i) hydration-induced swelling of the polymer nanoparticles followed by release through diffusion; (ii) enzymatic reaction resulting in rupture, cleavage or degradation of the polymer at the site of delivery, thereby releasing the drug from the entrapped inner core; and (iii) dissociation of the drug from the polymer and its de-adsorption/release from the swollen nanoparticles (Ghosh, 2000).

Methods for Preparation of Nanoparticles

The most commonly used technique is to prepare biodegradable nanoparticles from polylactic acid (PLA); poly-D-L-glycolide (PLG); poly-D-L-lactide-co-glycolide (PLGA) and poly-cyanoacrylate (PCA) through polymerization of monomers or from the preformed polymers. It can be used in several ways, as described below.

Solvent evaporation

In this technique the polymer is dissolved in an organic solvent (e.g. dichloromethane, chloroform or ethyl acetate). The drug is dissolved or dispersed in the preformed polymer solution followed by emulsification to form an oil/water (o/w) emulsion using appropriate surfactant/emulsifying agents (e.g. gelatin and polyvinyl alcohol). Later, the organic solvent is evaporated by increasing the temperature or pressure with continuous solution stirring (Fig. 6.2a). Particle size can be influenced by the type and concentrations of stabilizer, homogenizer speed and polymer concentration. To produce small particle size, high-speed homogenization or ultrasonication are often employed (Scholes *et al.*, 1993; Soppimath *et al.*, 2001; Pinto Reis *et al.*, 2006).

Spontaneous emulsification/ solvent diffusion

This is a modified solvent diffusion method where a water-miscible solvent (e.g. acetone

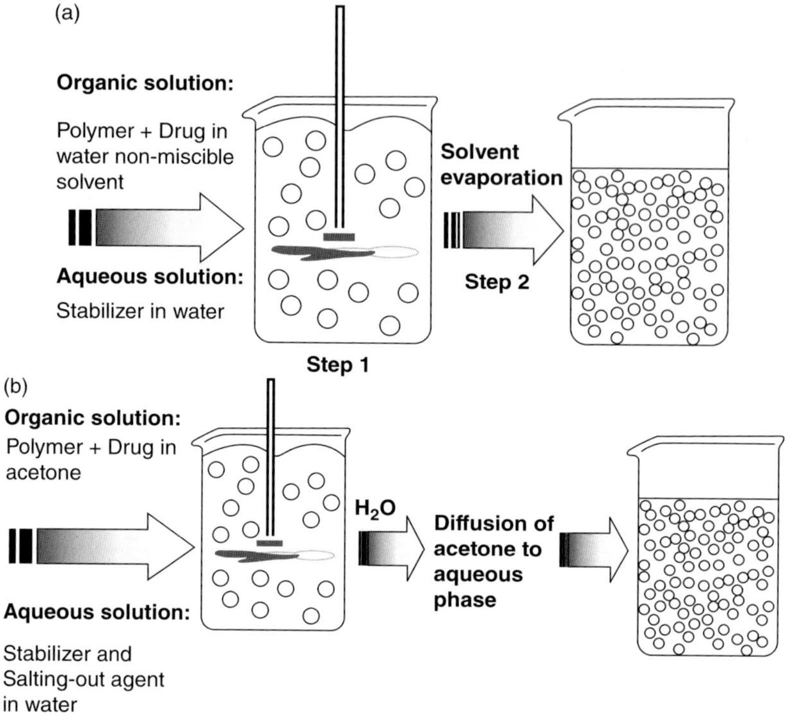

Fig. 6.2. (a) Schematic representation of the emulsification evaporation technique. Adapted from Pinto Reis *et al.*, 2006. © 2006 Elsevier; (b) schematic representation of the salting-out technique. Adapted from Pinto Reis *et al.*, 2006. © 2006 Elsevier.

or methanol) and a water-insoluble organic solvent (i.e. dichloromethane or chloroform) is used as an oil phase (Niwa *et al.*, 1993). Because of the spontaneous diffusion of solvents, an interfacial turbulence is created between the two phases leading to the formation of smaller particles. As the concentration of water-soluble solvent increases, smaller particle sizes of NPs can be achieved (Niwa *et al.*, 1993). This technique has several advantages, such as high encapsulation efficiencies (generally >70%), no need for homogenization, high batch-to-batch reproducibility, ease of scale-up, simplicity and narrow size distribution. However, high volumes of water to be eliminated from the suspension, leakage of water-soluble drug into the saturated–aqueous external phase during emulsification and reduced encapsulation efficiency (Pinto Reis *et al.*, 2006; Nagavarma *et al.*, 2012) are some of the concerns to be addressed.

Nanoprecipitation method/solvent displacement

This method involves the precipitation of a preformed polymer from an organic solution and the diffusion of the organic solvent in the aqueous medium in the presence or absence of a surfactant (Galindo-Rodriguez *et al.*, 2004; Ganachaud and Katz, 2005). Typically, it is used for hydrophobic drug entrapment, but has been adapted for hydrophilic drugs as well. Polymers and drugs are dissolved in a polar, water-miscible solvent (e.g. acetone, acetonitrile, ethanol or methanol). The solution is then poured (e.g. drop by drop) into an aqueous solution with surfactant. Nanoparticles are formed instantaneously by rapid solvent diffusion. Finally, the solvent is removed under reduced pressure (Govender *et al.*, 1999). This technique is well adapted for the incorporation of cyclosporin A, because of higher (98%) entrapment efficiencies (Allemann *et al.*, 1998).

Cyclosporin A is an immunosuppressant drug widely used in organ transplantation to prevent rejection. It reduces the activity of the immune system by interfering with the activity and growth of T cells.

Salting out

In this method, the polymer is dissolved in the organic phase, which should be water-miscible, such as acetone or tetrahydrofuran (THF). The organic phase is then emulsified in an aqueous phase under strong mechanical shear stress. The aqueous phase contains the emulsifier and a high concentration of salts, which are not soluble in the organic phase. Typically, the salts used are 60% w/w of magnesium chloride hexahydrate or magnesium acetate tetrahydrate in 1:3 polymer to salt ratio (Zweers *et al.*, 2004; Eley *et al.*, 2004). In contrast to the emulsion diffusion method, there is no diffusion of the solvent because of the presence of salts (Fig. 6.2b). The fast addition of pure water to the o/w emulsion under mild stirring reduces the ionic strength and leads to the migration of the water-soluble organic solvent to the aqueous phase, inducing nanosphere formation. The final step is purification of nanoparticles by cross-flow filtration or centrifugation to remove the salting-out agent (Zweers *et al.*, 2004; Eley *et al.*, 2004).

The main advantage is that it minimizes stress to protein encapsulants (Jung *et al.*, 2000). It does not require any temperature increase and therefore is useful for heat-sensitive substances. The exclusive application to lipophilic drugs and extensive nanoparticle washing steps (Couvreur *et al.*, 1995) are some of the limiting factors of this technique.

Biodegradable Polymeric Nanoparticles for Drug Delivery

Biodegradable nanoparticles have been used for site-specific delivery of drugs, vaccines and other biomolecules. A few of the most extensively used biodegradable polymer matrices for preparation of nanoparticles are described below.

Polylactic acid

Polylactic acid (PLA) is a biocompatible and biodegradable material that undergoes scission in the body to monomeric units of lactic acid as a natural intermediate product in carbohydrate metabolism. The most common route is the ring-opening polymerization of lactide with various metal catalysts or the direct condensation of lactic acid monomers (Fig. 6.3a). The use of PLA nanoparticles is therefore safe and devoid of any major toxicity issues. PLA nanoparticles have been prepared by the solvent evaporation, solvent displacement, salting out and solvent diffusion methodologies (Fessi *et al.*, 1989; Pinto Reis *et al.*, 2006).

Poly-ε-caprolactone

Poly-ε-caprolactone (PCL) is degraded, slower than the polylactide by hydrolysis of its ester linkages under normal physiological conditions in the human body, and has minimal or no toxicity. Therefore, it stands out as a candidate of choice for drug delivery and long-term implantable biomaterial. PCL is prepared by ring-opening polymerization of ε-caprolactone using a catalyst such as stannous octoate (Labet and Thielemans, 2009) (Fig. 6.3b). It has been used as the hydrophobic block of amphiphilic synthetic block copolymers in the vesicle membrane of polymersomes. PCL nanoparticles have been prepared mostly by nanoprecipitation, solvent displacement and solvent evaporation (Kumari *et al.*, 2010; Mahapatro and Singh, 2011).

Poly-D-L-lactide-co-glycolide

Poly-D-L-lactide-co-glycolide (PLGA) is one of the most successfully used biodegradable polymers (Fig. 6.3c). It is synthesized by means of ring-opening co-polymerization of two different biodegradable metabolite monomers, the cyclic dimers (1,4-dioxane-2,5-diones) of glycolic acid and lactic acid. These two monomers are normally found in the body and participate in a number of physiological and biochemical pathways, thus PLGA exerts very minimal systemic toxicity during drug delivery and biomaterial applications. PLGA NPs have been mostly prepared by the emulsification–diffusion, the solvent evaporation and the nanoprecipitation methods (Kumari *et al.*, 2010). They have been used to develop protein- and peptide-based nanomedicines, nanovaccines and genes containing nanoparticles for *in vivo* delivery systems.

Chitosan

Chitosan (Fig. 6.3d) is a modified natural carbohydrate polymer prepared by the partial N-deacetylation of crustacean-derived natural biopolymer chitin. There are at least four methods reported for the preparation of chitosan nanoparticles: ionotropic gelation, microemulsion, emulsification solvent diffusion and polyelectrolyte complex (Sinha *et al.*, 2004; Gan and Wang, 2007). Ionotropic gelation is based on electrostatic interaction between the amine group of chitosan and negatively charged polyanions such as tripolyphosphate (Kumari *et al.*, 2010). Chitosan and its derivatives such as trimethylchitosan (or quaternized chitosan) have been used in non-viral gene delivery and at approximately 50%, trimethylation, the derivative, is the most effective (Kean *et al.*, 2005). Oligomeric derivatives (3–6 kDa) are relatively non-toxic and display good gene delivery properties (Kean *et al.*, 2005). Trimethylchitosan has been shown to transfect breast cancer cells, and the cytotoxicity increases with the degree of trimethylation. Chitosan considerably enhances absorption of the peptides insulin and calcitonin across the nasal epithelium (Illum *et al.*, 1994).

Gelatin

Gelatin (Fig. 6.3e) is widely used in food applications, photographic plate and medical products, and is also attractive in controlled release due to its intrinsic non-toxic, biodegradable, bioactive and inexpensive properties (Kumari *et al.*, 2010). It is a polyampholyte, having both cationic and anionic groups as well as a hydrophilic group. Its mechanical properties, swelling behaviour and thermal properties depend significantly on the degree of cross-linking. Gelatin nanoparticles can be prepared by desolvation/coacervation and emulsion methods (Zillies *et al.*, 2008; Ofokansi *et al.*, 2010). In the pharmaceutical industry gelatin is used in the manufacture of capsules and tablets, and to prevent drug damage by air and light. Due to its good compatibility with human tissue, gelatin in sponge form is used to treat wounds and as a replacement agent for blood plasma.

Polymer-based Nanoparticles in Cancer Therapeutics

Nanotechnology has been regarded as one of the most promising approaches to combat cancer, the leading cause of death worldwide (Ferrari, 2005; Jemal *et al.*, 2009; Farokhzad and Langer, 2009). Nonetheless, there are only a few nanotechnology-based formulations that are approved for clinical application by the US FDA (Duncan, 2006). It is thus of paramount importance to develop smarter and more powerful drug formulations. Liposomes and polymeric nanoparticles are two of the most investigated anti-cancer drug carriers.

The concept of 'theranostics' is based on the 'find, fight and follow approach' that offers the possibility of early detection, disease targeting and treatment (Duro-Castano *et al.*, 2014). Nanotechnology is playing a pivotal role in providing new types of therapeutics for cancer. Most cancer patients die from drug-resistant and metastatic disease. These nanotherapeutics have the required potential with minimal side effects. It is expected that as diagnostic methods improve, treatment could be initiated at earlier stages of disease progression (Heath and Davis, 2008). A variety of imaging agents, such as nanoprobes for disease monitoring coupled with magnetic resonance imaging (MRI), positron emission tomography (PET) or optical imaging using near infra-red (NIR) fluorescent and luminescent probes are being explored in this regard.

The most important challenge in the successful formulation of polymeric drug delivery systems involves the preparation of carrier systems that are capable of encapsulating the preferred drug within its structure and effectively delivering it to the cancer cells. Control of particle size and polydispersity of NPs are important parameters, as they directly affect the physical stability,

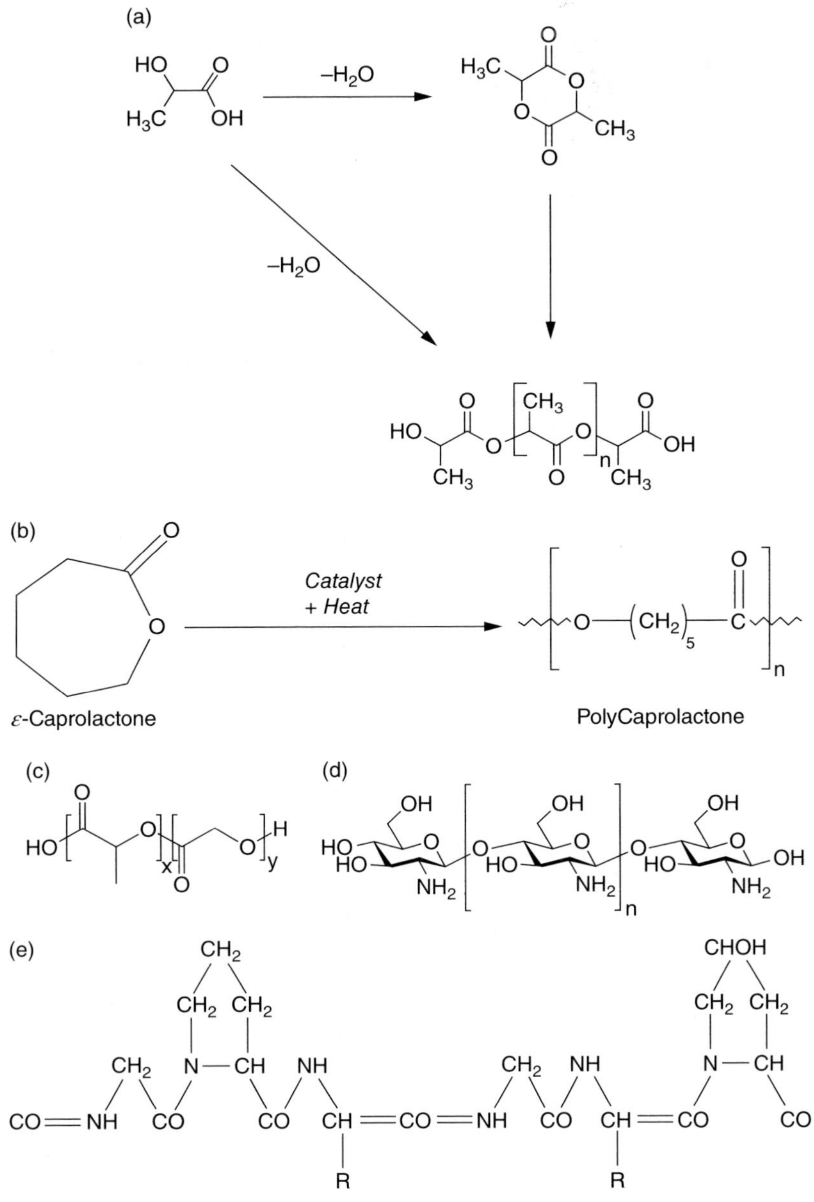

Fig. 6.3. (a) Polylactic acid (PLA) polymer; (b) ring-opening polymerization of ε-caprolactone to polycaprolactone; (c) chemical structure of poly-D-L-lactide-co-glycolide (PLGA); (d) chemical structure of chitosan; (e) chemical structure of gelatin.

cellular uptake, surface properties, biodistribution and drug release (Marin *et al.*, 2013).

Optimization and scaling up of PLGA nanoparticles encapsulating curcumin (natural plant-based material) by using solid–oil/water (S-O/W) through emulsification–solvent–evaporation/diffusion protocols (Ranjan *et al.*, 2012) yield interesting results. The *in vitro* studies revealed the sustained release of curcumin from the nanoparticles over a prolonged period of 10 days. Intracellular uptake and cell viability assays also demonstrated efficient uptake and action of the curcumin nanospheres in prostate cancer cell lines. PLGA nanospheres encapsulated with curcumin thus appear to be potential candidates for cancer therapy (Marin *et al.*, 2013). Curcumin (a yellow polyphenol, diferuloylmethane) is extracted from the rhizomes of turmeric. It possesses pharmacological activities that include antitumour, anti-amyloid, antioxidant and anti-inflammatory properties. The above-mentioned investigations into curcumin have established that free curcumin induces cell cycle arrest (apoptosis) and blocks nuclear factor kappa B (NF-κB) activation in various human cancer cell lines (e.g. LNCaP, PC3 and DU-145) (Duvoix *et al.*, 2005; Bisht *et al.*, 2007; Mukerjee and Vishwanatha, 2009).

The PLA- and PLGA-based nanoparticles can be rapidly cleared in the liver and captured by the reticuloendothelial system on their administration into the blood circulation (Fox *et al.*, 2009). These problems can be eliminated by the introduction of D-α-tocopheryl polyethylene glycol succinate (TPGS) into the hydrophobic PLA backbone. TPGS, a water-soluble derivative of D-α-tocopherol, is formed by esterification of vitamin E succinate with polyethylene glycol (Zhang *et al.*, 2012a), and it improves the aqueous solubility of drugs along with taxanes, antibiotics, cyclosporines and steroids. In addition, TPGS could serve as an excellent molecular biomaterial in developing targeted drug delivery systems as well as overcoming multidrug resistance. It is an inhibitor of P-glycoprotein and increases the cytotoxicity and oral bioavailability of antitumour agents (Zhang *et al.*, 2012b). A novel carrier system of cholic acid (CA), a star-shaped block copolymer (based on CA-PLA-TPGS nanoparticles) delivers paclitaxel in a sustained and controlled manner in breast cancer treatment. Most importantly, higher drug loading and entrapment efficiency, with faster drug release as well as higher cellular uptake and cytotoxicity, could be accomplished (Tang *et al.*, 2013). *In vivo* cell studies have indicated that the PTX-loaded star-shaped CA-PLA-TPGS nanoparticles showed significantly superior antitumour activity (Tang *et al.*, 2013).

Muthu *et al.* (2012) prepared TPGS-coated theranostic liposomes containing docetaxel and quantum dots with and without targeting moieties. Folic acid was used as the probe to target the folate receptor overexpressing MCF-7 breast cancer cell lines. Higher cellular uptake and cytotoxicity were observed for theranostic liposomes than for the non-targeted liposomes. Tao *et al.* (2013) demonstrated novel co-polymer docetaxel-loaded M-PLGA-TPGS NPs that are star-shaped and display higher cellular uptake in the MCF-7 breast cancer cells. Also found was a significantly higher level of cytotoxicity with docetaxel-loading compared with linear polymer. Subsequent *in vivo* data revealed significantly superior antitumour efficacy in treating breast cancer (Tao *et al.*, 2013). Paul *et al.* (2013) demonstrated the encapsulation of chelidonine in biodegradable PLGA polymers and evaluated the anti-cancer efficacy of nano-chelidonine (NC) vis-à-vis free chelidonine (FC) against HepG2 cells. NC exhibits rapid cellular uptake and a stronger apoptotic effect than FC, blocking HepG2 cells at the G2/M phase. p53, cyclin-D1, Bax, Bcl-2, cytochrome c, Apaf-1, caspase-9 and caspase-3 expressions also corroborated well, suggesting the greater anti-cancer potential of NC compared to free chelidonine (FC). *In vivo* studies revealed enhanced bioavailability of NC with better tissue distribution without any toxicity (Paul *et al.*, 2013). Thus, NC stands out as a possible candidate for therapeutic oncology.

After the introduction of chitosan, the zeta potential of the PLGA nanoparticle surface changes (altering the pH value) from negative to positive, giving the drug carrier more affinity with the cancer cells. PLGA NPs modified by chitosan revealed the versatility of the surface and a possible improvement in the efficacy of current PLGA-based drug delivery systems (Wang *et al.*, 2013).

There is another promising formulation, namely that of dendrimers. Dendrimers are synthetic polymeric macromolecules of nanometer dimensions possessing highly branched monomers with low polydispersity. They display a number of advantages as carriers or scaffolds for diagnosis and therapy. Their amphiphilic property and interior cavities could be used to encapsulate hydrophobic or hydrophilic drugs (Cho *et al.*, 2008; Tekade *et al.*, 2009; Hu *et al.*, 2011). The dendrimer–doxorubicin complex displays comparable *in vivo* antitumour efficacy as Doxil, an FDA-approved liposome-based doxorubicin delivery vehicle (Tong and Cheng, 2007). Compared to liposomes and micelles, dendrimer–drug conjugates may be more stable owing to their covalent attachment and uniform molecular structures. Hence, they are amenable for formulation, sterilization, transportation and storing (Lee *et al.*, 2005).

Nanotheranostics might be able offer a practical solution for the treatment of cancer and other fatal diseases, or at least treat these ailments in the early stages (Muthu *et al.*, 2014). However, some challenges need to be addressed, namely: (i) development of new molecular biomaterials with high performance in co-formulation of diagnostic and theranostic agents and identification of new biomarkers with high specificity; (ii) investigation of possible agonism and antagonism between the diagnostic and theranostic agents co-formulated in single nanocarriers; and (iii) intracellular autophagy and its inhabitation to nanotheranostics. None the less, the future for nanotheranostics appears to be bright.

Polymer-based therapeutics and magnetic resonance imaging

Magnetic resonance imaging (MRI) is a powerful tool for non-invasive exploration of internal body structures, and has excellent spatial and temporal resolution. Its power stems from the widely distributed, robust proton signalling ability of water in the body, which facilitates rapid and precise depiction of morphology with reduced imaging time (Suzuki *et al.*, 2014). The clinical efficacy of NPs in cancer imaging has already been demonstrated, especially in detecting liver cancer and staging lymph node metastasis non-invasively (Harris *et al.*, 2007). Polymeric NPs, such as superparamagnetic iron oxide NPs (SPIOs) and Gd-based compounds, have been shown to be effective as MRI contrasts (Andre *et al.*, 1999; Nitin *et al.*, 2004). For example, a mixture of SPIOs and doxorubicin (DOX) can be directly encapsulated by an amphiphilic block co-polymer composed of maleimide-PEG-poly (lactic acid), which self-assembles to form polymeric NP-based nanotheranostics, for both drug delivery and MRI contrast. Surface-presenting maleimide groups allow the conjugation of cRGD (arginine–glycine–aspartic acid) molecules or short peptides containing ten amino acids to target the $\alpha v\beta 3$ integrins (Nasongkla *et al.*, 2006, Guthi *et al.*, 2010). The resultant nanotheranostic particles allow the evaluation of pharmacokinetics by MRI in real-time. In tumour-bearing mice, targeted NPs showed a higher tumour accumulation compared to non-targeted NPs owing to the integrin-mediated endocytosis, which led to enhanced tumour retardation. Another pivotal function of integrins is the promotion of cell migration by virtue of their binding to extracellular matrix components. This phenomenon is responsible for the process of tumour proliferation, migration, invasion and metastasis (Moschos *et al.*, 2007).

Gadolinium compounds have been used to formulate polymeric NP-based theranostic formulations for real-time visualization of blood circulation and biodistribution, and to enhance magnetic resonance and tumour accumulation. They are also used in various ion channel electrophysiology experiments to block sodium leak channels and to stretch activated ion channels. Ye *et al.* (2006) conjugated gadolinium-(1,4,7,10-tetraazacyclododecane-1,4,7-trisacetic acid) (Gd-DO3A) to the side chains of poly(L-glutamic acid) (PGA) and subsequently fractionated the polymer-GD-DO3A conjugates into three groups with molecular weights of 87, 50 and 28 kDa (Fig. 6.4). Higher molecular weight conjugates (87 kDa) exhibited more prolonged blood circulation and increased tumour accumulation over the lower molecular weights.

Polymer-based therapeutics and chemical imaging

Photodynamic therapy (PDT) is a minimally invasive technique that kills target cancer cells in the presence of oxygen via the release of reactive oxygen species upon light activation of a photosensitizer. This methodology destroys the cancer cells through direct cellular damage, vascular shutdown and induction of the host immune response against the target cells (MacDonald and Dougherty, 2001).

By integrating fluorescence imaging techniques with polymeric NPs, image-guided drug delivery allows monitoring of the drug pharmacokinetics, intratumoural drug distribution and drug tumour accumulation in real time, and serves as the basis of specific photothermal injury to the targeted tissue (Kobayashi *et al.*, 2010). Peng *et al.* (2011) established multifunctional polymeric NPs composed of PEG-polycaprolactone di-block

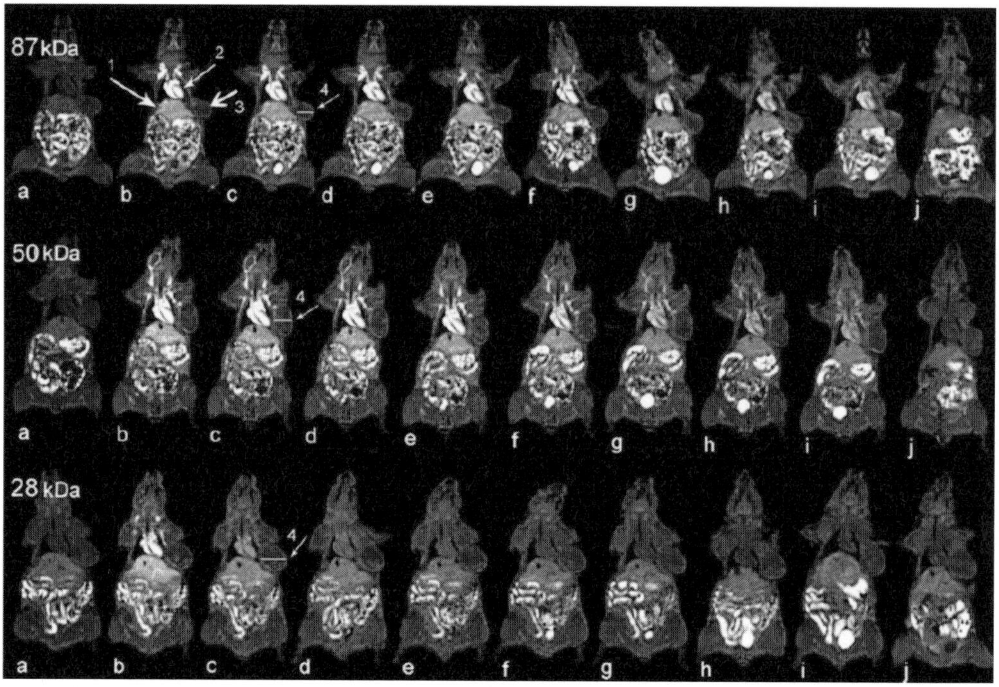

Fig. 6.4. Coronal MR images of tumour-bearing mice (a) before; and at (b) 1 min; (c) 11 min; (d) 20 min; (e) 30 min; (f) 60 min; (g) 120 min; (h) 180 min; (i) 240 min; and (j) 24 h after injection of PGA-1,6-hexanediamine-(Gd-DO3A) conjugates of different molecular weights. Arrows point to the (1) liver; (2) heart; and (3) tumour tissue; and (4) the cross section is for subsequent quantitative analysis (reprinted with permission from Ye *et al.*, 2006, © 2006 American Chemical Society).

co-polymer to load IR-780, a near-infrared (NIR) fluorescent dye, for both NIR imaging and photodynamic therapy (Fig. 6.5). The NPs were also labelled with radionuclide rhenium-188 (^{188}Re) for microSPECT-guided tumour imaging. In BALB/c athymic nude mice bearing HCT-116 colorectal carcinoma, these micellar nanoparticles demonstrated enhanced tumour accumulation and tumour growth inhibition. Chlorin e6 (Ce6) has also been commonly used as a photosensitizer in theranostic applications (Huang *et al.*, 2011).

Other chemical imaging techniques such as Raman imaging have been used recently in drug delivery systems (Garrett *et al.*, 2012). Raman spectroscopy, which provides chemical fingerprinting information, is a promising technique and has been used widely to study various pathological conditions (Haka *et al.*, 2005; Pandey *et al.*, 2015a,b). Raman imaging combines the chemical specificity of Raman spectroscopy with the imaging capability of an optical microscope, enabling imaging with chemical sensitivity.

Future Perspectives of Nanotherapeutics in Cancer

Nanoparticles provide opportunities for designing and fine-tuning properties that are not feasible with other types of therapeutic drugs, and their future as a new generation of cancer therapeutics is bright. For decades, researchers have been developing novel anti-cancer agents and formulations for delivering existing as well as newly developed drugs. The entry into clinical development of binary and ternary nanoparticles that combine synthetic polymers with proteins or drugs – as well as polymer micelles that incorporate covalently bound drugs – and their use in cancer therapeutics is expanding as their credibility has been established (Sinha *et al.*, 2006). The belief that nanoparticle–protein-bound technology could be a prospective way of developing anti-cancer agents was strengthened by the FDA approval of Abraxane to treat breast

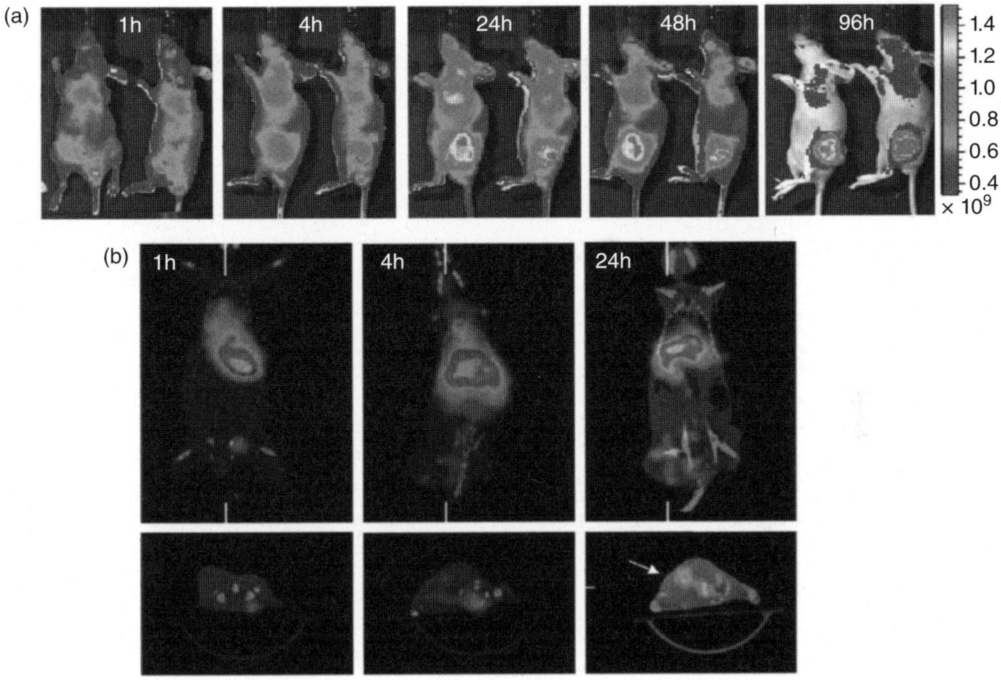

Fig. 6.5. Multifunctional PNPs composed of PEG-PCL di-block co-polymer were loaded with IR-780, an NIR florescent dye, for both NIR imaging and photodynamic therapy. The NPs were also labelled with 188Re for microSPECT-guided tumour imaging. (a) Time-lapse near-IR fluorescence images of mice bearing HCT-116 tumours after intravenous injections of PEG-PCL polymeric NPs containing IR-780. (b) MicroSPECT/CT images were obtained by first injecting 188 re-labelled IR-780 particles and then imaging at 1 h, 4 h and 24 h later, respectively (reprinted with permission from Peng *et al.*, 2011, © 2011 American Chemical Society).

cancer (2005), non-small cell lung cancer (2012) and late-stage pancreatic cancer (2013). The simple idea of eliminating cremophor from the taxol formulation and producing a compound with no hypersensitivity reactions and obviating the need for premedication has led to this new product being incorporated into various breast cancer adjuvant trials. Several binary molecules have been formulated and some of their pitfalls have led to the development of even more sophisticated 'ternary biomolecules' based on a thorough understanding of chemistry, biology and medicine. For specific targeting, the differences between cancerous cells and normal cells that include uncontrolled proliferation, insensitivity to negative growth regulation and antigrowth signals, angiogenesis and metastasis could be exploited. There is a growing body of knowledge of unique cancer markers, thanks to recent advances in proteomics and genomics, which form the basis of complex interactions between bioconjugated nanoparticles and cancer cells.

There is much synergy between imaging and nanotechnology in biomedical applications. Many of the principles used to target the delivery of drugs to cancer may also be applied to target imaging and diagnostic agents. Multifunctional nanoparticles that could serve as both diagnostic and therapeutic tools are currently being engineered. Researchers are continuing to look into nanoparticles that are conjugated to ligands targeting various receptors such as the folate receptor, epidermal growth factor receptor, estrogen and progesterone receptors and low-density lipoprotein receptors, as well as to ligands that target various antigens such as prostate-specific antigen. The use of more 'biologically friendly' carrier molecules is another prospective area in which more research is needed to help combat the side effects of some of the synthetic polymers.

The current continuous efforts, coupled with multidisciplinary team approaches, will indeed aid nanotechnology in shedding new light on diagnostics and therapeutics in cancer research.

Conclusions

Nanotechnology is a fast-expanding area of science. This area of research is anticipated to lead to the development of novel, sophisticated, multifunctional applications to recognize cancer cells, to deliver drugs to target tissue, to aid in reporting the outcome of therapy, to provide real-time assessment of therapeutic and surgical efficacy, and – most importantly – to monitor intracellular changes in preventing precancerous cells from becoming malignant.

Among different delivery systems, polymer-based nanocarriers possess favourable and useful properties such as being simple to synthesize, inexpensive, biocompatible, biodegradable, non-immunogenic, non-toxic and water-soluble. These polymer-based NPs have high motion in the smallest capillaries, allowing for effective uptake and selective drug accumulation at the target sites for competent cancer therapy. The use of polymers in NP formulations certainly provides greater multifunctional properties in the field of nanobiotechnology, NP-based drug delivery vehicles and efficient targeting in significant and effective anti-cancer therapeutics.

Acknowledgement

This work was supported by a UICC American Cancer Society Beginning Investigators Fellowship funded by the American Cancer Society.

References

Abhilash, M. (2010) Potential applications of nanoparticles. *International Journal of Pharma and Bio Sciences* 1, 1–12.

Allemann, E., Leroux, J.C. and Gurny, R. (1998) Polymeric nano-microparticles for the oral delivery of peptides and peptidomimetics. *Advanced Drug Delivery Reviews* 34, 171–189.

Andre, J.P., Toth, E., Fischer, H., Seelig, A., Macke, H.R. and Merbach, A.E. (1999) High relaxivity for monomeric Gd (DOTA)-based MRI contrast agents, thanks to micellar self-organization. *Chemistry – A European Journal* 5, 2977–2983.

Annick, L. (2005) The use of mucoadhesive polymers in ocular drug delivery. *Advanced Drug Delivery Reviews* 57, 1595–163.

Aynie, I.C, Vauthier, C., Fattal, E., Foulquier, M. and Couvreur, P. (1998) Alginate nanoparticles as a novel carrier for antisense oligonucleotide. In: Diederichs, J.E. and Muler, R. (eds) *Future Strategies of Drug Delivery with Particulate Systems.* Pharm Scientific Publisher, Stuttgart, pp. 5–10.

Aziz, N., Faraz, M., Pandey, R., Sakir, M, Fatma, T., Varma, A., Barman, I. and Prasad, R. (2015) Facile algae-derived route to biogenic silver nanoparticles: synthesis, antibacterial and photocatalytic properties. *Langmuir.* DOI: 10.1021/acs.langmuir.5b03081.

Babak K., Chiu, K., Black, K.L., Yamamoto, V., Khalsa, B., Ljubimova, J.Y., *et al.* (2011) Nanoplatforms for constructing new approaches to cancer treatment, imaging, and drug delivery: what should be the policy? *Neuro Image* 54, S106–S124.

Bisht, S., Feldmann, G., Soni, S., Ravi, R., Karikar, C. and Maitra, A. (2007) Polymeric nanoparticle-encapsulated curcumin ('nanocurcumin'): a novel strategy for human cancer therapy. *Journal of Nanobiotechnology* 5, 3.

Cho, K., Wang, X., Nie, S., Chen, Z.G. and Shin, D.M. (2008) Therapeutic nanoparticles for drug delivery in cancer. *Clinical Cancer Research* 14, 1310–1316.

Couvreur, P., Dubernet, C. and Puisieux, F. (1995) Controlled drug delivery with nanoparticles: current possibilities and future trends. *European Journal of Pharmaceutics and Biopharmaceutics* 41, 2–13.

Dinarvand, R., Sepehri, N., Manoochehri, S., Rouhani, H. and Atyabi, F. (2011) Polylactide-co-glycolide nanoparticles for controlled delivery of anticancer agents. *International Journal of Nanomedicine* 6, 877–895.

Dreaden, E.C., Austin, L.A., Mackey, M.A. and El-Sayed, M.A. (2012) Size matters: gold nanoparticles in targeted cancer drug delivery. *Therapeutic Delivery* 3(4), 457–478.

Duncan, R. (2006) Polymer conjugates as anticancer nanomedicines. *Nature Reviews. Cancer* 6(9), 688–701.
Duro-Castano, A., Conejos-Sánchez, I. and Vicent, M.J. (2014) Peptide-based polymer therapeutics. *Polymers* 6, 515–551.
Duvoix, A., Blasius, R., Delhalle, S, Schnekenburger, M., Morceau, F., Henry, E., Dicato, M. and Diederich, M. (2005) Chemopreventive and therapeutic effects of curcumin. *Cancer Letters* 223, 181–190.
Eley, J.G., Pujari, V.D. and McLane, J. (2004) Poly (lactide-co-glycolide) nanoparticles containing coumarin-6 for suppository delivery: in vitro release profile and in vivo tissue distribution, *Drug Delivery* 11(4), 255–261.
Farokhzad, O.C. and Langer, R. (2009) Impact of nanotechnology on drug delivery. *ACS Nano* 3, 16–20.
Farrugia, C.A. and Grover, M.J. (1999) Gelatin behaviour in dilute aqueous solution: designing a nanoparticulate formulation. *Journal of Pharmacy and Pharmacology* 51, 643–649.
Fernandez-Urrusuno, R., Calvo, P., Remunan-Lopez, C., Villa-Jato, J.L. and Alonso M.J. (1999) Enhancement of nasal absorption of insulin using chitosan nanopartilces. *Pharmaceutical Research* 16, 1576–1581.
Ferrari, M. (2005) Cancer nanotechnology: opportunities and challenges. *Nature Reviews Cancer* 5, 161–171.
Fessi, H., Puisieux, F., Devissaguet, J. Ph., Ammoury, N. and Benita, S. (1989) Nanocapsule formation by interfacial polymer deposition following solvent displacement. *International Journal of Pharmaceutics* 55, R1–R4.
Fox, M.E., Szoka, F.C. and Frechet, A.M.J. (2009) Soluble polymer carriers for the treatment of cancer: the importance of molecular architecture. *Accounts of Chemical Research* 8, 1141–1151.
Galindo-Rodriguez, S., Allemann, E., Fessi, H. and Doelker, E. (2004) Physicochemical parameters associated with nanoparticle formation in the salting-out, emulsification-diffusion, and nanoprecipitation methods. *Pharmaceutical Research* 21, 1428–1439.
Gan, Q. and Wang, T. (2007) Chitosan nanoparticle as protein delivery carrier – systematic examination of fabrication conditions for efficient loading and release. *Colloids and Surfaces B: Biointerfaces* 59(1), 24–34.
Ganachaud, F. and Katz, J.L. (2005) Nanoparticles and nanocapsules created using the ouzo effect: spontaneous emulsification as an alternative to ultrasonic and high-shear devices. *Physical Chemistry and Chemical Physics* 6, 209–216.
Garrett, N.L., Lalatsa, A., Begley, D., Mihoreanu, L., Uchegbu, I.F., Schätzlein, A.G. and Moger, J. (2012) Label-free imaging of polymeric nanomedicines using coherent anti-stokes Raman scattering microscopy. *Journal of Raman Spectroscopy* 43, 681–688.
Geckeler, K.E. and Nishide, H. (eds) (2010) *Advanced Nanomaterials*. Wiley-VCH, Weinheim, Germany.
Geckeler, K.E. and Rosenberg, E. (eds) (2006) *Functional Nanomaterials*. American Scientific, Valencia, California, USA.
Ghosh, P.K. (2000) Hydrophilic polymeric nanoparticles as drug carriers. *Indian Journal of Biochemistry and Biophysics* 37, 273–282.
Govender, T., Stolnik, S., Garnett, M.C., Illum, L. and Davis, S.S. (1999) PLGA nanoparticles prepared by nanoprecipitation: drug loading and release studies of a water soluble drug. *Journal of Controlled Release* 57(2), 171–185.
Gradishar, W.J. (2006) Albumin-bound paclitaxel: a next-generation taxane. *Expert Opinion on Pharmacotherapy* 7(8), 1041–1053.
Green, M., Manikhas, G., Orlov, S., *et al.* (2006) Abraxane, a novel Cremophor-free, albumin-bound particle form of paclitaxel for the treatment of advanced non-small-cell lung cancer. *Annals of Oncology* 17, 1263–1268.
Gref, R., Lück, M., Quellec, P., Marchand, M., Dellacherie, E., Harnisch, S., Blunk, T. and Müller, R.H. (2000) 'Stealth' corona-core nanoparticles surface modified by polyethylene glycol (PEG): influences of the corona (PEG chain length and surface density) and of the core composition on phagocytic uptake and plasma protein adsorption. *Colloids and Surfaces B: Biointerfaces* 18, 301–313.
Grund, S., Bauer, M. and Fischer, D. (2011) Polymers in drug delivery-state of the art and future trends. *Advanced Engineering Materials* 13, B61–B87.
Guthi, J.S., Yang, S.G., Huang, G., Li, S.Z., Khemtong, C., Kessinger, C.W., Peyton, M., Minna, J.D., Brown, K.C. and Gao, J.M. (2010) MRI-visible micellar nanomedicine for targeted drug delivery to lung cancer cells. *Molecular Pharmaceutics* 7, 32–40.
Haka, A.S., Shafer-Peltier, K.E., Fitzmaurice, M., Crowe, J., Dasari R.R. and Feld, M.S. (2005) Diagnosing breast cancer by using Raman spectroscopy. *Proceedings of the National Academy of Sciences* 102, 12371–12376.
Hamidi, M., Azadi, A. and Rafiei, P. (2008) Hydrogel nanoparticles in drug delivery. *Advanced Drug Delivery Reviews* 60(15), 1638–1649.

Harris, T.J., von Maltzahn, G. and Bhatia, S.N. (2007) Multifunctional nanoparticles for cancer therapy. In: Amiji, M.M. (ed.), *Nanotechnology for Cancer Therapy*. CRC Press, Boca Raton, pp.59–76.

Heath, J.R. and Davis, M.E. (2008) Nanotechnology and cancer. *Annual Review of Medicine* 59, 251–265.

Hu, J., Su, Y., Zhang, H., Xu, T. and Cheng, Y. (2011) Design of interior-functionalized fully acetylated dendrimers for anticancer drug delivery. *Biomaterials*, 32, 9950–9959.

Huang, P., Xu, C., Lin, J., Wang, C., Wang, X., Zhang, C., Zhou, X., Guo, S. and Cui, D. (2011) Folic acid-conjugated graphene oxide loaded with photosensitizers for targeting photodynamic therapy. *Theranostics* 1, 240–250.

Illum, L., Farraj, N.F. and Davis, S.S. (1994) Chitosan as novel nasal delivery system for peptide drugs. *Pharmaceutical Research* 11, 1186–1189.

Jahanshahi, M., Aghajani, H. and Ling, T. (2005) Assembly and purification of nanostructure bioproducts: protein nanoparticle characterizations and non-stick exterior coating adsorbents. *International Journal of Nanoscience and Nanotechnology* 1, 9–19.

Jemal, A., Siegel, R., Xu, J., and Ward, E. (2010) Cancer statistics. *CA: A Cancer Journal for Clinicians* 60(5), 277–300.

Jung, T., Kamm, W., Breitenbach, A., Kaiserling, E., Xiao, J.X. and Kissel, T. (2000) Biodegradable nanoparticles for oral delivery of peptides: is there a role for polymers to affect mucosal uptake? *European Journal of Pharmaceutics and Biopharmaceutics* 50, 147–160.

Kayser, O., Lemke, A. and Hernández-Trejo, N. (2005) The impact of nanobiotechnology on the development of new drug delivery systems. *Current Pharmaceutical Biotechnology* 6(1), 35.

Kean, T., Roth, S. and Thanou, M. (2005) Trimethylated chitosans as non-viral gene delivery vectors: Cytotoxicity and transfection efficiency. *Journal of Controlled Release* 103 (3), 643–653.

Kobayashi, H., Ogawa, M., Alford, R., Choyke, P.L. and Urano, Y. (2010) New strategies for fluorescent probe design in medical diagnostic imaging. *Chemical Reviews* 110, 2620–2640.

Kumaresh S.S., Aminabhavi, T.M., Kulkarni, A.R. and Rudzinski, W.E. (2001) Biodegradable polymeric nanoparticles as drug delivery devices. *Journal of Controlled Release* 70, 1–20.

Kumari, A., Yadav, S.K. and Yadav, S.C. (2010) Biodegradable polymeric nanoparticles based drug delivery systems. *Colloids and Surfaces B: Biointerfaces* 75, 1–18.

Kwon, H.Y., Lee, J.Y., Choi, S.W., Jang, Y. and Kim, J.H. (2001) Preparation of PLGA nanoparticles containing estrogen by emulsification-diffusion method. *Colloids and Surfaces A: Physicochemical and Engineering Aspects* 182, 123–130.

Labet, M. and Thielemans, W. (2009) Synthesis of polycaprolactone: a review. *Chemical Society Reviews* 38(12), 3484–3504.

Langer, R. and Folkman, J. (1976) Polymers for the sustained release of proteins and other macromolecules. *Nature* 263, 797–800.

Lee, C.C., MacKay, J.A., Frechet, J.M.J. and Szoka, F.C. (2005) Designing dendrimers for biological applications. *Nature Biotechnology* 23, 1517–1526.

Li, Y.P., Pei, Y.Y., Zhou, Z.H., Zhang, X.Y., Gu, Z.H., Ding, J., Zhou, J.J. and Gao, X.J. (2001) PEGylated polycyanoacrylate nanoparticles as tumor necrosis factor-alpha carriers. *Journal of Controlled Release* 71, 287–296.

Liechty, W.B., Kryscio, D.R., Slaughter, B.V. and Peppas, N.A. (2010) Polymers for drug delivery systems. *Annual Review of Chemical and Biomolecular Engineering* 1, 149–173.

Luppi, B., Bigucci, F., Corace, G., Delucca, A., Cerchiara, T., Sorrenti, M., *et al.* (2011) Albumin nanoparticles carrying cyclodextrins for nasal delivery of the anti-Alzheimer drug tarcine. *European Journal of Pharmaceutical Sciences* 44, 559–565.

MacDonald, I.J. and Dougherty, T.J. (2001) Basic principles of photodynamic therapy. *Journal of Porphyrins and Phthalocyanines* 5, 105–129.

Mahapatro, A. and Singh, D.K. (2011) Biodegradable nanoparticles are excellent vehicle for site directed in vivo delivery of drugs and vaccines. *Journal of Nanobiotechnology* 9, 55.

Mansour, H.M., Sohn, M.J., Al-Ghananeem, A. and DeLuca, P.P. (2010) Materials for pharmaceutical dosage forms: molecular pharmaceutics and controlled release drug delivery aspects. *International Journal of Molecular Sciences* 11, 3298–3322.

Marin, E., Briceño, M.I. and Caballero-George, C. (2013) Critical evaluation of biodegradable polymers used in nanodrugs. *International Journal of Nanomedicine* 8, 3071–3091.

Moghimi, S.M., Hunter, A.C. and Murray, J.C. (2001) Long-circulating and target specific nanoparticles: theory to practice. *Pharmacological Reviews* 53, 283–318.

Moschos, S.J., Drogowski, L.M., Reppert, S.L. and Kirkwood, J.M. (2007) Integrins and cancer. *Oncology* 21, 13–20.

Mu, L. and Feng, S.S. (2003) A novel controlled release formulation for the anticancer drug paclitaxel (Taxol(R)): PLGA nanoparticles containing vitamin E TPGS. *Journal of Controlled Release* 86, 33–48.
Mukerjee, A. and Vishwanatha, J.K. (2009) Formulation, characterization and evaluation of curcumin-loaded PLGA nanospheres for cancer therapy. *Anticancer Research* 29, 3867–3876.
Muthu, M.S., Kulkarni, S.A. and Raju, A., *et al.* (2012) Theranostic liposomes of TPGS coating for targeted co-delivery of docetaxel and quantum dots. *Biomaterials* 33(12), 3494–3501.
Muthu, M.S., Mei L. and Feng, S.S. (2014) Nanotheranostics: advanced nanomedicine or the integration of diagnosis and therapy. *Nanomedicine* 9, 1277–1280.
Nagavarma, B.V.N., Yadav, H.K.S., Ayaz, A., Vasudha, L.S. and Shivakumar, H.G. (2012) Different techniques for preparation of polymeric nanoparticles – a review. *Asian Journal of Pharmaceutical and Clinical Research* 5(3), 16–23.
Nasongkla, N., Erik, B., Jimin, R., Hua, A., Chalermchai, K., Setti, G.J., Shook-Fong, C., Dean, S.A., David, B.A. and Jinming, G. (2006) Multifunctional polymeric micelles as cancer-targeted, MRI-Ultrasensitive drug delivery systems. *Nano Letters* 6, 2427–2430.
Nitin, N., LaConte, L.E.W., Zurkiya, O., Hu, X. and Bao, G. (2004) Functionalization and peptide-based delivery of magnetic nanoparticles as an intracellular MRI contrast agent. *Journal of Biological Inorganic Chemistry* 9, 706–712.
Niwa, T., Takeuchi, H., Hino, T., Kunou, N. and Kawashima, Y. (1993) Preparations of biodegradable nanospheres of water-soluble and insoluble drugs with D, L-lactide/glycolide copolymer by a novel spontaneous emulsification solvent diffusion method, and the drug release behavior. *Journal of Controlled Release* 25(1–2), 89–98.
Ofokansi, K., Winter, G., Fricker, G. and Coester, C. (2010) Matrix-loaded biodegradable gelatin nanoparticles as new approach to improve drug loading and delivery. *European Journal of Pharmaceutics and Biopharmaceutics* 76(1), 1–9.
Pandey, R., Paidi, S.K., Kang, J.W., Spegazzini, N., Dasari, R.R., Valdez, T.A. and Barman I. (2015a) Discerning the differential molecular pathology of proliferative middle ear lesions using Raman spectroscopy. *Nature Scientific Reports* 5, 13305.
Pandey, R., Dingari, N.C., Spegazzini, N., Dasari, R.R., Horowitz, G.L. and Barman, I. (2015b) Emerging trend in optical sensing of alternate glycemic markers for diabetes monitoring. *Trends in Analytical Chemistry* 64, 100–108.
Patel, J.K. and Jivani, N.P. (2009) Chitosan based nanoparticles in drug delivery. *International Journal of Pharmaceutical Sciences and Nanotechnology* 2, 517–522.
Paul, A., Das, S., Das, J., Samadder, A. and Khuda-Bukhsh, A. (2013) Cytotoxicity and apoptotic signalling cascade induced by chelidonine-loaded PLGA nanoparticles in HepG2 cells in vitro and bioavailability of nano-chelidonine in mice in vivo. *Toxicology Letters*. 222, 17–22.
Peer, D., Karp, J.M., Hong, S., Farokhzad, O.C., Margalit, R. and Langer, R. (2007) Nanocarriers as an emerging platform for cancer therapy. *Nature Nanotechnology* 2, 761–770.
Peng, C.L., Shih, Y.H., Lee, P.C., Hsieh, T.M.H., Luo, T.Y. and Shieh, M.J. (2011) Multimodal image-guided photothermal therapy mediated by (188) Re-labeled micelles containing a cyanine-type photosensitizer. *Nano Letters* 5, 5594–607.
Peracchia, M.T., Harnisch, S., Pinto-Alphandary, H., Gulik, A., Dedieu, J.C., Desmaële, D., d'Angelo, J., Müller, R.H. and Couvreur, P. (1999) Visualization of in vitro protein-rejecting properties of PEGylated stealth polycyanoacrylate nanoparticles. *Biomaterials* 20(14), 1269–1275.
Pinto Reis, C., Neufeld, R.J., Ribeiro, A.J. and Veiga, F. (2006) Nanoencapsulation I. Methods for preparation of drug-loaded polymeric nanoparticles. *Nanomedicine: Nanotechnology, Biology and Medicine* 2, 8–21.
Prasad, R., Pandey, R. and Barman, I. (2015) Engineering tailored nanoparticles with microbes: quo vadis. *WIREs Nanomedicine & Nanobiotechnology*. DOI: 10.1002/wnan.1363.
Ranjan, A.P., Mukerjee, A., Helson, L. and Vishwanatha, J.K. (2012) Scale up, optimization and stability analysis of Curcumin C3 complex-loaded nanoparticles for cancer therapy. *Journal of Nanobiotechnology* 10, 38.
Rastogi, V., Yadav, P., Bhattacharya, S.S., Mishra, A.K., Verma, N., Verma, A. and Pandit, J.K. (2014) Carbon nanotubes: an emerging drug carrier for targeting cancer cells. *Journal of Drug Delivery*. DOI:10.1155/2014/670815.
Ravi Kumar, M.N., Bakowsky, U. and Lehr, C.M. (2004) Preparation and characterization of cationic PLGA nanospheres as DNA carriers. *Biomaterials* 25, 1771–1777.
Sahoo, S.K. and Labhasetwar, V. (2003) Nanotech approaches to drug delivery and imaging. *Drug Discovery Today* 8, 1112–1120.

Salouti, M. and Ahangari, A. (2014) Nanoparticle based drug delivery systems for treatment of infectious diseases, application of nanotechnology in drug delivery. In: Sezer, A.D. (ed.) *Application of Nanotechnology in Drug Delivery*. Intech, Rijeka.

Schmid, G. (2004) *Nanoparticles: From Theory to Applications*. Wiley-VCH, Weinheim, Germany.

Scholes, P.D., Coombes, A.G.A., Illum, L., Daviz, S.S., Vert, M. and Davies, M.C. (1993) The preparation of sub-200 nm poly(lactide-co-glycolide) microspheres for site-specific drug delivery. *Journal of Controlled Release* 25, 145–153.

Shaik, M.R., Korsapati, M. and Panati, D. (2012) Polymers in controlled drug delivery systems. *International Journal of Pharma Sciences* 2, 112–116.

Shokri, N., Akbari Javar, H., Fouladdel, Sh., Khalaj, A., Khoshayand, M.R., Dinarvand, R., Atyabi, F., Nomani, A. and Azizi, E. (2011) Preparation and evaluation of poly (caprolactone fumurate) nanoparticles containing Doxorubicin Hcl. *DARU* 19, 12–22.

Sinha, R., Kim G.J., Nie, S. and Shin D.M. (2006) Nanotechnology in cancer therapeutics: bioconjugated nanoparticles for drug delivery. *Molecular Cancer Therapeutics* 5(8), 1909–1917.

Sinha, V.R., Singla, A.K., Wadhawan, S., Kaushik, R., Kumria, R., Bansal, K. and Dhawan, S. (2004) Chitosan microspheres as a potential carrier for drugs. *International Journal of Pharmaceutics* 274(1–2), 1–33.

Soppimath, K.S., Aminabhavi, T.M., Kulkarni, A.R. and Rudzinski, W.E. (2001) Biodegradable polymeric nanoparticles as drug delivery devices. *Journal of Controlled Release* 70, 1–20.

Suzuki, Y., Iida, M., Miura, I., Inubushi, T. and Morikawa, S. (2014) A polymer-based magnetic resonance tracer for visualization of solid tumors by ^{13}C spectroscopic imaging. PLoS ONE 9(7), e102132. DOI:10.1371/journal.pone.0102132.

Tang, X., Cai, S., Zhang, R., Liu, P., Chen, H., *et al.* (2013) Paclitaxel-loaded nanoparticles of star-shaped cholic acid-core PLA-TPGS copolymer for breast cancer treatment. *Nanoscale Research Letters* 8, 420.

Tao, W., Zeng, X., Liu, T., Wang, Z., Xiong, Q., *et al.* (2013) Docetaxel-loaded nanoparticles based on star-shaped mannitol-core PLGA-TPGS diblock copolymer for breast cancer therapy. *Acta Biomaterialia* 9, 8916–8920.

Tekade, R.K., Kumar, P.V. and Jain, N.K. (2009) Dendrimers in oncology: an expanding horizon. *Chemical Reviews* 109, 49–87.

Tong, R. and Cheng, J. (2007) Anticancer polymeric nanomedicines. *Polymer Reviews* 47, 345–381.

Utreja, P., Jain, S. and Tiwary, A.K. (2010) Novel drug delivery systems for sustained and targeted delivery of anti-cancer drugs: current status and future prospects. *Current Drug Delivery* 7(2), 152–161.

Vila, A., Sanchez, A., Tobio, M., Calvo, P. and Alonso, M.J. (2002) Design of biodegradable particles for protein delivery. *Journal of Controlled Release* 78, 15–24.

Wang, X., Summers, C.J. and Wang, Z.L. (2004) Large scale hexagonal patterned growth of aligned ZnO nanorods for nano-optoelectronics and nanosensor arrays. *Nano Letters* 4, 423–426.

Wang, Y., Li, P. and Kong, L. (2013) Chitosan-modified PLGA nanoparticles with versatile surface for improved drug delivery. *American Association of Pharmaceutical Scientists* 14, 587–592.

Ye, F., Ke, T., Jeong, E.K., Wang, X., Sun, Y., Johnson, M. and Lu, Z.R. (2006) Noninvasive visualization of *in vivo* drug delivery of poly(L-glutamic acid) using contrast-enhanced MRI. *Molecular Pharmaceutics* 3, 507–515.

Zambaux, M.F., Bonneaux, F., Gref, R., Maincent, P., Dellacherie, E., Alonso, M.J., Labrude, P. and Vigneron, C. (1998) Influence of experimental parameters on the characteristics of poly(lac-tic acid) nanoparticles prepared by a double emulsion method. *Journal Controlled Release* 50, 31–40.

Zhang, Z.P., Mei, L. and Feng, S.S. (2012a) Vitamin E d-a-tocopheryl polyethylene glycol 1000 succinate-based nanomedicine. *Nanomedicine* 8, 1645–1647.

Zhang, Z.P., Tan, S.W. and Feng, S.S. (2012b) Vitamin E TPGS as a molecular biomaterial for drug delivery. *Biomaterials* 8, 4889–4906.

Zillies, J.C., Zwiorek, K., Hoffmann, F., Vollmar, A., Anchordoquy, T.J., Winter, G. and Coester, C. (2008) Formulation development of freeze-dried oligonucleotideloaded gelatin nanoparticles. *European Journal of Pharmaceutics and Biopharmaceutics* 70(2), 514–521.

Zweers, M.L.T., Engbers, G.H.M., Grijpma, D.W. and Feijen, J. (2004) In vitro degradation of nanoparticles prepared from polymers based on DL-lactide, glycolide and poly(ethylene oxide). *Journal of Controlled Release* 100(3), 347–356.

7 Polymer Nanocomposite-based Biosensors for Drug Delivery Applications

Monika Joshi*
Amity Institute of Nanotechnology, Amity University Uttar Pradesh, Noida, India

Abstract
Polymer nanocomposites (PNCs) have received much attention in various disciplines due to their high specific surface area, good compatibility, low density, high flexibility and improved functional properties. Recently, they have been explored as an emerging class of material in the biosensors due to their excellent sensitivity, selectivity, portability and lower cost. This chapter explores the properties and application of PNC material as a novel carrier in a drug delivery system. In this respect, the integration of biosensor and drug delivery systems is discussed in order to assess the challenges and future prospects. Different biosensors for drug delivery applications are also discussed.

Introduction

A sensor is a device that converts and displays a physical quantity in the form of an electrical signal. The physical quantity could be light, heat, motion, humidity, pressure, gas, etc. (Usher and Keating, 1996). A sensor should be selective to input but insensitive to other environmental factors. The input quantity for the sensor is non-electric and the device that converts it to an electrical signal is known as a transducer (Pallas-Areny and Webster, 1991). The converted signal could be electrical, mechanical, chemical, optical or thermal. Different transducers are used according to the type of actuation, such as electrochemical (pH probe), electromechanical (piezoelectric actuator, quartz, strain gauge), electroacoustic (gramophone, microphone), photoelectric (photodiode, solar cell), electromagnetic (antenna), magnetic (Hall-effect sensor), electrostatic (electrometer), thermoelectric (thermocouple) and electrical (capacitor, resistor).

There are several basic criteria for good and efficient sensing systems: (i) high sensitivity and selectivity; (ii) fast response time and recovery time; (iii) low analyst consumption; (iv) low operating temperature and temperature independence; and (v) stability in performance (Weppner, 1992). Several of their applications include medical diagnostics, medicinal kits, environmental monitoring, chemical industries and automobile industries (Miura *et al.*, 1995), and food and agriculture industries.

Recent advances in nanotechnology have resulted in novel classes of nanostructure thin films. Such films enhance chemical sensing properties via increased surface area to volume ratio, which in turn improves the active sensing area and interaction with the target molecules (Fritz and Roukes, 2002). Additionally, strong

*E-mail: mjoshi@amity.edu

photon and phonon quenching and amplification are also observed compared to the conventional bulk materials. Subsequent alterations in the optical and electronic properties can be tailored to suit the specific applications.

Types of Sensor

There are three broad categories:

Physical sensors

Physical sensors measure the aspects of indoor and outdoor environments. The change in these physical conditions can be measured by an electric instrument coupled with a sensor and input to a computer for analysis. Some examples include sensors for humidity, temperature, light, moisture, vibration, sound and pressure.

Chemical sensors

These sensors convert a chemical state into an electrical signal and provide real-time monitoring of many important types of input. They contain two functional units: a receptor and a transducer. The receptor part will be able to determine many input types simultaneously without separation and interference, while the transformer part converts that information into electrical signals. The type of sensing material and its size are important. For example, the sensitivity and selectivity of different sizes of the same material (bulk to nano) can be different. High response, low power consumption, stability, small size and low operating temperature are the key features of these sensors (Göpel and Schierbaum,1995). In contrast, real-time analysis and long-term stability are limiting factors. Some examples are explosive gas, air purifier, electrochemical, infrared (IR), optical and liquefied petroleum gas (LPG) sensors.

Biosensors

A biosensor is an analytical device used to detect isolated enzymes, antibodies, microorganisms, nucleic acids, immune systems, tissues, organelles and/or whole cells with the help of a physiochemical detector. It consists of electronic circuitry for the conversion and display of the analyte. This can be a reliable analytical device, able to perform quick and accurate analysis. To develop a sensitive, fast and stable biosensor, bio-components such as enzymes, antibodies, lactones, hormones, nucleic acids or tissues can be used as biological sensing elements. These bio-components provide an interface between the target species and the sensing material. Surface modification or an immobilization method can also be used to enhance the degree of sensitivity and selectivity of biosensors (Kissinger, 2005).

Characteristics of a Sensor

The characteristics of a sensor may be classified as being either static or dynamic. These parameters are essential in high fidelity mapping of output versus input. Static characteristics are those that can be measured after all transient effects have stabilized to their final or steady state. On the other hand, dynamic characteristics describe the sensor's transient properties.

SENSITIVITY. This is defined as the change in the output of the sensor to per unit change in the input. The sensitivity may be linear or nonlinear.

It is defined as:

$$S(\%) = (Rg - R_a/R_a) \times 100$$

where S is sensitivity, R_a is the resistance in the ambient atmosphere and Rg is the resistance in the presence of the measure and of the sensor.

SELECTIVITY. A sensor's response to measuring a single input in the presence of other inputs is known as selectivity. This characteristic is related to the discriminatory capacity of a device in front of a mixture of gases. For example, a glucose sensor that does not show a response to an enzyme may be considered to be selective.

REPEATABILITY. When all the ambient conditions are same, the repeatability of the sensor is to produce the same results for successive results of the same inputs in longer durations. To avoid the effects of no repeatability after repeated use, a thermal pre-treatment is given to nanomaterials, which decreases later material instabilities (Usher and Keating, 1996).

ACCURACY. Accuracy of a sensor is always comparative against a standard system. It defines how correctly the sensor output represents the true value. To assess it for a particular sensor, either the measurement should be benchmarked against a standard measurement, or the output should be compared with a system of known accuracy. For example, a humidity sensor that has a true value of 73% relative humidity (RH) is more accurate if it shows 73.1% than if it shows 74% RH.

REPRODUCIBILITY. This describes the sensor's ability to reproduce or replicate accurate results by someone else working independently. For example, after shutting down a sensing system and subsequently restarting it, a reproducible sensor will show the same response.

ERROR. The difference between the true value and that measured by the sensor is known as the error.

RESOLUTION. This signifies the smallest incremental change in the thing measured that could be detected in the output signal.

NOISE. Assuming there are no changes during a measurement process, if there are fluctuations in the output signal, this signifies the noise of the sensor. Ambient temperature, electromagnetic waves and mechanical vibrations are some examples of noise influencers.

STABILITY. Stability of the sensor is very important to produce identical output for the same inputs for a given time interval.

RESPONSE TIME. The time taken by a sensor to arrive at a stable value is the response time. This is generally expressed as the time at which the output reaches a certain percentage (for instance 95%) of its final value, in response to a stepped change of the input. The recovery time is defined in a similar way, but conversely.

HYSTERESIS. This is the difference between the output readings for the same measurement while increasing from the minimum value and later decreasing from the peak value.

Polymer Nanocomposite (PNC) Materials

Polymers are large organic molecules that consist of repeating units known as monomers, which are attached together through covalent bonds. There are two types: conducting and non-conducting polymers. Compared to other materials these are widely used as biosensors and in drug delivery applications due to their low fabrication cost, biocompatibility and biosensing ability at different temperatures. The polymers can participate directly as the sensitive layer, or be used as the media for immobilizing biomaterials on a surface.

Polymer nanocomposites are materials in which inorganic nanoparticles of 1–100 nm are uniformly dispersed in the polymeric matrix. The combination of an organic–inorganic matrix dramatically improves the performance and properties of the sensors (Ramanathan *et al.*, 2008; Caseri, 2009).

Polymers are easy to synthesize by means of a simple chemical process and their molecular chain structure can be modified by copolymerization or structural derivations. Polypyrrole (PPy), polyaniline (Pani), polythiophene (PTh) and their derivatives have been used as the active layers of sensors since the early 1980s. Most of the commercially available sensors are usually based on metal oxides and are operated at high temperatures. Sensors made of polymers, on the other hand, have many inherent advantages. Their nanocomposites are easily synthesized using simple and inexpensive chemical or electrochemical processes. Furthermore, polymers have good mechanical properties, allowing the sensors to be made easily. Nanomaterials of conjugated polymers are found to have superior performance relative to conventional materials because of their much larger exposed surface areas (Rajesh *et al.*, 2009).

Metals and their oxide nanoparticles such as titanium oxide (TiO_2), zinc oxide (ZnO), tin oxide (SnO_2), Ag, Au and metal oxide nanoparticles are routinely employed as sensing materials (Zheng *et al.*, 2000; Cukrov *et al.*, 2001; Chaabouni *et al.*, 2004). They have the merits of being low-cost, with good sensitivity and convenience in measuring the response. For example, systems based on carbon nanotubes (CNT) have been studied extensively due to their outstanding properties, which include electrical conductivity, high sensitivity, lower operating temperature and a faster response to a wider variety of input signals that may be detected, compared to other types of biosensors (Balasubramanian and Burghard, 2006). These are employed in the detection of cancer, glucose and other biological sensors. The large specific surface area improves the adsorption ability of the CNT–polymer conjugate, making it a promising material for drug delivery applications. However, to take advantage of these nanocomposites, more studies are still needed, especially those related to surface physics and chemistry.

Polymer Nanocomposite-based Biosensors

Development of nanotechnology has indeed created huge potential to build highly sensitive, low-cost, portable sensors with low power consumption. Of all known sensors, biosensors have attracted intensive research interest due to their widespread applications in biomedicine pharmaceutics and environmental monitoring (Dhand *et al.*, 2011).

Although there has been much research into electrical and analytical characterization, as well as into understanding the physical mechanisms that rule the sensing behaviour of nanomaterials, very little has been dedicated to polymer–nanocomposite materials (Tjong, 2012). A polymer nanocomposite is a hybrid material with a polymer and a nanomaterial, and at least one dimension on the nanoscale. In polymer nanocomposites, the nanomaterial could be amorphous or crystalline with various forms such as nanoparticles, nanorods, nanowires or quantum dots. The low density, facile nature, easy processing and superior chemical resistance of polymers results, when mixed with the nanomaterial, in high-performance polymer nanocomposites with greater surface to volume ratio and improved mechanical strength (Dos Santos *et al.*, 2003).

Nanoparticle- and nanowire-based sensors have been used for the detection of biological species (Shukla *et al.*, 2013). To develop a sensitive and highly stable biosensor, a suitable immobilization method must be used. The surface modification and functionalization of polymers using nanoparticles can also enhance the properties of biosensors (Ajayan *et al.*, 2003).

Proteins, catalysts and/or complementary DNA sequences with specific structures that are covalently or electrostatically bonded to the surface of the nanoparticles provide the selective sites for the species that needs to be detected. The sensor must be selective so that other proteins and molecules present do not attach to the surface of the nanocomposite. This type of approach could be invaluable for the treatment of cancers and tumours, where early detection is critical.

Biosensors for Drug Delivery Applications

Drug delivery is the process of transporting a medicine or pharmaceutical compound in the body (human or animal). Figure 7.1 shows the schematic of a delivery system. The process by which a drug is transported can have a significant effect on its efficacy. The concentration of drug and route provided for delivery also play an important role. Some drugs have a particular concentration range within which maximum benefit is achieved, and above or below this range could be toxic or

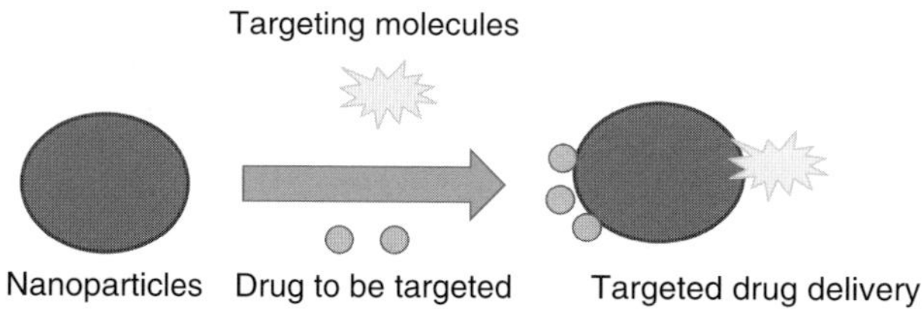

Fig. 7.1. Schematic of drug delivery system.

produce no therapeutic benefit. A smart drug delivery system consists of:

- Targeting moiety: The targeting moieties recognize the target. This includes proteins, lipoproteins, polysaccharides, antibodies and ligands with low molecular weight.
- Carrier: The system that carries an active drug molecule.
- Targeted drug: The drug that detects and performs specific therapeutic action at the specific site.

An ideal delivery system has many attributes including: (i) the prevention of harmful side-effects; (ii) a specific activity and increased bioavailability; (iii) targeted delivery or transport; (iv) low drug degradation; and (v) low drug dosages. Many systems are currently being developed to minimize degradation and loss, and to increase bioavailability and targeted delivery. Among these systems, considerable attention is being paid to biosensors. These systems could be polymers; micro- or nanoparticles made of insoluble, biodegradable natural and synthetic polymers; microcapsules; cells; cell ghosts; lipoproteins; liposomes; and micelles. The carriers could be made to be slowly degradable, stimuli-reactive (e.g. to pH or temperature) and even targeted, for example by conjugating them with specific antibodies against certain characteristic components in the area of interest (Sethi, 1994; Monošík *et al.*, 2012).

A biosensor-based drug delivery system is composed of a signal transducer, sensor interface, biological detector and an associate assay methodology, with each system component governed by its inherent operational considerations. These systems are generally classified into three categories:

- 1st-generation systems: microcapsules were used for the controlled released of the drug in the body, but lacked the ability to locate at a specific place.
- 2nd-generation systems: microcapsules were used with known and controlled pH and temperature. In these delivery systems the output signal could change with the preset values of pH or/and temperature.
- 3rd-generation systems: the delivery system was functionalized with specific bioreceptors for a specific target. That feature added an ideal delivery device that should integrate sensors and a memory element, and release components into a self-regulating smart device that could perform with minimal intervention.

The drug targeting in these systems was broadly done using two techniques: (i) passive; and (ii) active targeting. In passively targeted delivery, nanoparticles release their therapeutic action directly into the tumour milieu rather than within the cells. On the other hand, in an active targeted delivery, nanoparticles facilitate receptor-mediated endocytosis (RME), releasing therapeutic agents inside the target cells. Thus, active targeted drug delivery has higher therapeutic efficacy and lower toxicity.

Recent years have witnessed an unprecedented growth of research and applications in the area of nanoscience and nanotechnology. There is increasing optimism that nanotechnology, as applied to medicine, will bring significant advances in the diagnosis and treatment of diseases. Anticipated applications in medicine include drug delivery, both *in vitro* and *in vivo* diagnostics, nutraceuticals and production of improved biocompatible materials. Due to their small size (< 100 nm), high surface-to-volume ratio nanoparticles have a great potential as drug carriers. Different methods have been investigated for their synthesis, for example using natural biological substances such as albumin, gelatine, phospholipid polymers and solid metals containing nanoparticles. Their potential interaction with tissues and cells and subsequent toxicity, if any, greatly depends on the actual composition of the nanoparticle formulation (Fig. 7.2).

Nanomaterials have received considerable attention in the field of sensing because of their unique structural, electronic and chemical properties, large specific surface area, excellent conductivity and good biocompatibility. They do indeed serve as unique sensing tools. Their size is larger than typical molecules and smaller than viruses, but is similar in size to many proteins. They possess unique physical properties, which do not exist in bulk materials. These intrinsic properties allow them to operate well inside the cells. Drug delivery using nanosensors has unique physical characteristics, as the nanosensors deliver sensitivity that is orders of magnitude better than conventional devices, and they have performance advantages through their fast response and portability. For example, nanoshells

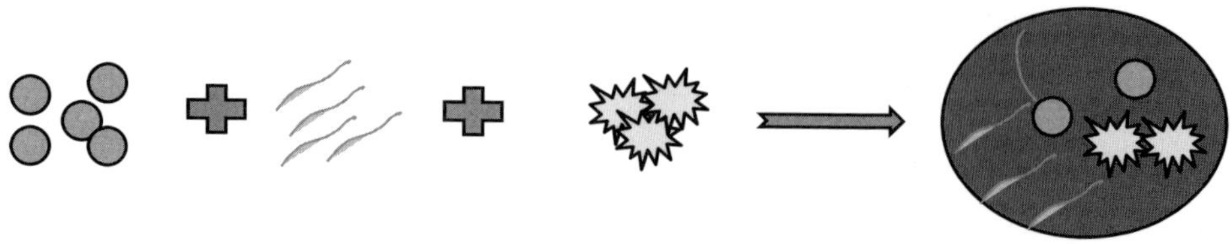

Fig. 7.2. Polymeric encapsulated drug delivery system.

(Kalele *et al.*, 2006) and nanorice (Zheng *et al.*, 2011) are made of a non-conducting core that is covered by a metallic shell. Nanoshells provide the opportunity to design all-optical nanoscale sensors – essentially new molecular-level diagnostic instruments – that could detect as little as a few molecules of a target substance.

Conclusion

The application of biosensors in drug delivery is an emerging and rapidly growing area. In this chapter, the basic characteristics and terminology related to sensors is discussed, followed by the properties of nanomaterials and the performance of biosensors in the presence of nanomaterials.

Acknowledgements

The author is thankful to Dr Ashok. K. Chauhan, Founder President, Ritanand Balved Educational Foundation for guidance and support; and also to Dr R.P. Singh, Distinguished Scientist, Amity institute of Nanotechnology and Dr L.M. Bharadwaj, Director, Amity Institute of Nanotechnology Amity University Uttar Pradesh, for their constant support and guidance.

References

Ajayan, P.M., Schadler, L.S. and Braun, P.V. (2003) *Nanocomposite Science and Technology*. Wiley-VCH, Verlag Gmbh, Weinheim, Germany.

Balasubramanian, K. and Burghard, M. (2006) Biosensors based on carbon nanotubes. *Analytical and Bioanalytical Chemistry* 385, 452–468.

Caseri, W. (2009) Inorganic nanoparticles as optically effective additives for polymers. *Chemical Engineering Communications* 196, 549–572.

Chaabouni, B., Abaab, M. and Rezig, B. (2004) Metrological characteristics of ZnO oxygen sensor at room temperature. *Sensors and Actuators B* 100, 200–204.

Cukrov, L.M., McCormick, P.G., Galatsis, K. and Wlodarski, W. (2001) Gas sensing properties of nanosized tin oxide synthesized by mechanochemical processing. *Sensors and Actuators B* 77, 491–495.

Dhand, C., Das, M., Datta, M. and Malhotra, B.D. (2011) Recent advances in polyaniline based biosensors. *Biosensors and Bioelectronics* 26(6), 2811–2821.

Dos Santos, D.S., Riul, S.A., Malmegrim, R.R., Fonseca, F.J., Oliveira, O.N. and Mattoso, L.H.C. (2003) A layer-by-layer film of chitosan in a taste sensor application. *Macromolecular Bioscience* 3, 591–595.

Fritz, S. and Roukes, M.L. (2002) *Understanding Nanotechnology*. Warner, New York.

Göpel, W. and Schierbaum, K.D. (1995) SnO_2 sensors: current status and future prospects. *Sensors and Actuators B: Chemical* 26(1–3), 1–12.

Kalele, S., Gosavi, S.W., Urban, J. and Kulkarni, S.K. (2006) Nanoshell particles: synthesis, properties and applications. *Current Science* 91(8), 1038–1052.

Kissinger, P.T. (2005) Biosensors—a prospective. *Biosensors and Bioelectronics* 20, 2512–2516.

Miura, N., Yan, Y., Nonaka, S. and Yamazoe, N. (1995) Sensing properties and mechanism of a planar carbon dioxide sensor using magnesia-stabilized zirconia and lithium carbonate auxiliary phase. *Journal of Material Chemistry* 5(9), 1391–1394.

Monošík, R., Streďanský, M. and Šturdík, E. (2012) Biosensors – classification, characterization and new trends. *Acta Chimica Slovaca* 5(1), 109–120.
Pallas-Areny, R. and Webster, J.G. (1991) *Sensors and Signal Conditioning*. Wiley, New York.
Rajesh, Ahuja, T. and Kumar D. (2009) Recent progress in the development of nano-structured conducting polymers/nanocomposites for sensor applications. *Sensors and Actuators B* 136, 275–286.
Ramanathan, T., Abdala, A., Stankovich, S., Dikin, D.A., Herrera-Alonso, M., Piner, R.D., Adamson, D.H., Schniepp, H.C., Chen, X., Ruoff, R.S., Nguyen, S.T., Aksay, I.A., Prud'Homme, R.K. and Brinson, L.C. (2008) Functionalized graphene sheets for polymer nanocomposites. *Nature Nanotechnology* 3(6), 327–331.
Sethi, R.S. (1994) Transducer aspects of biosensors. *Biosensors and Bioelectronics* 9, 243–264.
Shukla, S.K., Mishra, A.K., Arotiba, O.A. and Mamba, B.B. (2013) Chitosan-based nanomaterials: a state-of-the-art review. *International Journal of Biological Macromolecules* (59), 46–58.
Tjong, S.C. (2012) *Polymer Nanocomposites for Sensor Applications, in Polymer Composites with Carbonaceous Nanofillers, Properties and Applications*. Wiley-VCH Verlag GmbH. KGaA, Weinheim, Germany.
Usher, M.J. and Keating, D.A. (1996) *Sensors and Transducers: Characteristics, Applications, Instrumentation, Interfacing,* Second edition. Palgrave Macmillan, Basingstoke, United Kingdom.
Weppner, W. (1992) Advanced principles of sensors based on solid state ionics. *Materials Science and Engineering: B* 15, 48–55.
Zheng, L., Xu, M. and Xu, T. (2000) TiO_{2-x} thin films as oxygen sensor. *Sensors and Actuators B*, 66, 28–30.
Zheng, Y., Tao, J., Liu, H., Zeng, J., Yu, T., Ma, Y., Moran, C., Wu, L., Zhu, Y., Liu, J. and Xia, Y. (2011) Facile synthesis of gold nanorice enclosed by high-index facets and its application for CO oxidation. *Small* 7(16), 2307–2312.

8 Polymer–Drug Conjugates: Targeted Drug Delivery

Bhanu Malhotra[1], Harsha Kharkwal[2,*] and Amit Kumar Tyagi[3]
[1]*Amity Institute of Biotechnology and Amity Center for Carbohydrate Research, Amity University, Noida, India;* [2]*Amity Center for Carbohydrate Research and Amity Institute of Phytomedicine and Phytochemistry, Amity University Uttar Pradesh, Noida, India;* [3]*Department of Experimental Therapeutics, Division of Cancer Medicine, The University of Texas MD Anderson Cancer Center, Houston, Texas, USA*

Abstract

Polymer therapeutics is a promising area in medicine and has led to the recent development of enhanced, targeted drug delivery systems. The fast-growing field of polymeric drug conjugates, almost a dozen of which are close to the clinical trial stage, have demonstrated several advantages over the parent drugs. These include ease of drug administration with fewer side effects, improved patient compliance, enhanced therapeutic efficacy, concentration and absorption, improved pharmacokinetics and stability. This chapter considers the potential of polymer–drug conjugates, which are going beyond classical methodologies, and their utility for treating major human diseases and new targets.

Introduction

The polymer–drug conjugation approach is thought to improve therapeutic properties of peptides, proteins and oligonucleotides by offering various advantages including prolonged half-life, greater water solubility, increased stability and lower immunogenicity and antigenicity to the native parent drugs. The novel technology of polymeric drug conjugates not only protects the drug from premature degradation but also allows it to be specifically targeted to various cells and tissues in the body. The idea of utilizing polymers as therapeutic agents has been explored for several years. This technology was exploited for the first time by Helmut Ringsdorf in 1975 when he proposed that a polymer that was covalently bound to a drug molecule not only improved the solubility of the drug in water but also increased its bioavailability (Ringsdorf, 1975). These polymer–drug conjugates provide increased drug circulation times and allow slow and controlled release of drugs with time. They provide controlled and targeted release of drugs at the desired site of therapeutic need when there are changes in pH, enzyme concentration, and temperature (Gillies *et al.*, 2004; Torchilin, 2005). The Ringsdorf concept comprises a biocompatible polymeric backbone bound to a solubilizer serving hydrophilicity and increased water solubility; a drug, attached via a linker to the polymer backbone; and a targeting moiety which transports it to a desired physiological destination (Fig. 8.1).

*Corresponding author. E-mail: hkharkwal@amity.edu

Considerable interest is currently directed towards the development of anti-cancer polymer–drug conjugates. Since the well-known anti-cancer agents are generally limited by metabolic instability, poor water solubility and the limitation of dose-dependent toxicity, their role is curbed. The use of polymeric drug conjugates with anti-cancer agents focuses primarily on delivering the anti-cancer moiety at a dosage high enough to create cytotoxicity within the tumour cells. The therapeutic index of a drug is illustrated as the ratio between its toxic and therapeutic dose. The enhancement of the therapeutic index of the anti-cancer agents allows their availability to the tumour cells to be increased, while managing the possible side effects in cancer patients. The sheath effect of the polymer–drug conjugates limits the uptake by macrophages and the reticuloendothelial system (RES). These conjugates increase the therapeutic index of the drug in the tumour tissues by increasing the blood circulation time and utilizing their leaky vasculature and defective lymphatic drainage to advantage. This is referred to as the enhanced permeability and retention effect (EPR) (Matsumura and Maeda, 1986; Maeda, 2001) (Fig. 8.2).

Design of Polymer–Drug Conjugates

The potential of the technology has already been demonstrated, as a wide range of polymer–drug conjugates are available on the market for the treatment of various diseases. A polymeric prodrug is formed during the conjugation of the drug with the polymer. The initial requirement is to find a biocompatible polymer (natural or synthetic) that can be used *in vivo*. Preference is given to biodegradable, nonimmunogenic, stable and non-toxic polymers (Duncan *et al.*, 1992). The prodrug consists of a polymeric carrier, a polymer drug linker and an bioactive antitumour agent. Ligands are also important components that can facilitate active targeting of the drug. A perfect ideal drug targeting system should be able to overcome all the biological barriers which interfere with the drug, in order to target the diseased cells (Jain, 1998; Petrak, 2005).

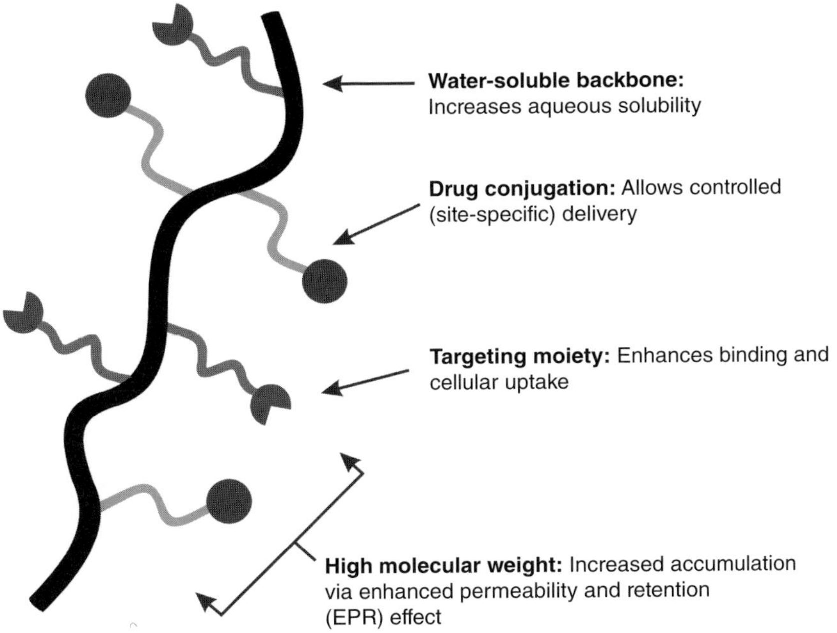

Fig. 8.1. Rationale for drug delivery via polymer–drug conjugates. (Reproduced, with permission, from Larson and Ghandehari, 2012.)

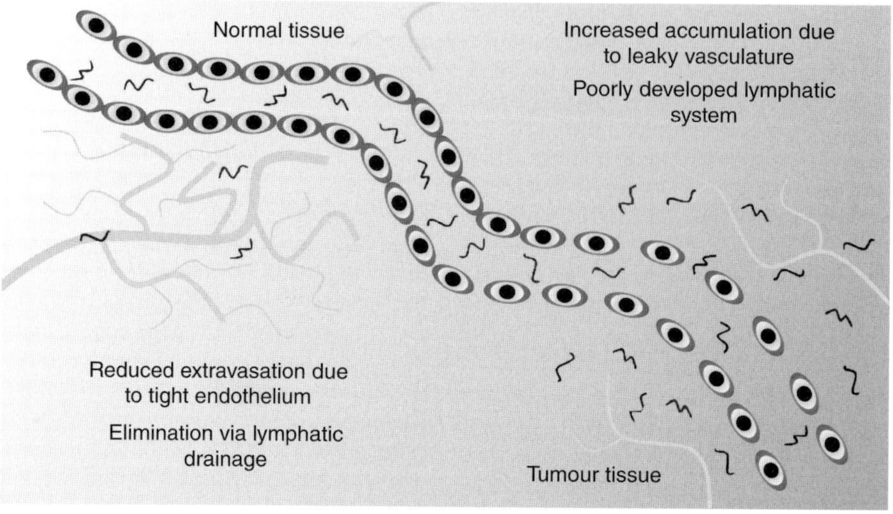

Fig. 8.2. The enhanced permeability and retention or 'EPR' effect; increased tumour accumulation of macromolecules occurs via a combination of increased extravasation and reduced lymphatic drainage in tumour tissues. (Reproduced, with permission, from Larson and Ghandehari, 2012.)

Polymers

Various polymers are used for drug conjugation, including N-(2-hydroxypropyl)methacrylamide (HPMA), polyethylene glycol (PEG), poly-L-glutamic acid (PG), polyvinyl pyrrolidone (PVP) and polyvinyl alcohol (PVA) (Yasukawa *et al.*, 1999; Kamada *et al.*, 2000; Kaneda *et al.*, 2004; Chipman *et al.*, 2006; Malugin *et al.*, 2007; Kopeček and Kopečková, 2010).

HPMA based N-(2-hydroxypropyl)-methacrylamide

HPMA copolymers are biocompatible hydrophilic drug carriers which have been widely used to target the asialoglycoprotein receptor of the hepatocytes. HPMA polymers were developed by Kopeček and his workers (Duncan, 2009; Duncan and Vicent, 2010; Kopeček and Kopečková, 2010; Lammers, 2010), however their increasing popularity initiated more research on them as biocompatible polymeric drug carriers. Hydrolytic stability is conferred by the presence of substituted α-carbon and an amide linkage in the side chain. Different functionalities can easily be incorporated into HPMA polymers using differently functionalized comonomers. In particular, drug side chains, reactive groups, imaging agents and targeting moieties can be used with ease. Certain oligopeptide sequences can be used as drug linkers in the development of HPMA-based copolymers (Šubr *et al.*, 1988). These oligiopeptide sequences ensure hydrolytic stability and lysosomal enzymatic cleavage directing cell internalization (Rejmanová *et al.*, 1985). Certain factors including drug loading, steric hindrance, drug structure, peptide length and structure are very important when considering drug release kinetics and stability (Kopeček, 1984; Putnam and Kopeček, 1995). A number of HPMA polymers that can be released in the presence of cathepsin B, a lysosomal enzyme, are identified with the isolation of the GFLG- glycyl phenylalanyl leucyl glycine tetrapetide sequence (Kasuya *et al.*, 2001; Vicent *et al.*, 2004; Hongrapipat *et al.*, 2008; Ray *et al.*, 2011). The first water-soluble polymeric drug conjugate investigated clinically was PK1 and was used in cancer therapy. It contains an anthracycline antibiotic called doxorubicin which is anti-cancerous and is attached using the tetrapeptide sequence GFLG to the HPMA polymer (Vasey *et al.*, 1999). The formulation of PK1 had 8.5% doxorubicin by weight and a molecular weight of 30 kDa. The tetrapetide linkage of doxorubicin demonstrated stability during intravenous administration, without the release of any free doxorubicin (Duncan *et al.*, 1992) (Fig. 8.3).

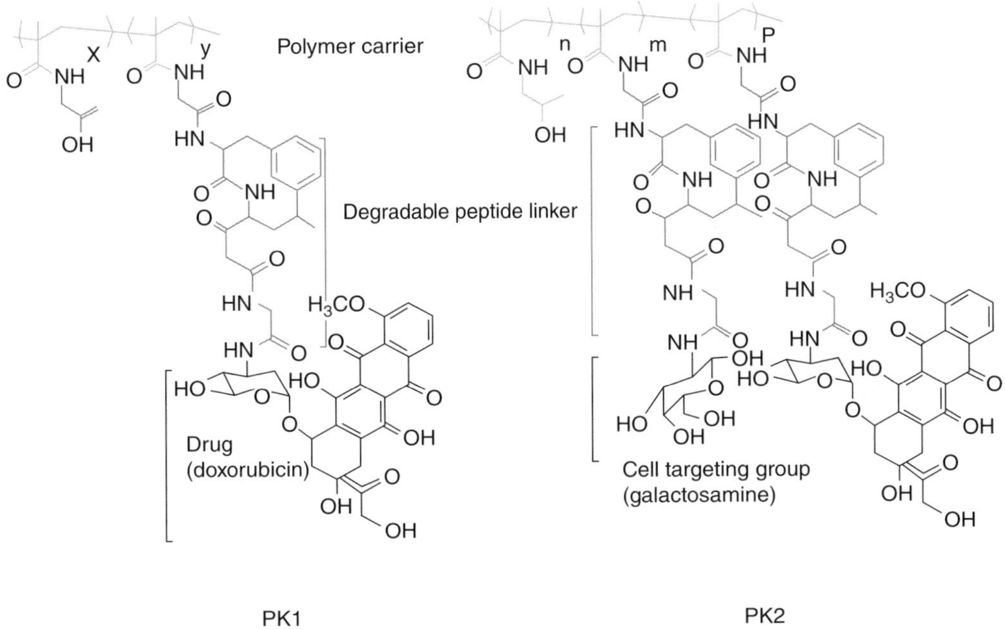

Fig. 8.3. PK1 and PK2. Structure of doxorubicin (PK1) and doxorubicin galactosamine (PK2) bound HPMA copolymer.

Recently, other HPMA copolymer–drug conjugates have been identified, illustrating pH-dependent release of the drug and suitable for use in pH-sensitive areas. Conjugates such as HPMA-based camptothecin (Caiolfa *et al.*, 2000; Gao *et al.*, 2009), taxanes (Caiolfa *et al.*, 2000; Gao *et al.*, 2009; Etrych *et al.*, 2010; Ray *et al.*, 2011), paclitaxel (Meerum Terwogt *et al.*, 2001) and geldanamycin (Borgman *et al.*, 2009; Greish *et al.*, 2011) have been investigated and used.

Poly ethylene glycol

Poly ethylene glycol (PEG) is a US Food and Drug Administration-approved commercial polymer. Conjugation of PEG can be done using structural modifications to the backbone. The result of modification not only enhances the half-life of the drug but also increases its plasma concentration (Pasut *et al.*, 2008). Protein–PEG conjugates have attracted considerable interest as their uptake by the RES is reduced. Examples include Oncaspar (PEG-asparaginase) (Graham, 2003), Pegasys (PEG-interferon α-2a) (Bailon *et al.*, 2001), Adagen (PEG-adensoine deaminase) (Levy *et al.*, 1988), Somavert (PEG-growth hormone receptor antagonist) (Yang, 2009), Neulasta (PEG-granulocyte colony-stimulating factor) (Roelfsema *et al.*, 2006). The use of certain esters or aldehydes, for example N-hydroxysuccinimide, confers conjugation with the amine group of lysine residues. A click chemistry approach has been used to create various functionalized PEGs in a highly specific reaction where byproducts are removed easily under non-toxic conditions (Nwe and Brechbiel, 2009; van Dijk *et al.*, 2009).

PEG-based drug conjugates offer the advantage of being biocompatible (Jevsevar *et al.*, 2010) and are widely synthesized on a large scale with different molecular weight specifications, adding to their potential for the treatment of various diseases. Since PEG is non-biodegradable its use is limited. It has been observed that drug conjugates of 3.5 nm can avoid the renal filtration path, but long-term exposure to them and accumulation over time can lead to toxic effects (Fang *et al.*, 2011). Another important disadvantage of using conventional PEG conjugates is achieving a very low drug loading due to limited conjugation at the end chain. To overcome this, a number of multiarm and branched PEGs have been developed (Zhao *et al.*, 2008; Nojima *et al.*, 2009).

In the process of drug development the safety of the polymer–drug conjugates is the main area of concern. A number of polymer–drug conjugates have progressed through specific clinical trials and are being used commercially. Table 8.1 shows examples of some of these (Pasut and Veronese, 2009; Galic *et al.*, 2011). Since the polymeric drug conjugates frequently exhibit different pharmacokinetic parameters and sometimes also show altered bio distribution, careful consideration must be given to their toxicity studies. The release of the drug from the polymer in a polymer drug conjugate is the key factor to exhibit the pharmacological effect (Malugin *et al.*, 2007; Larson *et al.*, 2010). Usually the drug is not released during systemic transport, but if it releases prematurely it can lead to undesired toxic effects in the body and the overall efficacy and safety profile of the polymeric drug conjugate will be poor (Meerum Terwogt *et al.*, 2001). Therefore, to achieve the desired efficacy along with safety in use, the ideal polymer–drug conjugate should exhibit a perfect balance between the drug–conjugate stability and drug release. A class of specific polymer–drug conjugates (i.e. those that can be used in temperature-sensitive and pH-sensitive systems, and are biodegradable) are also emerging and have been discussed further in this chapter.

Biodegradable polymers

Escape of the drug via kidney filtration and increase in the blood circulation time have been achieved using polymer–drug conjugates. When supplying anti-cancer agents to the tumour tissues these show greater accumulation in the tumour area due to the EPR effect. As discussed previously, drug conjugates of 3.5 nm can avoid the renal filtration path, but their long-term exposure and accumulation over time can lead to toxic effects hence the excretion of these carriers is a matter of concern (Fang *et al.*, 2011). It has been reported that if the conjugates are of a particular size they can not only escape kidney filtration, but can also help in the subsequent timely elimination and excretion from the body. This can only be achieved by using biodegradable polymers as drug carriers, because they have a slower rate of degradation which can facilitate the appropriate biodistribution and produce degradation products that are non-toxic. Biodegradation is generally achieved via enzymatic cleavage, hydrolysis and reductive degradation processes (Sun *et al.*, 2011). A number of biodegradable polymers have been investigated, particularly poly α-amino acids: polysialic acid (Pisal *et al.*, 2010), poly-L-lysine (Couffin-Hoarau

Table 8.1. The advent of polymeric drug conjugates in clinical trials. (Adapted, with permission, from Mohanty *et al.*, 2015.)

Polymeric drug conjugates	Stage of clinical trial	Target	Company
N-(2-hydroxypropyl)methacrylamide (HPMA) drug conjugates			
PK1 HPMA–doxorubicin (FCE28068)	Phase II	Breast and lung cancer	Pfizer UK
PK2 HPMA–doxorubicin galactosamine (FCE28069)	Phase I, II	Liver cell carcinoma	Pfizer UK
HPMA–palatinate (ProLindac) (AP5280)	Phase I, II	Colorectal and ovarian cancers	Access Pharmaceutical
HPMA–paclitaxel (PNU166945)	Phase I stopped	Tumours	Pfizer UK
Poly-L-glutamic acid (PG) drug conjugates			
PG–paclitaxel (Xyotax) (CT-2103)	Phase III	Ovarian, breast, lung and colorectal cancers	Cell Therapeutics
PG–camptothecin (CT2106)	Phase I	Ovarian, lung and colorectal cancers	Cell Therapeutics
Poly ethylene glycol (PEG) drug conjugates			
PEG–Camptothecin (Pegamotecan)	Phase I stopped	Tumours	Enzon Pharmaceuticals
PEG–SN38 (EZN–2208)	Phase I	Tumours	Enzon Pharmaceuticals

et al., 2009), dextrins (Baldwin and Kiick, 2010) and poly N-hydroxyalkyl glutamine (Metselaar *et al.*, 2003).

OPAXIO™, a biodegradable polymeric drug conjugate formerly marketed as XYOTAX, is under Phase III clinical trials (to expand its use in certain breast, ovarian, lung and colorectal cancers; Beer *et al.*, 2010). It is a conjugate of poly(L-glutamic acid) and the anticancer agent paclitaxel (Chipman *et al.*, 2006). Since poly (L-glutamic acid) is not excreted by renal clearance from the kidneys, Opaxio's water soluble paclitaxel has demonstrated a considerably higher maximum tolerated dose (MTD). Work is now focused on developing biodegradable derivatives of traditional polymers such as HPMA and PEG. A PEG-based multiarm biodegradable polymer, PEG 57(ENZ-2208), is currently in clinical trials. Also a biodegradable polymer contacting lower molecular weight blocks of PEG which are grouped via oligopeptide groups that can be cleaved enzymatically containing doxorubicin are formed (Pechar *et al.*, 2000).

Stimulus-sensitive polymers

The activity of anti-cancerous agents is limited because of insufficient cell internalization, severe toxic side effects, unsuitable physical and chemical properties, limited targeting and drug resistance. The focus is now to develop smart polymers which are stimuli specific and respond to changing environmental conditions such as temperature, pH, ionic strength, heat and ultrasound (Torchilin, 2009). These stimulus-specific polymers help in drug release, drug target and endosomal escape because of minute electrostatic and conformational changes.

It has often been noted that the pH of diseased and affected areas such as infarcts, tumours and sites of inflammation can lower slightly to 6.5, which is almost one unit away from the normal body blood pH of 7.4. This generally occurs in hypoxic conditions and cell death (Wike-Hooley *et al.*, 1984). It has also been found that the pH of late endosomes decreases to 5.0, thereby providing a gradient which can help release drugs (Bawa *et al.*, 2009). The polymer backbone can be engineered and made sensitive to small changes in pH, via acidic or basic groups that undergo rapid protonation or deprotonation in response to small changes in pH. Polymers reported with this type of chemistry are poly acrylamide, poly methacrylic acid, poly acrylic acid and poly(2-(dimethylamino)ethyl methacrylate (Bawa *et al.*, 2009).

The sensitivity to pH can be introduced primarily via pH-sensitive chemical bonds which help in the site-specific delivery of the drug. For polymeric-drug conjugation, pH sensitivity systems are a major breakthrough for facilitating pH sensitive release of drugs. For example, development of hydrazone bonds formed via the action of hydrazine on ketones or aldehydes exhibit hydrolysis under mildly acidic conditions (pH 5–6) such as that present in lysosomes while maintaining stability at pH values found in blood (pH 7.4); the underlining mechanism being the formation of chemical bonds which can result in site specific drug delivery (Howard *et al.*, 2011). Poly(acrylic acid)-based polymeric drug conjugates have been used for oral drug delivery, where they maintain their therapeutic effect in the stomach's acidic environment, and release the encapsulated drug in the alkaline environment of the small intestine because of swelling of the polymer matrix and ionization of the carboxylic acid groups.

It has been reported that diseased tissues and cells show considerably elevated temperature levels. Externally, hyperthermia can be used to trigger changes in the polymer structure that releases the drug. Polymers based on poly(N-isopropylacrylamide) can be used as aqueous temperature-sensitive polymers which offer a lower critical solution phase transition temperature. As a result of this, the collapse of polymeric chains occurs above the lower critical solution temperature (LCST) phase transition because of the dehydration of the hydrocarbon side chains (Chilkoti *et al.*, 2002). Genetic engineering has been used to create recombinant elastin such as polymers with defined composition and molecular weight properties (Trabbic-Carlson *et al.*, 2004). These contain mammalian tropoelastin motif tropoelastin (VPGXG)n, where X is any amino acid except proline (Tatham and Shewry, 2000). Modifications in composition and molecular weight can be used to create thermal-sensitive elastin-like polypeptides (ELPs). The ELP made with an LCST above 43°C, i.e. above the normal body temperature, has been used as an anticancer polymeric drug carrier generating localized hyperthermia in tumour cells (MacKay and Chilkoti, 2008).

Polymer drug linker

A linker is a component that facilitates specific temporal and spatial drug release in the tumour region and maintains the stability of the polymer–drug conjugate. The linker not only enhances the biodistribution of the drug, but also enables specific increase in the circulatory half-life of the prodrug from the polymeric backbone, until it receives a disease-specific trigger known as tissue-activated drug delivery (TADD) (Greish *et al.*, 2003). The enzymatic and hydrolytic cleavage of the linker occurs when the drug is to be released from it, which can be done by esterases. Therefore, a specific linker can be developed, which can stabilize the drug before its release in the tumour cells.

Ligands

The ligands function not only to increase the receptor-mediated binding of the polymeric drug, but also to help in internalization of the drug to the target cells. This leads to an increase in the efficacy and proficiency of a polymeric drug conjugate. The characteristic feature of tumour cells is over-expression of cell surface markers, which enable increased cellular uptake and active targeting. The ligand molecule can be a peptide, sugar moiety, hormone, antibody, etc. Considerable work has been done on ligand-based polymer–drug conjugates, and good results have been obtained by improvement in the anti-cancer activities of native drugs.

The ultimate goal of drug delivery research is to help patients by developing clinically useful formulations and targeted drug delivery to solid tumours; to minimize drug-originated systemic toxic effects is the main purpose of polymer drug conjugates. Drug targeting strategies have frequently been divided into categories of 'passive' and 'active'. These terms, however, do not represent what is really occurring *in vivo*, and tend to cause misunderstandings in defining a specific drug targeting strategy. So-called 'passive targeting' is based on drug accumulation in the areas around the tumours with leaky vasculature; commonly referred to as the enhanced permeation and retention (EPR) effect, while the term 'active targeting' simply means a specific 'ligand–receptor type interaction' for intracellular localization which occurs only after blood circulation and extravasation.

Dendrimers

Polymeric macromolecules known as dendrimers form a star-like structure which helps in the attachment and conjugation of the drug to the surface. Dendrimers help in maximizing the biological interaction potential and they can be synthesized with specificity of the monomeric units, core and surface functionality, which help in determining their physicochemical properties (Fig. 8.4). A dendrimer consists of an inner core bearing various generations which consist of exterior chemical groups (Fig. 8.5). Nanostructured dendrimer macromolecules have potential abilities in entrapment and conjugation of high molecular weight hydrophobic and hydrophilic moieties by the prodrug approach and host–guest interactions. The synthesis of poly(amido amine) dendrimer has also been reported (Tomalia *et al.*, 1986).

With respect to drug delivery, dendrimers have been associated with the formulation and encapsulation of various drugs (Gajbhiye *et al.*, 2009). Their extensive hyperbranching leads to open cavities between adjacent branches, enabling drug encapsulation (Bhadra *et al.*, 2003). This increases the solubility of poorly soluble drugs. Dendrimer-based polymeric drug conjugates have been administered orally as well as transdermally (Cheng *et al.*, 2007). Dendrimers can also behave as gene carriers as they can be complexed with DNA (Xu *et al.*, 2010). They have been used for drug delivery across biological membranes across the blood–brain barrier (Beg *et al.*, 2011), intestinal epithelia (Sadekar and Ghandehari, 2012), skin (Cheng *et al.*, 2007) and human placenta (Menjoge *et al.*, 2011). Various biodegradable dendrimers, asymmetric dendrimers and glycodendrimers have been identified (Cao *et al.*, 2010; Raghupathi *et al.*, 2011).

Despite much progress in drug delivery through dendrimers, their use is still debatable due to growing concerns of toxicity and biocompatibility. Dendrimers exhibit a considerably high affinity for peptides, amino acids, proteins, lipids and metal ions, therefore they cause

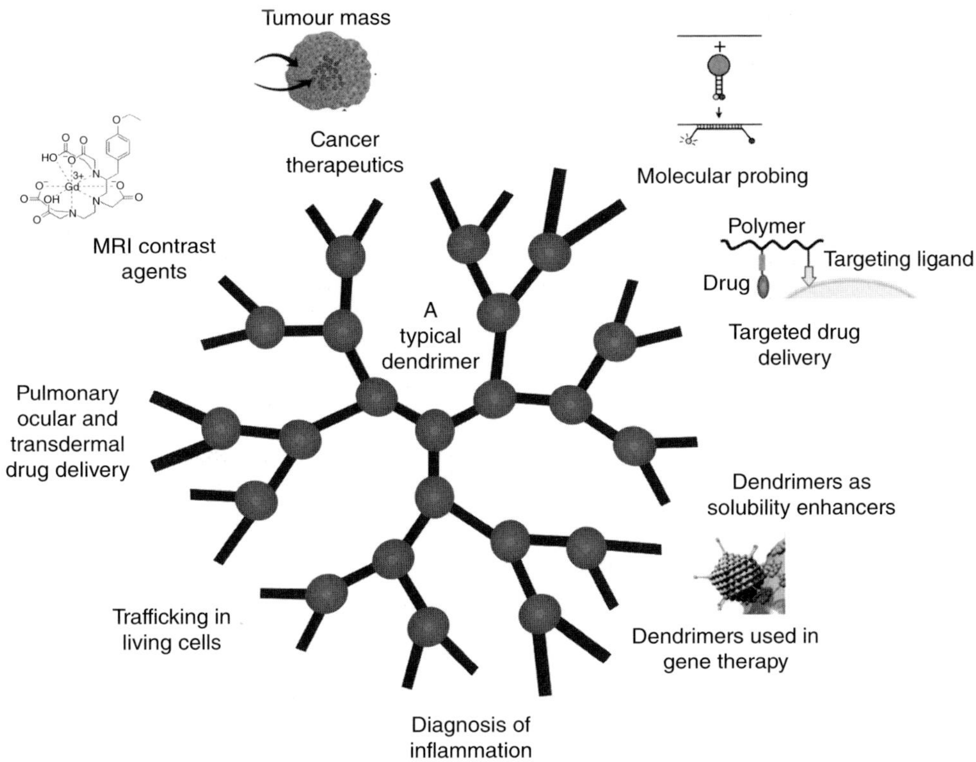

Fig. 8.4. A typical dendrimer. Nanostructured dendrimers.

toxicity by disrupting biological processes (Cheng *et al.*, 2011). Toxicities have been demonstrated both *in vivo* and in *vitro* (Ziemba *et al.*, 2011). There is, therefore, now a demand for biocompatible non-toxic dendrimer surface modifications which can be clinically trialled.

Targeting

Tremendous efforts are being made to develop therapeutic polymeric drug conjugates to overcome all possible limitations of this system. Physical properties of the polymer–drug conjugate system, such as surface charge, molecular weight, size, absorption, metabolism, excretion and elimination are important areas of concern to elicit proper targeting (Bae and Park, 2011). Drug conjugates of 3.5 nm can avoid the renal filtration path, but long-term exposure to them and accumulation over time can lead to toxic effects, hence the excretion of these carriers is a matter of concern. Conjugates with long blood circulation times increase the opportunity for greater interaction with their target. Targeting can be an active procedure which helps the conjugate to localize and internalize into target cells over smaller distances. Passive targeting, in contrast, is a concept that ensures the delivery of a polymeric drug conjugate at its specific, intended site of action. Future work should focus on the characteristics of an ideal target that is universally expressed by the tumour tissue. As the cancer cells overexpress a particular cell surface marker in comparison to normal cells, it is probable that polymeric drug conjugate binding to the tumour tissue is higher than in normal cells. The differences in the physiology of normal and cancerous tissues can, therefore, help in developing an ideal target polymeric drug conjugate for the latter. An ideal target should also help in endocytosis, helping the drug to exert its pharmacological action within the diseased cell (Danhier *et al.*, 2010).

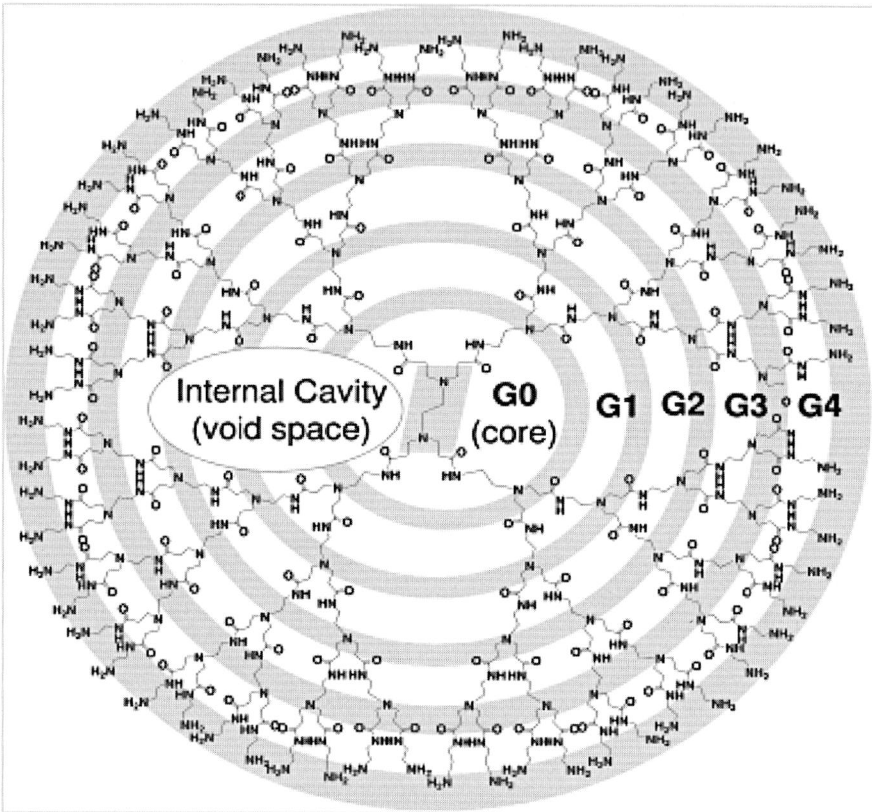

Fig. 8.5. Schematic diagram of a typical dendrimer. (Reproduced, with permission, from Wang *et al.*, 2012. Copyright Wiley-VCH Verlag GmbH & Co. KGaA.)

Hormones, sugars, proteins, growth factors, antibodies and their fragments can be utilized as targeting moieties (Torchilin, 2005). The target entity used should enable the stability of the polymeric drug conjugates during systemic circulation. Antibodies and their fragments show profound target selectivity and affinity, but the use of antibodies is limited due to immunogenicity concerns and the cost of production. However, the antibody fragments Fab and Fc can be synthesized in bacterial cultures and so decrease the cost. A number of factors should, therefore, be considered in selecting an ideal target for polymer–drug conjugates.

Nanotechnology-based Polymer–Drug Conjugates

Packaging small-molecular drugs into nanoparticles enhances bioavailability, compatibility and safety. Targeting particles to a diseased site can block a function there, deliver a drug to it and perform the necessary action. Nanocarriers used for targeting must make their way through blood–tissue barriers, with different strengths between organs and to highest in the brain, and finally to reach the destined target cells. They must enter the target cells to contact their cytoplasmic targets, if any.

The nanoparticles used for targeting are liposomes, dendrimers, protein nanoparticles and carbon nanoparticles such as fullerenes. Successful immunotargeting can also be achieved using monoclonal antibodies or chimeric antibodies/ humanized antibodies. This is currently being investigated in clinical trials.

Solid nanoparticles

Solid nanoparticles are mainly solid constructs prepared using techniques to reduce particle size and agglomerate molecules so they fall into the nano-size range. Agglomeration in particular is used to prepare nanocarriers of drugs. The reduction in size greatly facilitates the solubility of the drug due to an increase in its total surface area. This property is well exploited when the route of administration of the drug is oral, as it increases bioavailability through dissolution of the drug. One of the important concerns of nano-sized particles is that because of their increased surface area and high interfacial energy they tend to aggregate and then become unstable. This limitation can be overcome by use of surfactants during their preparation. Nanocrystals can be formulated into tablets or capsules, particularly for oral delivery.

Inorganic nanoparticles

Metal sulfides, calcium phosphates, carbon nanotubes and ceramics can be fabricated and used. Since they are not biodegradable, they generally have limited application. A metal oxide nanoparticle product known as Alcoscan is commercially available. This is administered orally and used in MRI bowel diagnostics because of its decreased toxicity and signal strength.

Liposomes

Liposomes are spherical vesicles consisting of bilayered membranes alternating with an aqueous core. These membranes are made up of phospholipids or amphiphilic lipids. Most research is directed towards them, and they are preferred because of their *in vivo* safety and versatility in vesicular transport of drugs. Various clinically approved liposomes are presented in Table 8.2.

Liposomes can carry water-soluble as well as lipophilic molecules, as the drugs soluble in water can be contained within the aqueous layer and the drugs soluble in lipids within the bilayer. Certain cationic liposomes can attract proteins and nucleic acids to their surface using electrostatic interactions. The size range of liposomes can extend from 30 nm to several micrometres. The classification of liposomes into small unilamellar vesicles, large unilamellar vesicles, multilamellar vesicles and multivesicular vesicles is done using the number of the lamellae and their size. Liposomes in cancer therapy can yield the best results. One example is the commercially available Myocet, a doxorubicin containing a liposome formulation. The Myocet

Table 8.2. Clinically approved liposomes. (Adapted, with permission, from Allen and Cullis, 2013.)

Commercial name	Formulation	Drug	Clinical manifestation
AmBisome	Intercalated amphotericin B within the liposome membrane with hydrogenated soy cholesterol, phosphatidylcholine, alpha tocopherol, disteroylphosphatidyl glycerol	Amphotericin B	Treatment of fungal infections administration is through the intravenous route
DaunoXome	Suv liposomes made up of cholesterol and distearylphosphatidylcholine, containing daunorubicin in the aqueous core of the liposome	Daunorubicin	Treatment of Kaposi's Sarcoma
Doxil	Intercalated doxorubicin within the aqueous core with hydrogenated soy cholesterol, phosphatidylcholine, ethanolamine, etc.	Doxorubicin	Breast cancer, ovarian cancers, AIDS caused Kaposi's Sarcoma
Epaxal	Bilayered vesicles made from purified influenza virus, cephalin, phosphatidylcholine	Inactivated Hepatitis A Virus	Vaccine for Hepatitis A

liposome is 180 nm and is made up of cholesterol and egg phosphatidylcholine. It creates a Mononuclear Phagocyte System (MPS) depot which helps a very slow drug release in the blood circulation. The future of polymer–drug conjugates involves various drugs at different stages of clinical trials, and work is still going on to develop the best formulations to enhance drug delivery.

The Future of Polymeric Drug Conjugates

A significant number of synthetic polymers such as HPMA, PG and PEG show great promise, yet considerable research is still needed on the availability and use of biodegradable polymers to enhance drug delivery to target tissues. In cases where loading is significantly low, ~10%, the pertaining of the polymeric backbone after treatment is a big disadvantage. Synthetic polymers cannot take full advantage of a long blood circulation time, and their molecular weight causes early renal elimination, so much more research is needed to address these concerns. There has also been a growing interest in the increased development of stimuli-sensitive polymers as advanced drug delivery vehicles to respond to all types of biologically relevant stimuli, which can exploit even minor changes in physiological signals and help in the timely release of the drug in more sustained fashion. The future of polymer therapeutics will result in the development of various drugs, following clinical trials, through the combined efforts of medicinal chemists, researchers, pharmaceutical scientists and clinicians. Synergistic collaborations will certainly further novel carrier technologies with increased pharmacological effects.

Acknowledgements

We thank Dr Ashok K Chauhan, Founder President, Ritanand Balved Educational Foundation, for support. The guidance from Shri. Atul Chauhan, Chancellor, Amity University Uttar Pradesh and Prof. Dr (Mrs) Balvinder Shukla, Vice Chancellor, Amity University Uttar Pradesh is greatly appreciated. Insights from articles cited in this manuscript have been of considerable assistance, and we are grateful to their authors.

References

Allen, T.M., and Cullis, P.R. (2013) Liposomal drug delivery systems: From concept to clinical application. *Advanced Drug Delivery Reviews: Perspectives and Prospects* 65(1), 36–48.

Bae, Y. and Park, K. (2011) Targeted drug delivery to tumors: Myths, reality and possibility. *Journal of Controlled Release* 153(3), 198–205.

Bailon, P., Palleroni, A., Schaffer, C., Spence, C., Fung, W., Porter, J., Ehrlich, G., Pan, W., Xu, Z., Modi, M., Farid, A., Berthold, W. and Graves, M. (2001) Rational design of a potent, long-lasting form of interferon: a 40 kda branched polyethylene glycol-conjugated interferon α-2a for the treatment of hepatitis C. *Bioconjugate Chemistry* 12(2), 195–202.

Baldwin, A. and Kiick, K. (2010) Polysaccharide-modified synthetic polymeric biomaterials. *Biopolymers* 94(1), 128–140.

Bawa, P., Pillay, V., Choonara, Y. and du Toit, L. (2009) Stimuli-responsive polymers and their applications in drug delivery. *Biomedical Materials* 4(2), 022001.

Beer, T., Ryan, C., Alumkal, J., Ryan, C., Sun, J. and Eilers, K. (2010) A phase II study of paclitaxel poliglumex in combination with transdermal estradiol for the treatment of metastatic castration-resistant prostate cancer after docetaxel chemotherapy. *Anti-Cancer Drugs* 21(4), 433–438.

Beg, S., Samad, A., Alam, M.I. and Nazish, I. (2011) Dendrimers as novel systems for delivery of neuropharmaceuticals to the brain. *CNS & Neurological Disorders – Drug Targets* 10(5), 576–588.

Bhadra, D., Bhadra, S., Jain, S. and Jain, N. (2003) A PEGylated dendritic nanoparticulate carrier of fluorouracil. *International Journal of Pharmaceutics* 257(1–2), 111–124.

Borgman, M., Ray, A., Kolhatkar, R., Sausville, E., Burger, A. and Ghandehari, H. (2009) Targetable HPMA copolymer–aminohexylgeldanamycin conjugates for prostate cancer therapy. *Pharmaceutical Research* 26(6), 1407–1418.

Caiolfa, V., Zamai, M., Fiorino, A., Frigerio, E., Pellizzoni, C., d'Argy, R., Ghiglieri, A., Castelli, M., Farao, M., Pesenti, E., Gigli, M., Angelucci, F. and Suarato, A. (2000) Polymer-bound camptothecin: initial biodistribution and antitumour activity studies. *Journal of Controlled Release* 65(1–2), 105–119.

Cao, W., Zhou, J., Wang, Y. and Zhu, L. (2010) Synthesis and in vitro cancer cell targeting of folate-functionalized biodegradable amphiphilic dendrimer-like star polymers. *Biomacromolecules* 11(12), 3680–3687.

Cheng, Y., Xu, Z., Ma, M. and Xu, T. (2007) Dendrimers as drug carriers: Applications in different routes of drug administration. *Journal of Pharmaceutical Sciences* 97(1), 123–143.

Cheng, Y., Zhao, L., Li, Y. and Xu, T. (2011) Design of biocompatible dendrimers for cancer diagnosis and therapy: current status and future perspectives. *Chemical Society Reviews* 40(5), 2673.

Chilkoti, A., Dreher, M., Meyer, D. and Raucher, D. (2002) Targeted drug delivery by thermally responsive polymers. *Advanced Drug Delivery Reviews* 54(5), 613–630.

Chipman, S., Oldham, F., Pezzoni, G. and Singer, J. (2006) Biological and clinical characterization of paclitaxel poliglumex (PPX, CT-2103), a macromolecular polymer–drug conjugate. *International Journal of Nanomedicine* 1(4), 375–383.

Couffin-Hoarau, A., Aubertin, A., Boustta, M., Schmidt, S., Fehrentz, J., Martinez, J. and Vert, M. (2009) Peptide–poly(l -lysine citramide) conjugates and their in vitro anti-HIV behavior. *Biomacromolecules* 10(4), 865–876.

Danhier, F., Feron, O. and Préat, V. (2010) To exploit the tumor microenvironment: passive and active tumor targeting of nanocarriers for anti-cancer drug delivery. *Journal of Controlled Release* 148(2), 135–146.

Duncan, R. (1992) Drug-polymer conjugates. *Anti-Cancer Drugs* 3(3), 175–210.

Duncan, R. (2009) Development of HPMA copolymer–anticancer conjugates: clinical experience and lessons learnt. *Advanced Drug Delivery Reviews* 61(13), 1131–1148.

Duncan, R. and Vicent, M. (2010) Do HPMA copolymer conjugates have a future as clinically useful nanomedicines? A critical overview of current status and future opportunities. *Advanced Drug Delivery Reviews* 62(2), 272–282.

Duncan, R., Seymour, L., O'Hare, K., Flanagan, P., Wedge, S., Hume, I., Ulbrich, K., Strohalm, J., Subr, V., Spreafico, F., Grandi, M., Ripamonti, M., Farao, M. and Suarato, A. (1992) Preclinical evaluation of polymer-bound doxorubicin. *Journal of Controlled Release* 19(1–3), 331–346.

Etrych, T., Šírová, M., Starovoytova, L., Říhová, B. and Ulbrich, K. (2010) HPMA copolymer conjugates of paclitaxel and docetaxel with pH-controlled drug release. *Molecular Pharmaceutics* 7(4), 1015–1026.

Fang, J., Nakamura, H. and Maeda, H. (2011) The EPR effect: unique features of tumor blood vessels for drug delivery, factors involved, and limitations and augmentation of the effect. *Advanced Drug Delivery Reviews* 63(3), 136–151.

Gajbhiye, V., Palanirajan, V., Tekade, R. and Jain, N. (2009) Dendrimers as therapeutic agents: a systematic review. *Journal of Pharmacy and Pharmacology* 61(8), 989–100.

Galic, V., Wright, J., Lewin, S. and Herzog, T. (2011) Paclitaxel poliglumex for ovarian cancer. *Expert Opinion on Investigational Drugs* 20(6), 813–821.

Gao, S., Sun, Y., Kopečková, P., Peterson, C. and Kopeček, J. (2009) Antitumor efficacy of colon-specific hpma copolymer/9-aminocamptothecin conjugates in mice bearing human-colon carcinoma xenografts. *Macromolecular Bioscience* 9(11), 1135–1142.

Gillies, E., Goodwin, A. and Fréchet, J. (2004) Acetals as pH-sensitive linkages for drug delivery. *Bioconjugate Chemistry* 15(6), 1254–1263.

Graham, M. (2003) Pegaspargase: a review of clinical studies. *Advanced Drug Delivery Reviews* 55(10), 1293–1302.

Greish, K., Fang, J., Inutsuka, T., Nagamitsu, A. and Maeda, H. (2003) Macromolecular therapeutics. *Clinical Pharmacokinetics* 42(13), 1089–1105.

Greish, K., Ray, A., Bauer, H., Larson, N., Malugin, A., Pike, D., Haider, M. and Ghandehari, H. (2011) Anticancer and antiangiogenic activity of HPMA copolymer-aminohexylgeldanamycin-RGDfK conjugates for prostate cancer therapy. *Journal of Controlled Release* 151(3), 263–270.

Hongrapipat, J., Kopeckova, P., Prakongpan, S. and Kopeček, J. (2008) Enhanced antitumor activity of combinations of free and HPMA copolymer-bound drugs. *International Journal of Pharmaceutics* 351(1–2), 259–270.

Howard, M., Ponta, A., Eckman, A., Jay, M. and Bae, Y. (2011) Polymer micelles with hydrazone-ester dual linkers for tunable release of dexamethasone. *Pharmaceutical Research* 28(10), 2435–2446.

Jain, R. (1998) The next frontier of molecular medicine: delivery of therapeutics. *Nature Medicine* 4(6), 655–657.

Jevsevar, S., Kunstelj, M. and Porekar, V. (2010) PEGylation of therapeutic proteins. *Biotechnology Journal* 5(1), 113–128.

Kamada, H., Tsutsumi, Y., Yamamoto, Y., Kihira, T., Kaneda, Y., Mu, Y., Kodaira, H., Tsunoda, S.I., Nakagawa, S. and Mayumi, T. (2000) Antitumor activity of tumor necrosis factor-alpha conjugated with polyvinylpyrrolidone on solid tumors in mice. *Cancer Research* 60(22), 6416–6420.

Kaneda, Y., Tsutsumi, Y., Yoshioka, Y., Kamada, H., Yamamoto, Y., Kodaira, H., Tsunoda, S., Okamoto, T., Mukai, Y., Shibata, H., Nakagawa, S. and Mayumi, T. (2004) The use of PVP as a polymeric carrier to improve the plasma half-life of drugs. *Biomaterials*, 25(16), 3259–3266.

Kasuya, Y., Lu, Z., Kopečková, P., Minko, T., Tabibi, S. and Kopeček, J. (2001) Synthesis and characterization of HPMA copolymer–aminopropylgeldanamycin conjugates. *Journal of Controlled Release* 74(1–3), 203–211.

Kopeček, J. (1984) Controlled biodegradability of polymers – a key to drug delivery systems. *Biomaterials* 5(1), 19–25.

Kopeček, J. and Kopečková, P. (2010) HPMA copolymers: origins, early developments, present, and future. *Advanced Drug Delivery Reviews* 62(2), 122–149.

Lammers, T. (2010) Improving the efficacy of combined modality anticancer therapy using HPMA copolymer-based nanomedicine formulations. *Advanced Drug Delivery Reviews* 62(2), 203–230.

Larson, N. and Ghandehari, H. (2012) Polymeric conjugates for drug delivery. *Chemistry of Materials* 24(5), 840–853. DOI:10.1021/cm2031569.

Larson, N., Ray, A., Malugin, A., Pike, D. and Ghandehari, H. (2010) HPMA copolymer-aminohexylgeldanamycin conjugates targeting cell surface expressed GRP78 in prostate cancer. *Pharmaceutical Research* 27(12), 2683–2693.

Levy, Y., Hershfield, M., Fernandez-Mejia, C., Polmar, S., Scudiery, D., Berger, M. and Sorensen, R. (1988) Adenosine deaminase deficiency with late onset of recurrent infections: response to treatment with polyethylene glycol-modified adenosine deaminase. *The Journal of Pediatrics* 113(2), 312–317.

Maeda, H. (2001) The enhanced permeability and retention (EPR) effect in tumor vasculature: the key role of tumor-selective macromolecular drug targeting. *Advances in Enzyme Regulation* 41(1), 189–207.

MacKay, J.A. and Chilkoti, A. (2008) Temperature sensitive peptides: engineering hyperthermia-directed therapeutics. *International Journal of Hyperthermia* 24(6), 483–495.

Malugin, A., Kopečková, P. and Kopeček, J. (2007) Liberation of doxorubicin from HPMA copolymer conjugate is essential for the induction of cell cycle arrest and nuclear fragmentation in ovarian carcinoma cells. *Journal of Controlled Release* 124(1–2), 6–10.

Matsumura, Y. and Maeda, H. (1986) A new concept for macromolecular therapeutics in cancer chemotherapy: mechanism of tumoritropic accumulation of proteins and the antitumor agent smancs. *Cancer Research* 46(12 Pt 1), 6387–6392.

Meerum Terwogt, J., ten Bokkel Huinink, W., Schellens, J., Schot, M., Mandjes, I., Zurlo, M., Rocchetti, M., Rosing, H., Koopman, F. and Beijnen, J. (2001) Phase I clinical and pharmacokinetic study of PNU166945, a novel water-soluble polymer-conjugated prodrug of paclitaxel. *Anti-Cancer Drugs* 12(4), 315–323.

Menjoge, A., Rinderknecht, A., Navath, R., Faridnia, M., Kim, C., Romero, R., Miller, R. and Kannan, R. (2011) Transfer of PAMAM dendrimers across human placenta: prospects of its use as drug carrier during pregnancy. *Journal of Controlled Release* 150(3), 326–338.

Metselaar, J., Bruin, P., de Boer, L., de Vringer, T., Snel, C., Oussoren, C., Wauben, M., Crommelin, D., Storm, G. and Hennink, W. (2003) A novel family of l-amino acid-based biodegradable polymer–lipid conjugates for the development of long-circulating liposomes with effective drug-targeting capacity. *Bioconjugate Chemistry* 14(6), 1156–1164.

Mohanty, A.K., Dilnawaz, F., Prasad Mohanta, G. and Kumar Sahoo, S. (2015) Polymer–drug conjugates for targeted drug delivery. In: Devarajan, P.V. and Jain, S. (eds.) *Targeted Drug Delivery: Concepts and Design.* Springer, Heidelberg, Germany, pp. 389–407.

Nojima, Y., Suzuki, Y., Yoshida, K., Abe, F., Shiga, T., Takeuchi, T., Sugiyama, A., Shimizu, H. and Sato, A. (2009) Lactoferrin conjugated with 40-kDa branched poly(ethylene glycol) has an improved circulating half-life. *Pharmaceutical Research* 26(9), 2125–2132.

Nwe, K. and Brechbiel, M. (2009) Growing applications of "click chemistry" for bioconjugation in contemporary biomedical research. *Cancer Biotherapy & Radiopharmaceuticals* 24(3), 289–302.

Pasut, G. and Veronese, F. (2009) PEG conjugates in clinical development or use as anticancer agents: An overview. *Advanced Drug Delivery Reviews* 61(13), 1177–1188.

Pasut, G., Sergi, M. and Veronese, F. (2008) Anti-cancer PEG-enzymes: 30 years old, but still a current approach. *Advanced Drug Delivery Reviews* 60(1), 69–78.

Pechar, M., Ulbrich, K., Šubr, V., Seymour, L. and Schacht, E. (2000) Poly(ethylene glycol) multiblock copolymer as a carrier of anti-cancer drug doxorubicin. *Bioconjugate Chemistry* 11(2), 131–139.

Petrak, K. (2005) Essential properties of drug-targeting delivery systems. *Drug Discovery Today* 10(23–24), 1667–1673.

Pisal, D., Kosloski, M. and Balu-Iyer, S. (2010) Delivery of therapeutic proteins. *Journal of Pharmaceutical Sciences* 99(6), 2557–2575.

Putnam, D. and Kopeček, J. (1995) Enantioselective release of 5-fluorouracil from n-(2-hydroxypropyl)methacrylamide-based copolymers via lysosomal enzymes. *Bioconjugate Chemistry* 6(4), 483–492.

Raghupathi, K., Azagarsamy, M. and Thayumanavan, S. (2011) Guest-release control in enzyme-sensitive, amphiphilic-dendrimer-based nanoparticles through photochemical crosslinking. *Chemistry – A European Journal* 17(42), 11752–11760.

Ray, A., Larson, N., Pike, D., Grüner, M., Naik, S., Bauer, H., Malugin, A., Greish, K. and Ghandehari, H. (2011) Comparison of active and passive targeting of docetaxel for prostate cancer therapy by hpma copolymer–rgDfK conjugates. *Molecular Pharmaceutics* 8(4), 1090–1099.

Rejmanová, P., Kopeček, J., Duncan, R. and Lloyd, J. (1985) Stability in rat plasma and serum of lysosomally degradable oligopeptide sequences in N-(2-hydroxypropyl) methacrylamide copolymers. *Biomaterials* 6(1), 45–48.

Ringsdorf, H. (1975) Structure and properties of pharmacologically active polymers. *Journal of Polymer Science: Polymer Symposia* 51(1), 135–153.

Roelfsema, F., Biermasz, N., Pereira, A. and Romijn, J. (2006) Nanomedicines in the treatment of acromegaly: focus on pegvisomant. *International Journal of Nanomedicine* 1(4), 385–398.

Sadekar, S. and Ghandehari, H. (2012) Transepithelial transport and toxicity of PAMAM dendrimers: implications for oral drug delivery. *Advanced Drug Delivery Reviews* 64(6), 571–588.

Šubr, V., Kopeček, J., Pohl, J., Baudyš, M. and Kostka, V. (1988) Cleavage of oligopeptide side-chains in N-(2-hydroxypropyl)meth-acrylamide copolymers by mixtures of lysosomal enzymes. *Journal of Controlled Release* 8(2), 133–140.

Sun, H., Meng, F., Dias, A., Hendriks, M., Feijen, J. and Zhong, Z. (2011) α-Amino acid containing degradable polymers as functional biomaterials: rational design, synthetic pathway, and biomedical applications. *Biomacromolecules* 12(6), 1937–1955.

Tatham, A. and Shewry, P. (2000) Elastomeric proteins: biological roles, structures and mechanisms. *Trends in Biochemical Sciences* 25(11), 567–571.

Tomalia, D., Baker, H., Dewald, J., Hall, M., Kallos, G., Martin, S., Roeck, J., Ryder, J. and Smith, P. (1986) Dendritic macromolecules: synthesis of starburst dendrimers. *Macromolecules* 19(9), 2466–2468.

Torchilin, V. (2005) Recent advances with liposomes as pharmaceutical carriers. *Nature Reviews Drug Discovery* 4(2), 145–160.

Torchilin, V. (2009) Multifunctional and stimuli-sensitive pharmaceutical nanocarriers. *European Journal of Pharmaceutics and Biopharmaceutics* 71(3), 431–444.

Trabbic-Carlson, K., Liu, L., Kim, B. and Chilkoti, A. (2004) Expression and purification of recombinant proteins from *Escherichia coli*: comparison of an elastin-like polypeptide fusion with an oligohistidine fusion. *Protein Science* 13(12), 3274–3284.

van Dijk, M., Rijkers, D., Liskamp, R., van Nostrum, C. and Hennink, W. (2009) Synthesis and applications of biomedical and pharmaceutical polymers via click chemistry methodologies. *Bioconjugate Chemistry* 20(11), 2001–2016.

Vasey, P.A., Kaye, S.B., Morrison, R., Twelves, C., Wilson, P., Duncan, R., *et al.* (1999) Phase I clinical and pharmacokinetic study of PK1 [*N*-(2-hydroxypropyl)methacrylamide copolymer doxorubicin]: first member of a new class of chemotherapeutic agents – drug-polymer conjugates. *Clinical Cancer Research* 5, 83–94.

Vicent, M., Manzanaro, S., de la Fuente, J. and Duncan, R. (2004) HPMA copolymer-1,5-diazaanthraquinone conjugates as novel anticancer therapeutics. *Journal of Drug Targeting* 12(8), 503–515.

Wang, X., Guerrand, L., Wu, B., Li, X., Boldon, L., Chen, W.-R. and Liu, L. (2012) Characterizations of polyamidoamine dendrimers with scattering techniques. *Polymers* 4(1), 600–616.

Wike-Hooley, J., Haveman, J. and Reinhold, H. (1984) The relevance of tumour pH to the treatment of malignant disease. *Radiotherapy and Oncology* 2(4), 343–366.

Xu, Q., Wang, C. and Wayne Pack, D. (2010) Polymeric carriers for gene delivery: chitosan and poly(amidoamine) dendrimers. *Current Pharmaceutical Design* 16(21), 2350–2368.

Yang, B. (2009) Drug Profile: Somavert. Available at: www.discoverymedicine.com/Benjamin-Yang/2009/05/21/drug-profile-somavert (accessed 29 April 2016).

Yasukawa, T., Kimura, H., Tabata, Y., Miyamoto, H., Honda, Y., Ikada, Y. and Ogura, Y. (1999) Targeted delivery of anti-angiogenic agent TNP-470 using water-soluble polymer in the treatment of choroidal neovascularization. *Investigative Ophthalmology & Visual Science* 40(11), 2690–2696.

Zhao, H., Rubio, B., Sapra, P., Wu, D., Reddy, P., Sai, P., Martinez, A., Gao, Y., Lozanguiez, Y., Longley, C., Greenberger, L. and Horak, I. (2008) Novel prodrugs of sn38 using multiarm poly(ethylene glycol) linkers. *Bioconjugate Chemistry* 19(4), 849–859.

Ziemba, B., Janaszewska, A., Ciepluch, K., Krotewicz, M., Fogel, W., Appelhans, D., Voit, B., Bryszewska, M. and Klajnert, B. (2011) In vivo toxicity of poly(propyleneimine) dendrimers. *Journal of Biomedical Materials Research Part A* 99A(2), 261–268.

9 Protein–Drug Conjugates: A New Class of Biotherapeutics

Deepshikha Pande Katare[1,*], Savita Mishra[1], Harsha Kharkwal[2] and S.K. Jain[3]
[1]Centre for Medical Biotechnology, Amity Institute of Biotechnology, Amity University Uttar Pradesh, Noida, India; [2]Amity Center for Carbohydrate Research and Amity Institute of Phytomedicine and Phytochemistry, Amity University Uttar Pradesh, Noida, India; [3]Hamdard Institute of Medical Sciences and Research, Hamdard University, New Delhi, India

Abstract

There is an increasing need for a novel drug delivery system in the current clinical scenario. Over the past few decades recombinant human proteins, enzymes, monoclonal antibodies and drug conjugates (ADCs) have changed the pharmaceutical industry. This chapter highlights current and emerging methods for the development of stable and effective antibody–drug conjugates that provide target-specific therapy for various life-threatening diseases such as cancer.

Introduction

Over past decades considerable efforts have been made to develop drug delivery systems (DDS). In the human body any drug or active pharmaceutical ingedient (API) needs to be in its desirable concentration; that is, its therapeutic dose. A major feature of a DDS is both the amount and duration of drug, and the type of dosage and route of administration. To regulate the pharmacokinetics, dynamics and efficacy of drugs, several new approaches have been established, collectively known as novel DDS (Charman *et al.*, 1999; Kopeček, 2003). Collaboration between pharmaceuticals and other disciplines such as molecular biology and macromolecular science have led to the development of conjugation systems or medical devices that can modify not only the bioavailability and bioequivalence of drugs but also improve their effectiveness and safety, as well as the convenience and compliance of the patient (Niculescu-Duvaz and Springer, 1997; Santini *et al.*, 2000; Torchilin, 2001).

Recent approaches to administration of medication such as proteins, peptides, disease-specific antibodies, gene-based therapies or cytotoxic drugs for cancer treatment may not use the classical routes (oral, topical, intraperitoneal, mucosal, nasal, inhalation route, etc.) because of their susceptibility to enzymatic degradation and very low absorbance into the circulatory system due to molecular size or ionic charge issues (Wang and Von Recum, 2011). To enhance the bioavailability, decrease toxicity and minimize drug degradation, drug-specific systems are being formulated. These systems use carrier-based technology to increase the effect of drugs in the human body. Most drug carriers are made up of biopolymers, nanoparticles, microcapsules, liposomes, micelles, emulsions, fast-dissolving

*Corresponding author. E-mail: dpkatare@amity.edu

films, protein-based conjugates or lysosomes. These carriers slow degradation and drug release, and can be pH- and temperature-sensitive (Fischer *et al.*, 2002; Brannon-Peppas and Blanchette, 2012; Zhang *et al.*, 2014). In recent years, antibody-mediated targeting using immunotoxins, immunoliposomes and other conjugates have been used to target specific cells.

In recent treatments, protein–drug conjugates have been found to be a novel and effective delivery system. In this method small molecules (such as cytotoxic drugs) are combined with proteins (mostly antibodies) to form a complex protein–drug conjugate (PDC) structure that could target specific tissues or receptors. These therapeutic antibodies can recognize specific antigens due to high sensitivity towards them (Perez *et al.*, 2013; Hallam, 2013; Kopeček, 2013). Cytotoxic drugs bind to the antibody through linkers that are recognized by the specific cells.

This chapter highlights current therapeutics; the components of antibody–drug conjugates (ADCs) and their mechanism of action; and linkage between the drugs and carrier antibody and other PDCs.

Novel Drug Delivery Systems

Since 2000, the worldwide market for novel drug delivery systems (NDDS) has increased remarkably. More specially, in the area of cancer therapy, it is projected to reach up to USD15.9 billion by 2020 (Transparency Market Research, 2015). According to BCC Research (Wellesley, Maryland, USA, www.bccresearch.com) by 2018 the market for advanced DDS will rise to USD174 billion with a compound annual growth rate (CAGR) of 2.8% (BCC research pharmaceutical report, PHM006J, www.bccresearch.com/market-research/pharmaceuticals/advanced-drug-delivery-markets-phm006j.html, January 2014). Since completion of the Human Genome Project in 2000, it has been established that around 30,000 drug targets are available in the human body. This has, in turn, altered the projected market value of DDS (Fig. 9.1). Individual therapy, controlled bioavailability, managing side effects and sustained release of formulations /API are major requirements of a novel drug delivery system (NDDS).

NDDS can be based on physical and biochemical mechanisms. The physical mechanism uses some specific methods such as drug osmosis, desorption of surface-bound compounds, diffusion through the carrier matrix or carrier wall, electro-transport and the combined erosion process. On the other hand, biochemical mechanisms are based on biochemical processes such as gene therapy, drug–liposome/ micelle formulation, fast-dissolving films/tablets and polypeptide drug adducts, monoclonal antibody/protein–drug conjugates and various other carrier molecules for controlled release and target specificity. The drug-carrying molecules have the ability to

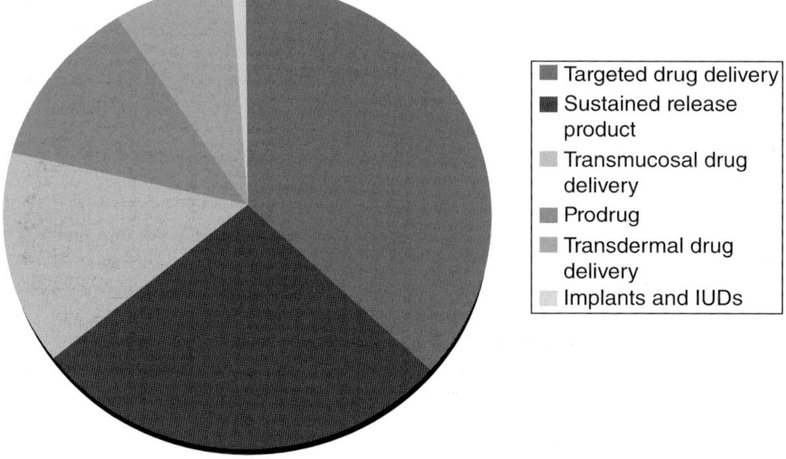

Fig. 9.1. Summary of global sales of novel drug delivery products during 2014, adapted from BCC research.

respond to specific stimuli such as light exposure, pH alteration or temperature change (Kaparissides *et al.*, 2006; Allen and Cullis, 2013; Hehir and Cameron, 2014). Targeted release of PDCs into the systems can reduce the occurrence of drug dosage and also the oscillation in levels of drug in the circulating system. However, the only disadvantage is high expenditure, which makes production and dosage adjustment difficult.

Targeted therapies (also termed biological therapy) are used to identify and attack specific cancer cells with a lesser amount of damage to the normal cells (Ferrara and Kerbel, 2005). Small-molecule, antibody-bound drugs or bioconjugated drugs are the most used mediators for this type of therapy (Fig. 9.2). Targeted therapy either blocks certain enzymatic actions, downregulates the action of proteins involved in growth of cancer cells, or delivers cytotoxic substances directly to the cancer cells and kills them, rather than simply interfering in cell proliferation as in traditional chemotherapy. Targeted therapies also have lesser adverse effects than other types of cancer therapy such as radiation or chemotherapy (Swapnil *et al.*, 2013; Zolot *et al.*, 2013). Traditional chemotherapeutic agents can attack the cancer cell as well as healthy cells. After monoclonal antibodies have been combined with the cancer-killing cytotoxic drugs, antibody–drug conjugates can act as sensitive identifiers between the healthy and diseased cells. In recent years, a significant approach has been to target only all hurriedly dividing cells rather than target tumour tissues, thereby killing the tumour cells but reducing damage to the healthy tissues (Beck *et al.*, 2010; Beck, 2014).

Bioconjugation for Targeted Drug Delivery

Bioconjugation is an approach in which biomolecules (drugs) are linked with different substrates (ligands). Synthetically modified biomolecules display varied functionalities such as cell cycle tracing, enzymatic function, protein biodistribution, biomarker detection and targeted drug delivery

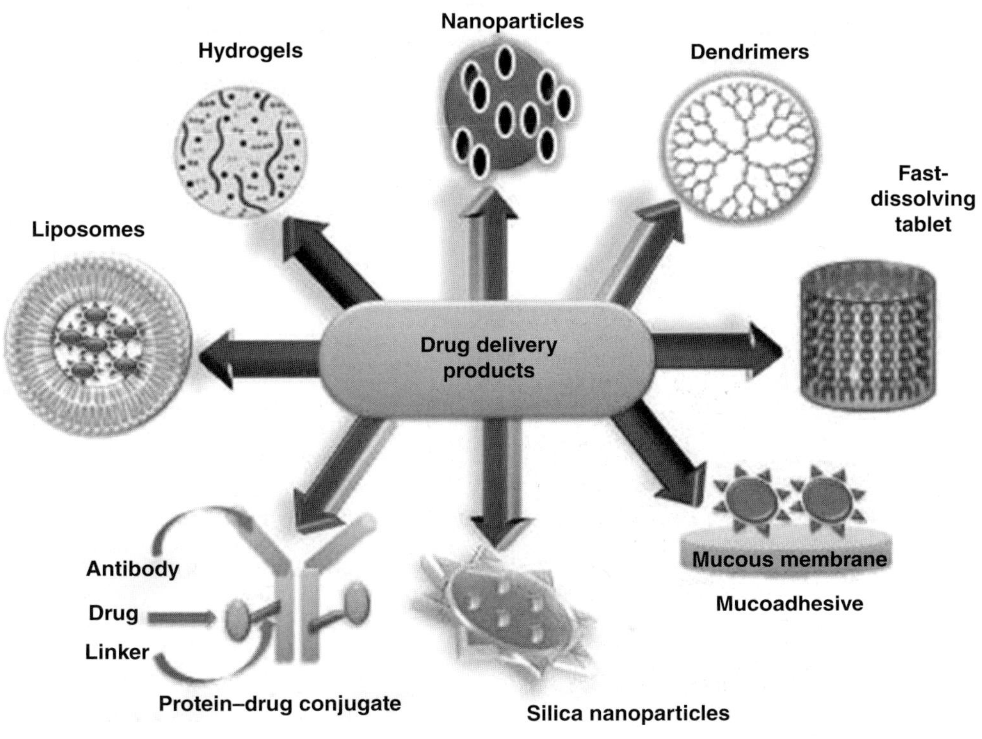

Fig. 9.2. Various drug delivery products.

in cancer treatment (Stephanopoulos and Francis, 2011. For the development of bioconjugates, chemical reactions such as amino acid residue coupling, modification in the amino acid residues, and N- and C-terminus modification are routinely employed (Francis and Carrico, 2010). Various anti-cancer bioconjugates such as aptamer–nanoparticle bioconjugates, carbon nanotube-based conjugates, aptamer–dendrimer bioconjugates and liposome-based conjugates are used for killing specific tumour cells (Zhou and Rossi, 2014). Some of the challenges with bioconjugation / bioconjugate drugs are: enhancing specificity and stability; improving bioavailability and reducing toxicity of carrier molecules. Despite these challenges, bioconjugation is by far the most constructive technique for target-based treatment. Protein–drug conjugates (PDCs) are one of the latest technologies for cancer treatment (Hermanson, 2013).

Protein–Drug Conjugates/ Antibody–Drug Conjugates

Historical background

The phenomenon of ADCs was introduced in 1956 by the German scientist Paul Ehrich, who invented the concept of selective targeting of tumour cells through antibodies and delivering the cytotoxic drugs to those particular tumour cells (Ehrlich, 1956). He used the term 'magic bullet' to describe the idea, but it took nearly 50 years for it to actually take shape and to produce the first ADCs (Trail, 2013). MTX was linked with the antibody 791T/36 to target leukaemia cells (Affleck and Embleton, 1992). Prior to this invention, polyclonal antibodies were being used as the targeting agents but the emergence of ADCs has altered the market scenario and certainly aided in their advancement. Since 1990, a high number of drugs have been approved by the FDA and in the year 2012 alone, 39 drugs were approved. Market analysts predict that up to 2016, the number of approvals could be around 35 drugs per year (Berggren *et al.*, 2012; Kitson *et al.*, 2013).

The immunotherapeutic use of target-based antibody–drug conjugates may be the next major drug discovery in cancer treatment (Fig. 9.3). The idea behind PDCs is to avoid harm to the normal cells and to eliminate the tumour cells effectively. Individually, cytotoxic drugs are very harmful to cells, but the toxicity is manageable when combined with a targeted antibody. The linkage between cytotoxic drug and antibody is stable, but chemical or linker dependent (explained later in the chapter), and goes directly to

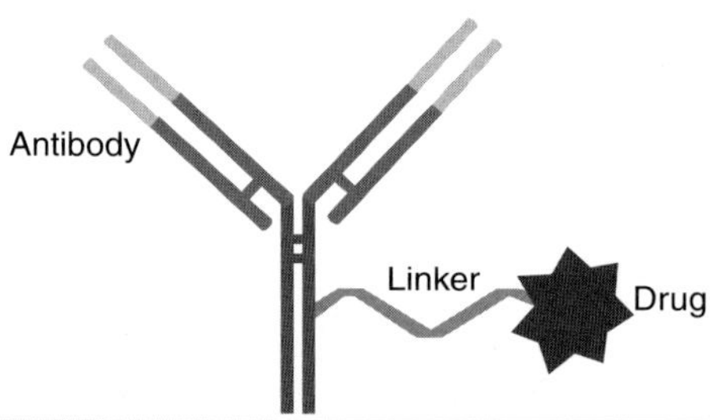

Antibody	Linker	Cytotoxic drug
• Selective; higher expression in tumour cells • Limited expression in normal cells • Stable conjugation site and should not interfere in binding of antigen antibody • Stable at body pH	• Cleavable or non-cleavable linker • Circulation stable • Specific intracellular delivery such as enzymatic cleavage or monoclonal antibody degradation should be selective • Limited heterogeneity	• Highly efficacious • Capable for structural modification so that linker can attach to it • Circulation stable • Stable in lysis activity • Should have defined mechanism of action • Bystander effect

Fig. 9.3. Schematic diagram of antibody–drug conjugate.

Table 9.1. Market-approved antibodies and drug conjugates.

Drug name	Trade name	Approval year	Cytotoxic compound	Linker	Target cell	Reference
Gemtuzumab ozogamicin	Mylotarg	June 2000 but withdrawn from the market in June 2010 because of lower patient survival	Calicheamicin	4-(4-acetylphenoxy) butanoic acid	Leukemic blast cells in acute myelogenous leukaemia (AML)	Ricart, 2011; Bross *et al.*, 2001
Brentuximab vedotin	Adcetris	2011	Monomethyl auristatin E (MMAE)	Para-aminobenzylcarbamate	Anaplastic large cell lymphoma (sALCL) and Hodgkin's lymphoma (HL)	Younes *et al.*, 2012b
Trastuzumab emtansine	Kadcyla	2013	Trastuzumab molecule bound to several molecules of mertansine	Succinimidyl trans-4-(maleimidylmethyl) cyclohexane-1-carboxylate	HER2-positive breast cancer	Burris, 2011; Verma *et al.*, 2012

the targeted tumour cells and releases the drug within them (Kovtun and Goldmacher, 2007). The first ADC was produced by the pharmaceutical company Wyeth with the trade name Mylotarg (gemtuzumab ozogamicin), for the treatment of acute myelogenous leukaemia. However, it was withdrawn from the market due to the inefficiency in maintaining the survival rate (Reichert, 2013). In 2011 the FDA approved another drug, brentuximab vedotin (Adcetris), a chimeric monoclonal antibody, for the treatment of Hodgkin's and anaplastic large cell lymphoma. This was one of the initial ADCs in which the antibody target cell membrane protein C30 was linked with the drug (cathepsin) through a cleavable linker (valine–citrulline) (Lambert, 2013; Sievers and Senter, 2013). In 2013 ado-trastuzumab emtansine (Kadcyla; Genentech/Roche) medication for late-stage HER2-positive breast cancer was approved by the FDA. In this medication, trastuzumab was combined with the maytansinoid DM1 via a linker nonreducible thioether (Phillips *et al.*, 2008). Currently more than 100 clinical trials are in the pipeline for the validation of several ADCs for cancer treatment (Table 9.1 and 9.4; Mullard, 2013).

Characteristics of ADC components

Target antigen and antibody

Expression of tumour antigens is very important in the development of ADCs. Malignant cells, tumour endothelial cells or other tumour microenvironment-associated antigens are highly expressed in diseased patients and such expression makes ADCs suitable for the targeting of those cells (Beverly, 2014). However, expression of these tumour-specific antigens can vary throughout the body (and in normal cells as well) and even within the tumour cells, and this affects the overall effectiveness of the ADCs. Since angiogenesis is a hallmark of solid tumour cells, the angiogenesis-related receptors (epidermal growth factor receptor (EGFR), human epidermal growth factor receptor 2 (HER2) and vascular growth factor receptor (VEGFR), etc.) are candidates for the ADC target. B- and T-cell receptors are widely used because of their over-expression in haematological malignancies (Lartigue, 2012). The specificity of the antibody towards antigen-specific binding means that the antibody targets only those cells with tumourigenic properties. In the first generation of ADCs the lack of selectivity of cysteine or lysine residues of antibody caused more side effects, and selection of normal and tumour cells was often inadequate, leading to disappointing results and consequently to withdrawal of some drug conjugates (US FDA, 2012). However, second and third generation FDA-approved ADCs such as Kadcyla have shown promising results with good efficacy in Phase III trials (Table 9.2).

Linkers

The linker between the antibody and the drug plays a crucial role in providing the appropriate function and stability for the ADCs. Most linkers are small molecules, a chemical compound or a peptide, which are covalently coupled to the antibody and drug molecule. Minimum harm to the normal cells and maximum exposure of drug to the tumour cells, coupled with the stability of ADCs, are key features of a good linker. The process of designing a linker relates directly to the mechanism of ADCs (Feng *et al.*, 2014). Linker molecules possess functional groups such as disulfides, thioethers or hydrazones and aid in controlling the distribution and release of the cytotoxic compound to the target cells. On the basis of bond splicing, the linker could be cleavable or non-cleavable. After a non-cleavable linker has internalized the ADCs into the tumour cell, it degrades completely and releases the attached drug. More importantly, chemical modification of the drug structure does not affect the efficacy of the drug. An example of a non-cleavable linker is the thioether in various ADCs of immunogen bond; whereas a cleavable linker releases both the drug and MAb into the target cell and escapes from the cell surface, a process known as 'bystander killing' (Kovtun and Goldmacher, 2007). On the basis of their mechanisms, cleavable linkers can be divided into three major classes:

- Enzyme activity sensitive: These types of linkers are recognized by some lysosomal proteases, are cleaved at the dideptide bond and release the drug to the tumour cell. The valine–citrulline dipeptide linker is an example of this class that is utilized in many ADCs such as Adcetris (Perez *et al.*, 2013).

Table 9.2. Target cancer cell antigen and its ADCs.

Target cancer	Antigen	ADC for treatment	Reference
Non-Hodgkin's lymphomas (NHLs)	CD30, B-cell activation antigens, CD70, CD19, CD37, CD20, CD72, CD79 and CD10, CD21	SGN-75 (Phase I study)	Thompson *et al.*, 2011; Perez *et al.*, 2013)
Anaplastic large cell lymphoma (AML)	CD33	Adcetris (FDA approved 2011)	Perez *et al.*, 2013; 36, 42
Hodgkin's lymphoma (HL)	CD30	Adcetris (FDA approved 2011)	Perez *et al.*, 2013; Younes *et al.*, 2012a; Senter and Sievers, 2012
Multiple myeloma (MM)	CD74, CD56, endothelin B receptor, CD138	hLL1-DOX (Phase II trial)	Perez *et al.*, 2013; Feng *et al.*, 2014
Non-small cell lung cancer	NaPi2b	DNIB0600A (Phase I study)	Gordon *et al.*, 2013
Breast cancer	ErbB2, ErbB4, ErbB3	Trastuzumab-DM1 (FDA approved 2013)	Perez *et al.*, 2013; Gordon *et al.*, 2013
Pancreatic cancer	MUC16, MUC-1	DMUC5754A (Phase I trial)	Liu *et al.*, 2013
Ovarian cancer	MUC16, IM-1	DMUC5754A (Phase I trial)	Liu *et al.*, 2013
Prostate cancer	CD56, CD33 and PSMA, STEAP-, CD44v6 and TENB2		Perez *et al.*, 2013
Kidney cancer	CAIX and TIM-1 (CDX-014)	SGN-75(Phase I study)	Thompson *et al.*, 2011

- Low pH sensitive: During circulation ADCs are stable because of their high pH but, in the lysosomal compartment and due to low pH conditions, hydrolysis of the acidic groups occurs. In turn this causes drug release (e.g. the hydrazone linker in Mylotarg. However, there are stability concerns about these linkers and hence the corresponding ADCs have been withdrawn from the market (Perez *et al.*, 2013).
- Glutathione sensitive: Linkers with the disulfide bond are glutathione sensitive. Anti-CD19-maytansinoid conjugate, lorvotuzumab mertansine and SAR3419 are known examples (Perez *et al.*, 2013).

In recent years, scientists have focused on various methods in developing site-specific conjugation to improve the stability and therapeutic value of ADCs. Stability of the linker can directly or indirectly affect the toxicity, pharmacokinetic properties and therapeutic index of the ADCs (Bąchor *et al.*, 2013). The efficacy of the ADCs can be increased by delivering more drugs through MAb. The high ratio of drug over antibody such as in the case of doxorubicin conjugates as 1:25 can elevate the effect of particular ADCs (Fig. 9.4). To achieve this, branched linker or polymeric carriers are used (Ricart and Tolcher, 2007; Sapra *et al.*, 2011; Perez *et al.*, 2013).

Cytotoxic drugs

Cytotoxic agents cause apoptosis in the tumour cells. High potency, chemical traceability, stability and ability to induce apoptosis are major characteristics of a good cytotoxic agent. Two main categories are used in ADCs (Table 9.3; Ricart and Tolcher, 2007; Perez *et al.*, 2013; Trail, 2013):

- Microtubule-disrupting agents that hinder the cell cycle via inhibition of microtubule formation, (e.g. maytansinoids and dolastatin analogues).
- DNA-disrupting agents that have properties of binding the minor grooves of the DNA or

acting on protein related to replication and degradation of the DNA double strand. Drugs such as duocarmycins and calicheamicins bind to the minor grooves of the DNA but some, such as amatoxin analogues and SN-38, attack RNA polymerase II/III and topoisomerase I, respectively, and break the DNA strand.

Tumour cells are resistant to various drugs and sensitivity of a toxin towards multidrug resistance (MDR) is a major factor in the optimization of any drugs. Interestingly, MDR proteins such as MRP1/3 and P-glycoprotein (Pgp) can assist the drug to overcome this problem. Alteration in expression of MDR protein can also help in the apoptosis of cancer cells (Benderra *et al.*, 2005).

ADC mechanism of action

Although the concept of designing ADCs is quite simple, their successful development depends on optimization of the building blocks:

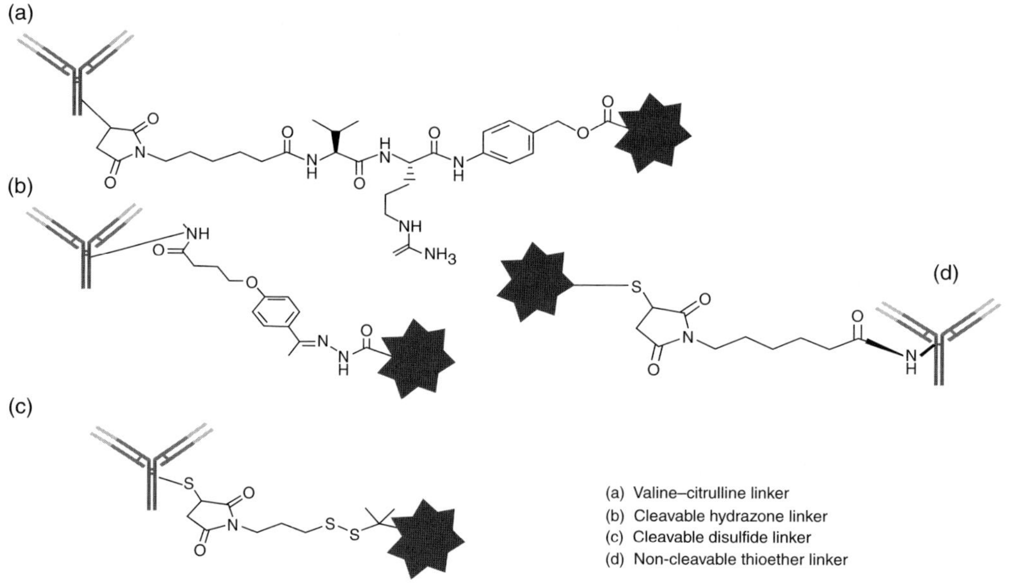

Fig. 9.4. Different types of linkers.

Table 9.3. Cytotoxic drugs and their action (Kovtun and Goldmacher, 2007; Ricart and Tolcher, 2007; Perez *et al.*, 2013).

Drug	Target molecule	Mechanism of action
α-Amanitin 7	RNA polymerase II	Inhibits RNA polymerase II action hinders transcription
Derivatives of doxorubicin	Topoisomerase II	Nucleophilic attack on DNA double stand
Maytansinoids	Tubulin	Microtubule inhibitor
Pyrrolobenzodiazepine (PBD)	Minor groove of DNA	Cross-linking of opposing strands of DNA
Auristatins	Tubulin	Microtubule inhibitor
Calicheamicin and its analogues	Specific DNA minor groove	DNA breaks
Duocarmycins	Specific DNA minor groove	Adenine alkylation (adenine-N3) DNA
Anthracyclines	DNA- & RNA-related molecules	Replication and transcription inhibition

Table 9.4. Major ADCs in clinical trials as novel drug delivery systems (NDDS).

Drug name	Target antigen	Cytotoxic drug	Linker	Mode of action	Disease and clinical trial phase	Reference
Inotuzumab ozogamicin (CMC-544)	CD22	Calicheamicin	4-(4-acetylphe-noxy)butanoic acid	DNA damage	Phase II for NHL Phase I for ALL	Li *et al.*, 2013; Yardley *et al.*, 2012
DCDT2980S	CD22	Monomethyl auristatin E (MMAE),	Valine–citrulline	Microtubule inhibitor (G2/M arrest)	Phase II for NHL	Lambert, 2013
Glembatumumab vedotin (CDX-011)	GPNMB	MMA-E	Protease-sensitive peptide linker	Microtubule inhibitor (G2/M arrest)	Phase II for breast cancer	Whiteman *et al.*, 2014
SAR3419	CD19	Maytansine derivate	Cleavable disulfide linker	Microtubule inhibitor (G2/M arrest)	Phase II for relapsed/ refractory B-cell NHL	Tannir *et al.*, 2014; Minotti *et al.*, 2004
Lorvotuzumab mertansine (IMGN901)	CD56	Maytansinoid DM1	Cleavable disulfide linker	Microtubule inhibitor (G2/M arrest)	Phase II for SCLC and multiple myeloma	Mathur and Weiner, 2013
SGN 75	CD70	MMA-F	Plasma-stable linker	Microtubule inhibitor (G2/M arrest)	Phase I for NHL and renal cancer	Younes *et al.*, 2012a
MDX-1203	CD70	Duocarmycin	Dipeptide non-cleavable	DNA damage	Phase I NHL	Trail, 2013
BAY94-9343	CD37, folate receptor 1, CD56	Maytansine derivative DM4		Microtubule inhibitor (G2/M arrest)	Phase I for solid tumours	Minotti *et al.*, 2004; Govindan and Goldenberg, 2012
Brentuximab Vedotin (SGN-35)	CD30	MMAE	Protease-cleavable linker		Stage II-IV Hodgkin's lymphoma	Mullard, 2013
IMMU132	TROP-2	SN-38, the active metabolite of irinotecan			Phase I/II study for epithelial cancers	Minotti *et al.*, 2004
Labetuzumab-SN38 IMMU-130	CEACAM5	SN-38, the active metabolite of irinotecan	Cleavable, Phe-Lys		Phase II for metastatic colorectal cancer	Ducry and Stump, 2009; Perez *et al.*, 2013
RG-7596	CD79b	MMAE			Phase II for DLBCL and follicular non-Hodgkin's lymphoma	Minotti *et al.*, 2004
Pinatuzumab vedotin (RG-7593)	CD 22	MMAE			Phase II for DLBCL and follicular non-Hodgkin's lymphoma	Minotti *et al.*, 2004
ASG-5ME	SLC44A4				Metastatic pancreatic ductal adenocarcinoma (PDA)	Perez *et al.*, 2013

ALL, acute lymphoblastic leukaemia; NHL, Non-Hodgkins lymphoma; DLBCL, diffuse large B-cell lymphoma; SCLC, small cell lung cancer antibody, linker and cytotoxic drugs. The target antigen should be characterized and highly expressed in the cancer cells and low at the normal cells for maximum efficacy and minimum toxicity of conjugates. The linker must have a bifunctional attachment site to bind both antibody and toxin, as well as being stable during circulation. More importantly, cytotoxic drugs must have high potency for killing tumour cells and should not be cleaved before reaching the target cell. The mode of action of the ADC complex is given below.

Blood circulation

In the blood circulation, the ADC complex acts as a single antibody, and other parts of the complex do not react and are stable in the bloodstream. Because of these properties, the whole complex can move easily from administration site to target site. Linkers play an important role in the stability of the complex and limits damage to normal tissue. Degradation of the ADC complex may lead to premature delivery of the cytotoxic drug, which is very harmful to normal tissues and should be prevented as far as possible (Perez *et al.*, 2013).

Target identification and binding

The antibody recognizes tumour-specific receptors (the antigens), which are highly expressed on the surface of cancer cells and bind to them. For antigen–antibody binding, high immunoaffinity and stability of the ADC complex are basic requirements. The accessibility of the ADC to the target cell may also reduce due to fewer uptakes of the ADC by the endothelial stromal and epithelial barrier (Fig. 9.5). Hence, ADC design relies on the selection and development of stable and compatible monoclonal antibodies (Perez *et al.*, 2013).

Internalization

Internalization is the process of insertion of ADC into the tumour cell. This is achieved through receptor-mediated endocytosis and, because of the inefficiency of the antibody–antigen complex, it is one of the most challenging steps. Although this process is inadequately understood, factors such as the epitope, the intracellular trafficking pattern and antigen–antibody interactions play major roles. Endocytosis of ADC could be either immune cell-mediated (by natural killer cell or macrophages) through or intracellular lysis. However, most ADCs follow endocytosis via direct intracellular lysis (Perez *et al.*, 2013).

Drug release mechanism

Once internalization of ADCs is accomplished, ADC undergoes lysosomal-mediated degradation to release the drug into the cytoplasm (Fig. 9.6).

Drug activity and cell death

A cytotoxic drug stimulates various signalling pathways for cancer cell apoptosis. Most drugs act either as microtubule inhibitors to arrest cell growth (G2/M arrest) or cause DNA damage through crosslinking with a DNA strand (Ricart and Tolcher, 2007; Trail, 2013).

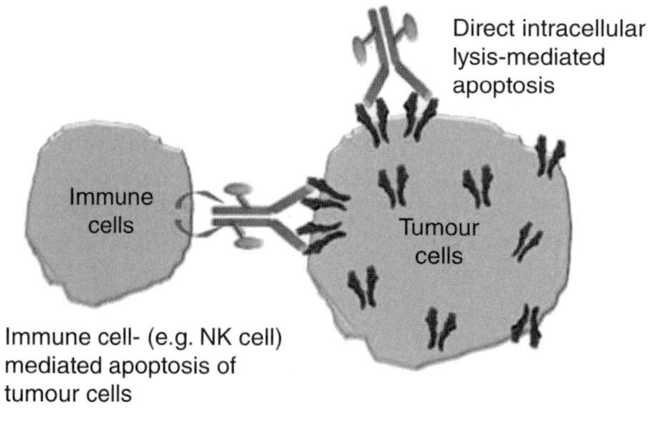

Fig. 9.5. Target binding of ADCs.

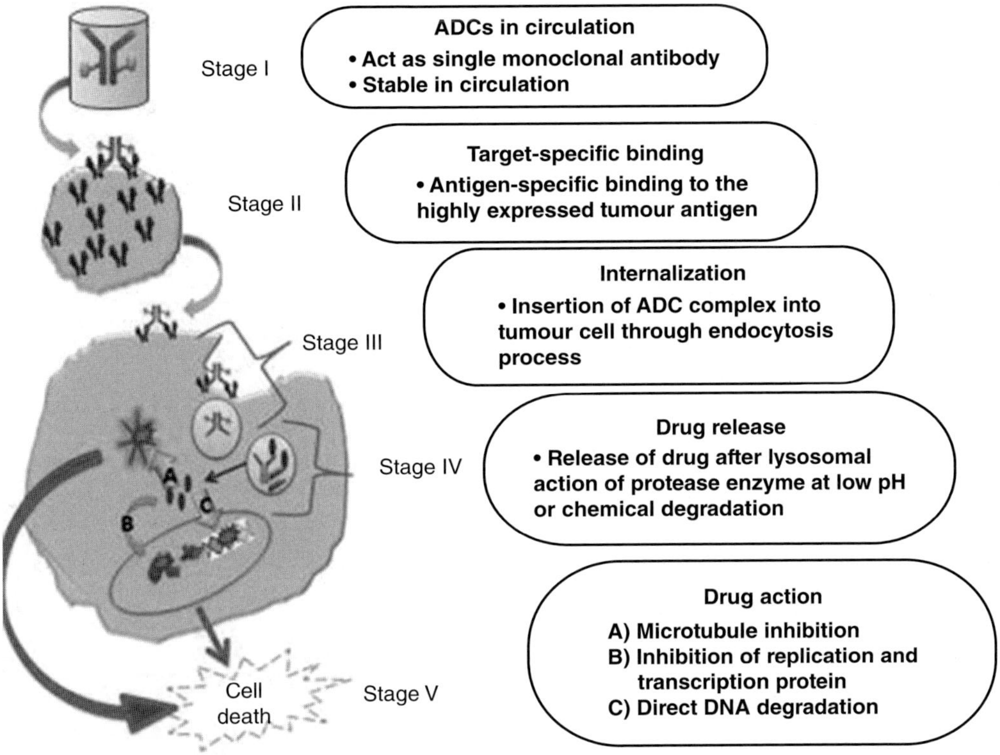

Fig. 9.6. ADC mechanism of action.

Challenges

Although ADC/PDC development seems a very simple process there are various challenges to be addressed. In the early stages, linker instability is a very big concern. Linkers are often either too stable to release the drug or unstable during circulation, causing low efficacy, poor targeting and high toxicity (Ducry and Stump, 2009; Mitragotri *et al.*, 2014). Lower expression of tumour antigen, insufficient potency and unproductive internalization of the cytotoxic drug are some other problems that make ADC production more challenging. Larger size and degradation susceptibility are also major issues for the formulation of PDCs.

Conclusions and Future Prospects

Systemic cytotoxic therapy for malignancies, as well as other life-threatening diseases, can be addressed by ADCs and PDCs. Proteins are a commercially very well-established class of therapeutics and certainly have a bright future in the design and delivery of drug carriers. Conjugates such as immunotoxin and immunoliposome have not only displayed antitumour efficacies in clinical trials, but are also less toxic compared to conventional therapies. Overall, current progress in the development of PDCs is promising and it is likely that these DDS will play a major role in the near future.

References

Affleck, K. and Embleton, M.J. (1992) Monoclonal antibody targeting of methotrexate (MTX) against MTX-resistant tumour cell lines. *British Journal of Cancer* 65(6), 838–844.

Allen, T.M. and Cullis, P.R. (2013) Liposomal drug delivery systems: from concept to clinical applications. *Advanced Drug Delivery Reviews* 65(1), 36–48.

Bąchor, R, Kluczyk, A, Stefanowicz, P. and Szewczuk, Z. (2013) New method of peptide cleavage based on Edman degradation. *Molecular Diversity* 17(3), 605–611.

BCC Research Pharmaceuticals Report (2014) Global Markets and Technologies for Advanced Drug Delivery Systems. PHM006J. BCC Research, Wellesley, USA.

Beck, A., Haeuw, J.F., Wurch, T., Goetsch, L., Bailly, C. and Corvaïa, N. (2010) The next generation of antibody-drug conjugates comes of age. *Discovery Medicine* 10(53), 329–339.

Beck, B.A. (2014) Antibody drug conjugates. *Current Opinion in Oncology* 26(5), 476–483

Benderra, Z., Faussat, A.M., Sayada, L., Perrot, J.Y., Tang, R., Chaoui, D. and Legrand, O. (2005) MRP3, BCRP, and P-glycoprotein activities are prognostic factors in adult acute myeloid leukemia. *Clinical Cancer Research* 11(21), 7764–7772.

Berggren, R., Møller, M., Moss, R., Poda, P. and Smietana, K. (2012) Outlook for the next 5 years in drug innovation. *Nature Reviews Drug Discovery* 11(6), 435–436.

Beverly A.T. (2014) Antibody drug conjugates. *Current Opinion in Oncology* 26, 476–483.

Brannon-Peppas, L. and Blanchette, J.O. (2012) Nanoparticle and targeted systems for cancer therapy. *Drug Delivery Reviews* 64, 206–212.

Bross, P.F., Beitz, J., Chen, G., Chen, X.H., Duffy, E., Kieffer, L. *et al.* (2001) Approval summary: gemtuzumab ozogamicin in relapsed acute myeloid leukemia. *Clinical Cancer Research* 7(6), 1490–1496.

Burris III, H.A. (2011) Trastuzumab emtansine: a novel antibody-drug conjugate for HER2-positive breast cancer. *Expert Opinion on Biological Therapy* 11(6), 807–819.

Charman, W.N., Chan, H.K., Finnin, B.C. and Charman, S.A. (1999) Drug delivery: a key factor in realising the full therapeutic potential of drugs. *Drug Development Research* (3–4), 316–327.

Ducry, L. and Stump, B. (2009) Antibody–drug conjugates: linking cytotoxic payloads to monoclonal antibodies. *Bioconjugate Chemistry* 21(1), 5–13.

Ehrlich, P. (1956) The relationship existing between chemical constitution, distribution, and pharmacological action. *The Collected Papers of Paul Ehrlich* 1, 596–618.

Feng, Y., Zhu, Z., Chen, W., Prabakaran, P., Lin, K. and Dimitrov, D.S. (2014) Conjugates of small molecule drugs with antibodies and other proteins. *Biomedicines* 2(1), 1–13.

Ferrara, N. and Kerbel, R.S. (2005) Angiogenesis as a therapeutic target. *Nature* 438(7070), 967–974.

Fischer, P.M., Wang, S., and Zhelev, N. (2002) US Patent No. 6,472,507. Washington, DC: US Patent and Trademark Office.

Francis, M.B. and Carrico, I.S. (2010) New frontiers in protein bioconjugation. *Opinion in Chemical Biology* 14(6), 771–773.

Gordon, M.S., Gerber, D.E., Infante, J.R., Xu, J., Shames, D.S., Choi, Y. and Burris, H.A. (2013) A phase I study of the safety and pharmacokinetics of DNIB0600A, an anti-NaPi2b antibody–drug-conjugate (ADC), in patients (pts) with non-small cell lung cancer (NSCLC) and platinum-resistant ovarian cancer (OC). *Journal of Clinical Oncology* 31(15), 2507.

Govindan, S.V. and Goldenberg, D.M. (2012) Designing immunoconjugates for cancer therapy. *Expert Opinion on Biological Therapy* 12(7), 873–890.

Hallam, T. (2013) Antibody-drug conjugates and cancer treatment: making "smart bombs" smarter. *Scientific American.* Available at: blogs.scientificamerican.com/guest-blog/antibody-drug-conjugates-and-cancer-treatment-making-smart-bombs-smarter (accessed 18 April 2016).

Hehir, S. and Cameron, N.R. (2014) Recent advances in drug delivery systems based on polypeptides prepared from N-carboxyanhydrides. *Polymer International* 63(6), 943–954.

Hermanson, G.T. (2013) *Bioconjugate Techniques,* Third Edition. Academic Press, London.

Kaparissides, C., Alexandridou, S., Kotti, K. and Chaitidou, S. (2006) Recent advances in novel drug delivery systems. *Journal of Nanotechnology Online* 2, 1–11.

Kitson, S.L., Quinn, D.J., Moody, T.S., Speed, D., Watters, W. and Rozzell, D. (2013) Antibody–drug conjugates (ADCs) – Biotherapeutic bullets. *Chimica Oggi-Chemistry Today* 31(4), 29–35.

Kopeček, J. (2003) Smart and genetically engineered biomaterials and drug delivery systems. *European Journal of Pharmaceutical Sciences* 20(1), 1–16.

Kopeček, J. (2013) Polymer–drug conjugates: origins, progress to date and future directions. *Advanced Drug Delivery Reviews* 65(1), 49–59.

Kovtun, Y.V. and Goldmacher, V.S. (2007) Cell killing by antibody–drug conjugates. *Cancer Letters* 255(2), 232–240.

Lambert, J.M. (2013) Drug-conjugated antibodies for the treatment of cancer. *British Journal of Clinical Pharmacology*, 76(2), 248–262.

Lartigue, J.D. (2012) Antibody–drug conjugates: guided missiles deployed against cancerous cells. *Oncology Live.* Available at: global.onclive.com/publications/Oncology-live/2012/June-2012/Antibody-Drug-Conjugates-Guided-Missiles-Deployed-Against-Cancerous-Cells (accessed 18 April 2016).

Li, D., Poon, K.A., Yu, S.F., Dere, R., Go, M., Lau, J. and Polson, A.G (2013) DCDT2980S, an anti-CD22-monomethyl auristatin E antibody–drug conjugate, is a potential treatment for non-Hodgkin lymphoma. *Molecular Cancer Therapeutics* 12(7), 1255–1265.

Liu, J., Moore, K., Birrer, M., Berlin, S., Matulonis, U., Infante, J. and Burris, H. (2013) Abstract LB-290: Targeting MUC16 with the antibody–drug conjugate (ADC) DMUC5754A in patients with platinum-resistant ovarian cancer: a phase I study of safety and pharmacokinetics. *Cancer Research* 73(8 Supplement), LB-290.

Mathur, R.B.A. and Weiner, G.J. (2013) Picking the optimal target for antibody-drug conjugates. *American Society of Clinical Oncology* 17–132, DOI: 10.1200/EdBook_AM.2013.33.e103.

Minotti, G., Menna, P., Salvatorelli, E., Cairo, G. and Gianni, L (2004) Anthracyclines: molecular advances and pharmacologic developments in antitumor activity and cardiotoxicity. *Pharmacological Reviews* 56(2), 185–229.

Mitragotri, S., Burke, P.A. and Langer, R. (2014) Overcoming the challenges in administering biopharmaceuticals: formulation and delivery strategies. *Nature Reviews Drug Discovery* 13(9), 655–672.

Mullard, A. (2013) Maturing antibody-drug conjugate pipeline hits 30. *Nature Reviews Drug Discovery* 12(5), 329–332.

Niculescu-Duvaz, I. and Springer, C.J. (1997) Antibody-directed enzyme prodrug therapy (ADEPT): a review. *Advanced Drug Delivery Reviews* 26(2), 151–172.

Perez, H.L., Cardarelli, P.M., Deshpande, S., Gangwar, S., Schroeder, G.M., Vite, G.D. and Borzilleri, R.M. (2013) Antibody–drug conjugates: current status and future directions. *Drug Discovery Today* 19(7), 869–881.

Phillips, G.D.L., Li, G., Dugger, D.L., Crocker, L.M., Parsons, K.L., Mai, E. and Sliwkowski, M.X. (2008) Targeting HER2-positive breast cancer with trastuzumab-DM1, an antibody–cytotoxic drug conjugate. *Cancer Research* 68(22), 9280–9290.

Reichert, J.M. (2013) Antibodies to watch in 2013: mid-year update. In MAbs. *Landes Bioscience* 5(4), 513–517.

Ricart, A.D. (2011) Antibody-drug conjugates of calicheamicin derivative: gemtuzumab ozogamicin and inotuzumab ozogamicin. *Clinical Cancer Research* 17(20), 6417–6427.

Ricart, A.D. and Tolcher, A.W. (2007) Technology insight: cytotoxic drug immunoconjugates for cancer therapy. *Nature Clinical Practice Oncology* 4, 245–255.

Santini Jr, J.T., Richards, A.C., Scheidt, R., Cima, M.J. and Langer, R. (2000) Microchips as controlled drug-delivery devices. *Angewandte Chemie International Edition* 39(14), 2396–2407.

Sapra, P., Hooper, A.T., O'Donnell, C.J. and Gerber, H.P. (2011) Investigational antibody drug conjugates for solid tumors. *Expert Opinion on Investigational Drugs* 20(8), 1131–1149.

Senter, P.D. and Sievers, E.L. (2012) The discovery and development of brentuximab vedotin for use in relapsed Hodgkin lymphoma and systemic anaplastic large cell lymphoma. *Nature Biotechnology* 30(7), 631–637.

Sievers, E.L. and Senter, P.D. (2013) Antibody-drug conjugates in cancer therapy. *Annual Review of Medicine*, 64, 15–29.

Stephanopoulos, N. and Francis, M.B. (2011) Choosing an effective protein bioconjugation strategy. *Nature Chemical Biology* 7(12), 876–884.

Swapnil, K., Vijay, S. and Chandrakant, M. (2013) Targeted drug delivery: a backbone for cancer therapy. *Asian Journal of Pharmaceutical Research* 3(1), 40–46.

Tannir, N.M., Forero-Torres, A., Ramchandren, R., Pal, S.K., Ansell, S.M., Infante, J.R. and Thompson, J.A. (2014) Phase I dose-escalation study of SGN-75 in patients with CD70-positive relapsed/refractory non-Hodgkin lymphoma or metastatic renal cell carcinoma. *Investigational New Drugs* 32(6), 1246–1257.

Thompson, J.A., Forero-Torres, A., Heath, E.I., Ansell, S.M., Pal, S.K., Infante, J.R. (2011) The effect of SGN-75, a novel antibody–drug conjugate (ADC), in treatment of patients with renal cell carcinoma (RCC) or non-Hodgkin lymphoma (NHL): a phase I study. *Journal of Clinical Oncology* 29, 3071.

Torchilin, V.P. (2001) Structure and design of polymeric surfactant-based drug delivery systems. *Journal of Controlled Release* 73(2), 137–172.

Trail, P.A. (2013) Antibody drug conjugates as cancer therapeutics. *Antibodies* 2(1), 113–129.

Transparency Market Research (2015) Novel drug delivery systems (NDDS) in cancer therapy market to reach an estimated value of USD 15.9 billion in 2020. *PRNewswire* 19 August 2015, Albany, New York, USA.

US FDA (Food and Drug Administration) (2012) FDA: Pfizer voluntarily withdraws cancer treatment Mylotarg from US market. Available at: www.fda.gov/Safety/MedWatch/SafetyInformation/SafetyAlertsforHumanMedicalProducts/ucm216458.htm (accessed 18 April 2016).

Verma, S., Miles, D., Gianni, L., Krop, I.E., Welslau, M., Baselga, and Blackwell, K. (2012) Trastuzumab emtansine for HER2-positive advanced breast cancer. *New England Journal of Medicine* 367(19), 1783–1791.

Wang, N.X., and Von Recum, H.A. (2011) Affinity-based drug delivery. *Macromolecular Bioscience* 11(3), 321–332.

Whiteman, K.R., Johnson, H.A., Mayo, M.F., Audette, C.A., Carrigan, C.N., LaBelle, A., and Lutz, R.J. (2014) Lorvotuzumab mertansine, a CD56-targeting antibody-drug conjugate with potent antitumor activity against small cell lung cancer in human xenograft models. In MAbs. *Landes Bioscience* 6(2), 90–89.

Yardley, D.A., Weaver, R., Melisko, M.E., Saleh, M.N., Arena, F.P., Forero, A. and Vahdat, L.T. (2012) Abstract P6-10-01: a randomized phase 2 study of the antibody-drug conjugate CDX-011 in advanced GPNMB-overexpressing breast cancer: The EMERGE study. *Cancer Research* 72(24), 6–10.

Younes, A., Gopal, A.K., Smith, S.E., Ansell, S.M., Rosenblatt, J.D., Savage, K.J. and Chen, R. (2012a) Results of a pivotal phase II study of brentuximab vedotin for patients with relapsed or refractory Hodgkin's lymphoma. *Journal of Clinical Oncology* 30(18), 2183–2189.

Younes, A., Yasothan, U. and Kirkpatrick, P. (2012b) Brentuximab vedotin. *Nature Reviews Drug Discovery* 11(1), 19–20.

Zhang, Q., Wang, X., Li, P.Z., Nguyen, K.T., Wang, X.J., Luo, Z. and Zhao, Y. (2014) Biocompatible, uniform, and redispersible mesoporous silica nanoparticles for cancer-targeted drug delivery in vivo. *Advanced Functional Materials* 24(17), 2450–2461.

Zhou, J. and Rossi, J.J. (2014) Cell-type-specific, aptamer-functionalized agents for targeted disease therapy. *Molecular Therapy – Nucleic Acids* 3(6), e169.

Zolot, R.S., Basu, S. and Million, R.P. (2013) Antibody–drug conjugates. *Nature Reviews Drug Discovery* 12(4), 259–260.

10 Microencapsulation for Controlled Gastrointestinal Delivery of Probiotics and Prebiotics

Preeti Panthari[1,*] and Harsha Kharkwal[2]

[1]*Amity Institute of Phytochemistry and Phytomedicine, Amity University, Noida, India;*
[2]*Amity Center for Carbohydrate Research and Amity Institute of Phytomedicine and Phytochemistry, Amity University Uttar Pradesh, Noida, India*

Abstract

Microencapsulation of bioactive compounds (such as antioxidants, vitamins, minerals, omega-3 lipids and probiotics) has been increasingly studied extensively due to interest in nutraceutical components and functional foods. The main objective of this technique is to protect the bioactive compounds from diminished functionality due to environmental conditions such as oxygen, pH, humidity, light or temperature. Among the different microencapsulation processes, spray drying produces a final powder product with good-quality properties for distribution, transportation and storage. In this regard, a wide variety of encapsulation agents have been studied for increasing the viability of the bioactive compounds and to promote an additional functionality in the final product as well, such as prebiotics. Prebiotics are soluble carbohydrates that humans are unable to digest, which selectively enhance *Bifidobacterium* and *Lactobacillus* growth (microorganisms commonly present in the human gut). Some examples include inulin, fructans (fructo-oligosaccharides) and galacto-saccharides. In addition, several microorganisms (probiotics) have demonstrated beneficial effects in humans, and these have been attributed to lactic acid and short-chain fatty acid production, as well as to a reduction in the pH of the colon, which causes a decrease in the survival of pathogenic bacteria. This chapter considers the enhanced efficacy of probiotics and prebiotics through microencapsulation in addressing gastrointestinal diseases.

Introduction

Microencapsulation involves the packing of a core (of a bioactive or functional ingredient) within a secondary material (encapsulant) to form a microcapsule (Mooranian *et al.*, 2015). The encapsulant provides a protective matrix or shell around the core, thus isolating it from the outer environment until release is triggered by environmental changes. This ultimately saves the bioactive or functional ingredient from degradation or unnecessary interactions with surrounding materials, for example other food components in any food application. Before designing the microcapsulated ingredient its end use should be kept in mind (Furuike *et al.*, 2015). This requires knowledge of: (i) the core; (ii) the encapsulant materials; (iii) interactions between the core, matrix and the environment; (iv) the stability of the microencapsulated ingredient in storage and when incorporated into the food matrix; and (v) the mechanisms that control the

*Corresponding author. E-mail: preeti.panthari81@gmail.com

release of the core (Turino *et al.*, 2015). As the molecular structure of the core is known, information on the core's interactions with other food components, its fate upon consumption, its target site for action and – in the case of a bioactive core – its function in the body after ingestion, may also be known. Different encapsulated materials are in use, for example proteins (sodium caseinate, whey protein, isolated wheat proteins, soy proteins, gelatins, zein and albumin), carbohydrates (native starches, modified starches, resistant starches, maltodextrins, dried glucose syrups, gum acacia, alginates, pectins, carrageenan, chitosan, cellulosic materials, sugars and derivatives), lipids and waxes (vegetable fats and oils, hydrogenated fats, palm stearin, carnauba wax, beeswax, shellac, polyethylene glycol), alone or in combination (Neish, 2009).

A successful oral delivery system depends on the encapsulation technique, capsule material, capsule size, capsule stability, mass transfer, polymer coating, release mechanism, etc. Another aspect is immunogenicity of the capsule material in oral delivery. In addition, delivery of the viable probiotic at the host intestine without hampering the physiological properties of the encapsulated bacteria is one of the primary concerns. The major limitation of oral consumption is the complexity of delivering the 'live bacteria'. Poor survival of probiotics during gastric transit in the host's gastrointestinal (GI) tract has been a serious concern, as non-viable bacteria may not be very effective in conferring the potential therapeutic effects. Better immunomodulatory effects are exhibited by the viable probiotics in comparison to heat-killed bacteria or bacterial components (Ditu *et al.*, 2014).

Microencapsulation Processes

Microencapsulation depends on the core properties, the encapsulant materials and the need of the target food application. Traditionally, a range of microencapsulates are used for food ingredients. Many mechanical processes are in use, such as emulsification, spray drying, fluidized-bed coating, centrifugal extrusion, spinning disk, pressure extrusion and hot-melt extrusion. Some chemical processes used include ionotropic gelation, simple coacervation, complex coacervation, solvent evaporation, liposomes and cyclodextrin complexation.

Spray drying

The most common technique is spray drying as it is economical, flexible and produces good-quality products (Deng *et al.*, 2014). The core material is dispersed into the polymer solution via emulsion or dispersion. In the drying chamber, homogenization and atomization of the mixture take place. This leads to solvent evaporation and formation of microcapsules. Although it can be operated on a continuous basis, one of the basic requirements, high temperature, is not suitable for the encapsulation of probiotic bacteria (Gharsallaoui *et al.*, 2007).

Extrusion and emulsion

Extrusion involves projection of the emulsion core and coating material through a nozzle at high pressure. Extrusion of polymers has been reported on a laboratory scale using syringes (Gombotz and Siow, 2012). Prilling is a technique that involves formation of droplets in a controlled manner. Multi-nozzle systems are used for mass production. However, the most extensively used methodology is emulsion.

Microencapsulation laboratory procedures

There is plenty of literature available on microencapsulation of probiotic bacteria used at small scale where it involves emulsion, extrusion and/or coacervation. (Zuidam and Eyal, 2010). Of the entrapment procedures, the most common are gelled biopolymer systems having kappa–carrageenan or calcium–alginate matrices (Kumar *et al.*, 2009). An alternative microencapsulation technique that involves a single-step process is interfacial polymerization. Here, the emulsion is formed in a discontinuous phase (aqueous suspension of cells) as well as a continuous phase (organic solvent). The droplets contain the cell and the reaction is initiated when a biocompatible reagent, soluble in the continuous organic phase, is added to the emulsion (Ikem *et al.*, 2010).

Prebiotics and probiotics

The use of prebiotics and probiotics in dietary practices has increased worldwide in recent years. A prebiotic is a selectively fermented ingredient that results in specific changes in the composition and/or activity of the gastrointestinal microbiota, which in turn provide the health benefit(s) (Saier and Mansour, 2005).

Probiotics are live microorganisms, which when ingested in adequate quantities offer one or more demonstrated health benefits such as increasing immunity levels in lactose intolerance, food allergies, diarrhoea and colon cancer (Gill *et al.*, 2001; Reid, 2001). According to a report by the Food and Agriculture Organization/World Health Organization (FAO/WHO), probiotics are 'live microorganisms which when administered in adequate amounts confer a health benefit on the host'.

A number of microorganisms with probiotic potential are used in food products or as supplements. Lactic acid bacteria (*Lactobacillus* spp.) and bifidobacteria (*Bifidobacterium* spp.) have the most potential. They are saccharolytic bacteria and thus metabolize sugars. The presence of probiotics in the human small and large intestines leads to many health benefits, although the literature varies on this point. They do indeed benefit health by producing nutrients and cofactors, competing with pathogens for binding sites and stimulating the host immune response (Gøbel *et al.*, 2010). Many reports have demonstrated positive health benefits in the treatment of inflammatory bowel disease, irritable bowel syndrome, *Helicobactor pylori* infections and antibiotic-associated diarrhoea (Johnston *et al.*, 2011). Most of them have supporting clinical data (Chávarri *et al.*, 2010).

One important point is the viability of the probiotic during use. In many cases it has been found that viability is lost as the probiotic passes through the low pH of the stomach and high bile salt condition of the intestine. However, viability can be sustained by microencapsulation of the bacteria in a polymer matrix (Capela *et al.*, 2006; Champagne and Kailasapathy, 2008). The encapsulate should be non-cytotoxic and non-antimicrobial to ensure the host and bacteria are not harmed (Kailasapathy, 2002, 2005). Release of the bacteria needs to be targeted in the GI tract in order to gain the full health benefits (You *et al.*, 2004). The naturally available polysaccharides and proteins are good alternatives for microencapsulation as they do not impart any cytotoxicity (Boh, 2006). Figure 10.1 highlights the therapeutic effects of probiotics.

Probiotics: food products

The probiotics can be consumed in three ways:

- Culture concentrate (dried or deep-frozen) to be added to a food for industrial or home uses.
- Food products (fermented/non-fermented).
- Dietary supplements (drug products in powder, capsule or tablet forms) (Tannis, 2008).

The most convenient way to consume probiotics is in the form of food products. Table 10.1 highlights the probiotic products available in the market. According to the reports of Statista, sales of probiotics worldwide were projected to grow to USD8 billion by 2014 (www.statista.com/statistics/492591/dollar-sales-probiotic-supplements-united-states; Prajapati and Senan, 2014). Currently, the most developed markets are in Europe and Japan, accounting for more than half of total sales (McGee *et al.*, 2011). The needs of the consumer have developed during last few decades as there is much more awareness of health issues (McGee *et al.*, 2011). Probiotic bacteria are readily available in the form of fermented milks, ice cream, different types of cheese, baby food, powdered milk, whey-based beverages, sour cream, buttermilk, normal and flavoured liquid milk, and concentrated milk (Iyer and Kailasapathy, 2005). A variety of non-dairy products are also available, such as vegetarian foods, cereal, confectionery, soya-based products, fruit juice, breakfast cereals and baby foods. As a large part of the population turns to vegetarianism, so considerable development in probiotic food products has taken place. Dairy products have a large share of the probiotic food available in the market. Cargill (2009) reported worldwide sale of probiotic through yogurt to be 78%, so developing dairy products containing

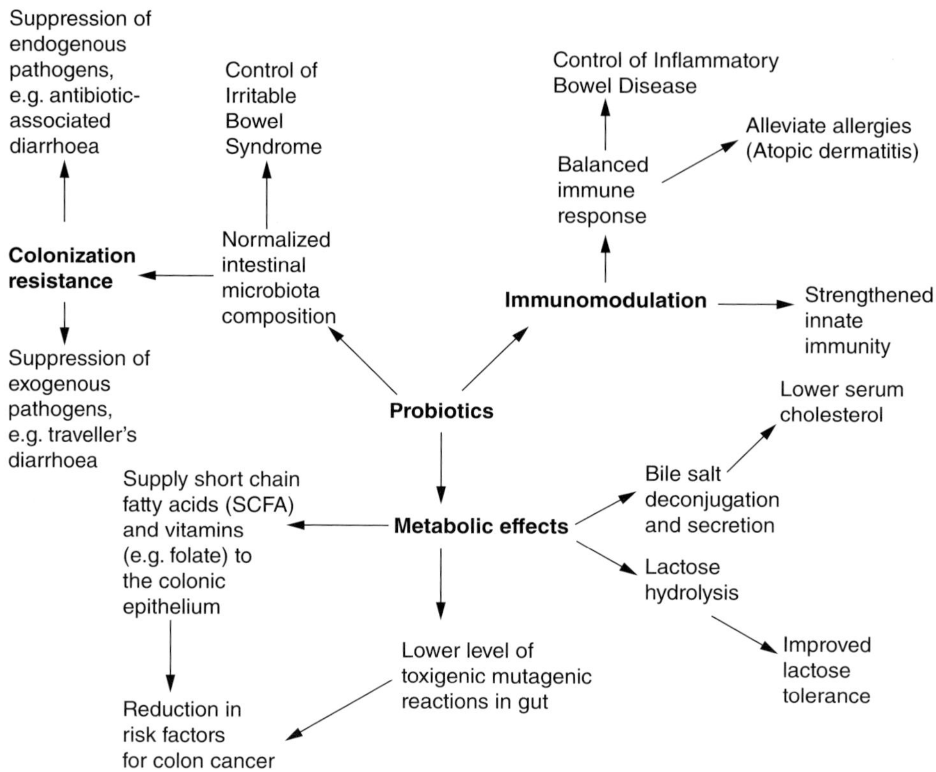

Fig. 10.1. Therapeutic benefits of probiotics (adapted from Sarrela *et al.*, 2002).

probiotic bacteria is commercially important for the manufacturers. Table 10.2 lists the micro-organisms used as probiotics.

Probiotics Delivery and the GI Tract

The gastrointestinal tract possesses the highest complexity in the human body as the immune tissues constitute the largest and most complex fraction of the human immune system. There is considerable variation in bacterial numbers and populations between the stomach, small intestine and colon. Much research has been done in the field of gut microbial ecology, but current knowledge is insufficient to understand this ecosystem and certainly demands more research (Augustin *et al.*, 2011). Microbial colonization in the GI tract starts immediately after birth. Although a sterile gut is found in the newborn baby, with ageing gradual changes take place in the gut leading to an adult-type microflora (Fig. 10.2)

Many factors, such as age, health, hormonal functions, stress level, drug use and diet modulate the intestinal microflora (Bhujbal *et al.*, 2014; Cortes-Rojas *et al.*, 2014; Mahdavi *et al.*, 2014; Xiao *et al.*, 2014).

Gut microbiota

Weaning solid foods change the composition and complexity of gut microbiota. Changes in diets during adulthood are also greatly responsible for its composition (Wang *et al.*, 2008). Development of 16S ribosomal RNA (rRNA) gene sequence-based metagenomic methods has led to major advances in defining the total gut microbial population. Results show that 90% of the bacteria belong to Bacteroidetes and Firmicutes (Urbanska *et al.*, 2007). The most important roles of gut microbiota in maintaining health are summarized below (Carroll *et al.*, 2007).

Table 10.1. Probiotic products available worldwide (adapted from Mortazavian *et al.*, 2012)

Dairy products	Non-dairy products
Regular full-fat yogurts	Vegetable-based drinks
Iranian yogurt drink (Doogh)	Fermented banana
Acidophilus milk drink	Tomato-based drink
Stirred fruit yogurts	Many dried fruits
Fermented dairy beverage	Green coconut water
Whey protein-based drinks	Peanut milk
Symbiotic acidophilus milk	Cranberry, pineapple and orange juices
Cheddar cheese	Ginger juice
Feta cheese	Grape and passion fruit juices
Cheese from caprine milk	Cabbage juice
Semi-hard reduced-fat cheese	Carrot juice
White-brined cheese	Noni juice
Minas Fresco cheese	Onion
Cottage cheese	Non-fermented fruit juice beverages
Canestrato Pugliese hard cheese	Non-fermented soy-based frozen desserts
Argentine Fresco cheese	Fermented soymilk drink
Goat semi-solid cheese	Soy-based stirred yogurt
Turkish Beyaz cheese	Rice-based yogurt
Iranian white-brined cheese	Oat-based drink
White-brined cheese	Oat-based products
Minas fresh cheese	Oat bran products
Symbiotic ice cream	Fermented maize beverage
Fermented goat's milk	Wheat fermented malt-based drink
Probiotic ice cream	Millet or sorghum flour fermented probiotic beverage
Low-fat ice cream	Starch-saccharified probiotic drink
Mango soy fortified probiotic yogurt	Meat products
Frozen yogurt	Tempeh (based on soybean)
Frozen symbiotic dessert	Chocolate
Acidophilus butter	
Traditional Greek yogurt	
Frozen dairy dessert	
Corn milk yogurt	
High pressure homogenized probiotic fermented milk	
Banana-based yogurt	
Mango soy fortified probiotic yogurt	
Yog-ice cream	

Structure and histological function

Microbiota define the intestinal structure and function of the tract. Mucin secretion and degradation is regulated by the mucus layer that maintains the required balance. It produces an obstacle to pro-inflammatory compounds and uptake of antigens.

The microflora define development in cells and tissues. Fatty acid (butyrate) secreted by colonic microbes regulates cell growth and differentiation, and hence inhibits cell growth transformation.

Metabolic functions

The gut bacteria are responsible for producing a number of vitamins. Biotransformation of bile by microbial enzymes results in the metabolism of glucose and cholesterol. Different pathways are provided by the microbiome for the fermentation of non-digestible substrates such as fibres and endogenous mucus. The growth of the microbes is dependent on the fermentation/metabolism of non-digestible substrates. The major fatty acids produced are acetate, propionate and butyrate, followed by lactate, ethanol, succinate,

Table 10.2. Microorganisms used as probiotics (adapted from Stefan *et al.*, 2013)

Microorganisms	Genus	Species
Lactic acid bacteria	*Lactobacillus* spp.	*L. acidophilus*
		L. amylovorus
		L. brevis
		L. bulgaricus
		L. casei
		L. crispatus
		L. delbrueckii spp. *bulgaricus*[b]
		L. fermentum
		L. gallinarum[a]
		L. gasseri
		L. helveticus
		L. johnsonii
		L. lactis
		L. paracasei
		L. plantarum
		L. reuteri
		L. rhamnosus
		L. salivarius
	Streptococcus spp.	*S. salivarius* subsp. *thermophilus*[b]
	Lactococcus spp.	*L. lactis*
	Leuconostoc spp.	*Lc. mesenteroides*
	Pediococcus spp.	*P. pentosaceus*
		P. acidilactici
Bifidobacteria	*Bifidobacterium* spp.	*B. adolescentis*
		B. animalis[c]
		B. bifidum
		B. breve
		B. essensis
		B. infantis
		B. laterosprous
		B. longum
Propionibacteria	*Propionibacterium* spp.	*P. acidipropionici*
		P. freudenreichii
		P. jensenii
		P. thoenii
Enterobacteria	*Enterococcus* spp.	*E. faecalis*[a]
		E. faecium
Sporulated bacteria	*Bacillus* spp.	*B. alcalophilus*
		B. cereus
		B. clausii
		B. coagulans
		B. subtilis
Other bacteria	*Escherichia* spp.	*E. coli*
	Sporolactobacillus spp.	*S. inulinus*[a]
Yeasts	*Saccharomyces* spp.	*S. cerevisiae (boulardii)*[d]

[a]Mainly for animals; [b]*L. delbrueckii* subsp. *bulgaricus* is typically used as a starter culture for yogurt; mainly in pharmaceutical preparations; [c]Some *Bifidobacterium animalis* strains are commonly referred to on commercial labels as *B. lactis*; [d]Probiotic *Saccharomyces cerevisiae* strain is marketed as *S. boluardii*.

formate, valerate, caproate, isobutyrate, 2-methylbutyrate and isovalerate. In the cecum and colon, bacterial formation takes place as the short-chain fatty acids absorb salts and water. These bacteria show a protective effect on the intestinal epithelium. The main source of energy is butyrate, which is preferred by colonic bacteria (Adrio and Demain, 2014).

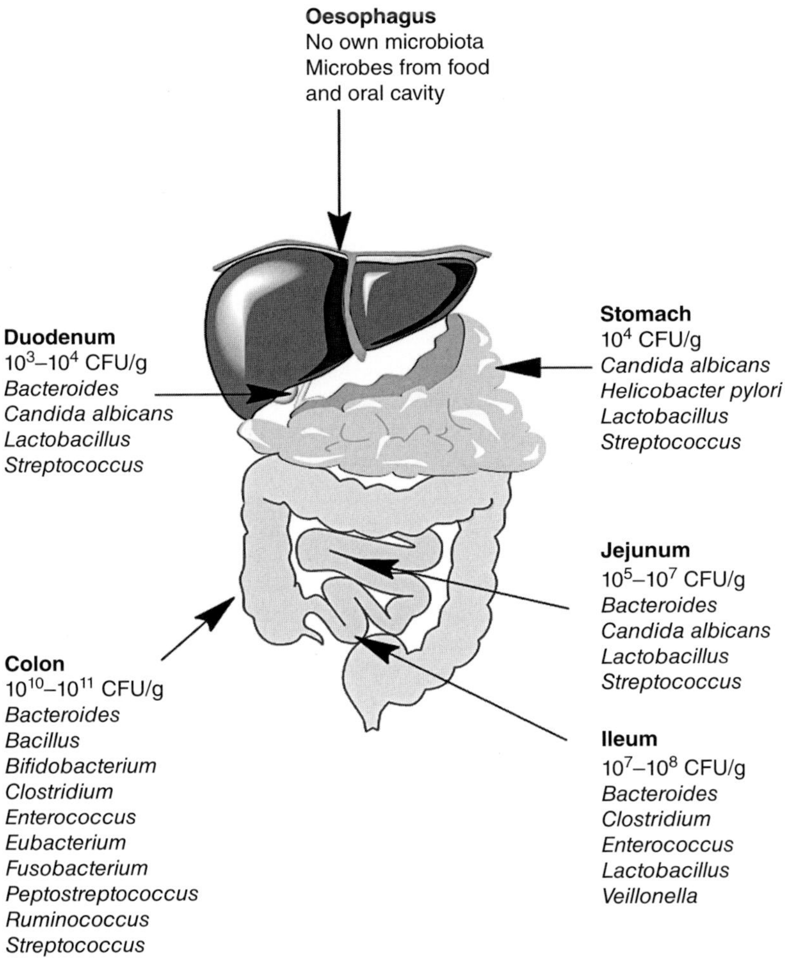

Fig. 10.2. Microbial colonization of human gastrointestinal tract (adapted from Isolauri *et al*., 2004; Vyas and Ranganathan, 2012). CFU, colony forming units.

Protective functions

The commensal organisms produce antimicrobial compounds and compete for the nutrients and sites of attachment in the gut lining; hence colonization by pathogens is limited (Fig. 10.3). This results in a reduction in lipopolysaccharide and peptidoglycan production. The microflora are also responsible for the development of the immune system (Sanchez and Demain, 2011). The commensal bacteria of the GI tract, therefore, play active roles in the development and homeostasis of the immune system.

After ingestion, the microencapsulated probiotic passes quickly from the oesophagus and reaches the stomach where, due to high levels of acid, the greatest loss of bacteria is expected (Wandrey *et al.*, 2010). The gastric emptying time is found to be between 5 min and 2 h; the half gastric emptying time is estimated to be 80.5 min (Cook *et al.*, 2012). The maximum fluid capacity of the stomach is 1.5 l, but it contains only 50 ml in a fasted state. The presence of pepsin is responsible for the breaking down of proteins, produced in the low pH of the stomach. The next stage is entry into the small intestine. In two separate studies transit has been reported to be 3.2±1.6 h (Davis *et al.*, 1986) and ranging from 0.5 to ~ 9.5 h (Coupe *et al.*, 1991). From the small intestine the microcapsule moves to the large intestine where the pH lowers to 5.26–6.72 in the ascending colon and 5.20–7.02

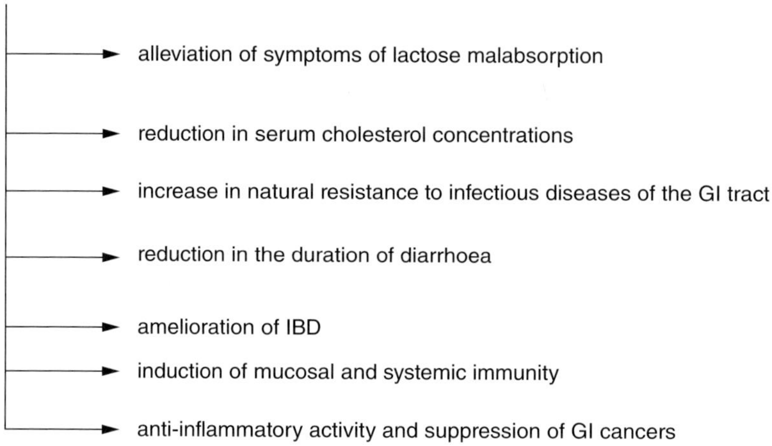

Fig. 10.3. Therapeutic effects of probiotics. GI, gastrointestinal; IBD, inflammatory bowel disease.

in the descending colon. Later in the GI tract the encapsulate encounters low levels of liquid as the contents compact into faeces (Mariat *et al.*, 2009).

Conclusion and Future Prospects

In the future there will clearly be a huge demand for probiotic bacteria and bioactive prebiotic substances in the development of innovative health-based products. This will lead to a cost-effective fermentation technology for maximizing the symbiotic health properties of newer foods. Current technological development has led to precise microcapsule formation from simple immobilization. The use of nutraceuticals and food ingredients has been a boon for advancement in this field. In the near future, if we are successful in achieving the delivery of microencapsulated probiotic bacteria, the concept of a healthy gut may become a reality. Microencapsulation, coupled with developing nanoencapsulation protocols, will be vital for the preservation and storage of the bacteria in unfavourable environmental conditions, leading to novel food, pharmaceutical and medical applications.

Acknowledgements

We thank Dr Ashok K. Chauhan, Founder President, Ritanand Balved Educational Foundation, for support. The guidance from Shri. Atul Chauhan, Chancellor, Amity University Uttar Pradesh, and Prof. Dr (Mrs) Balvinder Shukla, Vice Chancellor, Amity University Uttar Pradesh, is greatly appreciated. Insights from articles cited in this chapter have been of considerable assistance, and we are grateful to their authors.

References

Adrio, J.L. and Demain, A.L. (2014) Microbial enzymes: tools for biotechnological processes. *Biomolecules* 4, 117–139.

Augustin, M.A., Abeywardena, M.Y., Patten, G., Head, R., Lockett, T., De Luca, A. *et al.* (2011) Effects of microencapsulation on the gastrointestinal transit and tissue distribution of a bioactive mixture of fish oil, tributyrin and resveratrol. *Journal of Functional Foods* 3(1), 25–37.

Bhujbal, S.V., de Vos, P. and Niclou, S.P. (2014) Drug and cell encapsulation: Alternative delivery options for the treatment of malignant brain tumors. *Advanced Drug Delivery Reviews* 67–68, 142–53.

Boh, B. (2006) Developements et applications industrielles des microcapsules. In: Capela, P., Hay, T.K.C. and Shah, N.P., Effect of cryoprotectants, prebiotics and microencapsulation on survival of probiotic organisms in yoghurt and freeze-dried yoghurt. *Food Research International* 39(2), 203–211.

Capela, P., Hay, T.K.C. and Shah, N.P. (2006) Effect of cryoprotectants, prebiotics and microencapsulation on survival of probiotic organisms in yoghurt and freeze-dried yoghurt. *Food Research International* 39(2), 203–211.

Cargill (2009) Cargill beverage concepts will address consumer demands for health, taste and texture at IFT 2008. Available at: www.cargill.com/newscenter/news-releases/2008/NA3007612.jsp (accessed 20 July 2009).

Carroll, I.M., Andrus, J.M., Bruno-Bárcena, J.M., Klaenhammer, T.R., Hassan, H.M. and Threadgill, D.S. (2007) Anti-inflammatory properties of *Lactobacillus gasseri* expressing manganese superoxide dismutase using the interleukin 10-deficient mouse model of colitis. *The American Journal of Physiology: Gastrointestinal and Liver Physiology* 293, G729–G738.

Champagne, C.P. and Kailasapathy, K. (2008) Encapsulation of probiotics. In: Garti, N. (ed.) *Delivery and Controlled Release of Bioactives in Foods and Nutraceuticals*. Woodhead Publishing, Cambridge, UK, pp. 344–369.

Chávarri, M., Marañón, I., Ares, R., Ibáñez, F.C., Marzo, F. and Villarán, M.D.C. (2010) Microencapsulation of a probiotic and prebiotic in alginate-chitosan capsules improves survival in simulated gastro-intestinal conditions. *International Journal of Food Microbiology* 142, 185–189.

Cook, M.T., Tzortzis, G., Charalampopoulos, D. and Khutoryanskiy, V.V. (2012) Microencapsulation of probiotics for gastrointestinal delivery. *Journal of Controlled Release* 162(1), 56–67.

Cortes-Rojas, D.F., Souza, C.R.F. and Oliveira, W.P. (2014) Encapsulation of eugenol rich clove extract in solid lipid carriers. *Journal of Food Engineering* 127, 34–42.

Coupe, A., Davis, S. and Wilding, I. (1991) Variation in gastrointestinal transit of pharmaceutical dosage forms in healthy subjects. *Pharmaceutical Research* 8, 360–364.

Davis, S.S., Hardy, J.G. and Fara, J.W. (1986) Transit of pharmaceutical dosage forms through the small intestine. *Gut* 27, 886–892.

Deng, X.X., Chen, Z., Huang, Q., Fu, X. and Tang, C.H. (2014) Spray-drying microencapsulation of beta-carotene by soy protein isolate and/or OSA-Modified starch. *Journal of Applied Polymer Science*, 131(12), 40399.

Ditu, L.M., Chifiriuc, M.C., Bezirtzoglou, E., Marutescu, L., Bleotu, C., Pelinescu, D., Mihaescu, G. and Lazar, V. (2014) Immunomodulatory effect of non-viable components of probiotic culture stimulated with heat-inactivated *Escherichia coli* and *Bacillus cereus* on holoxenic mice. *Microbial Ecology in Health and Disease*, 25, 23239.

Furuike, T., Nasu, K. and Tamura, H. (2015) Synthesis of sulfated cyclodextrin amphiphiles with liposomal encapsulation properties. *Journal of Encapsulation and Adsorption Sciences* 5, 144–154.

Gharsallaoui, A., Roudaut, G., Chambin, O., Voilley, A. and Saurel, R. (2007) Applications of spray-drying in microencapsulation of food ingredients: an overview. *Food Research International* 40, 1107–1121.

Gill, H.S., Rutherfurd, K.J., Cross, M.L. and Gopal, P.K. (2001) Enhancement of immunity in the elderly by dietary supplementation with the probiotic Bifidobacterium lactis HN019. *The American Journal of Clinical Nutrition*, 74(6), 833–9.

Gøbel, R., Larsen, N., Mølgaard, C., Jakobsen, M. and Michaelsen, K.F. (2010) Probiotics to young children with atopic dermatitis: a randomized placebo-controlled trial. *International Journal of Probiotics and Prebiotics* 5(2), 53–60.

Gombotz, W.R. and Siow, F.W. (2012) Protein release from alginate matrices. *Advanced Drug Delivery Reviews* 64, 194–205.

Ikem, V.O., Menner, A., Horozov, T.S. and Bismarck, A. (2010) Highly permeable macroporous polymers synthesized from pickering medium and high internal phase emulsion templates. *Advanced Materials* 22(32), 3588–3592.

Isolauri, E., Salminen, S. and Ouwehand, A.C. (2004) Microbial-gut interactions in health and disease. Probiotics. *Best Practice & Research Clinical Gastroenterology*, 18, 299–313.

Iyer, C. and Kailasapathy, K. (2005) Effect of co-encapsulation of probiotics with prebiotics on increasing the viability of encapsulated bacteria in simulated gastrointestinal conditions and in yoghurt. *Journal of Food Science* 70(1), M18–M23.

Johnston, B.C., Goldenberg, J.Z., Vandvik, P.O., Sun, X. and Guyatt, G.H. (2011) Probiotics for the prevention of pediatric antibiotic-associated diarrhea. *Cochrane Database of Systematic Reviews 2011 Nov 9; (11).*

Kailasapathy, K. (2002) Microencapsulation of probiotic bacteria: technology and potential applications. *Current Issues in Intestinal Microbiology* 3, 39–48.
Kailasapathy, K. (2005) Survival of free and encapsulated probiotic bacteria and effect on the sensory properties of yoghurt. *Food Science and Technology* 1, 1–2.
Kumar, S., Dwevedi, A. and Kayastha, A.M. (2009) Immobilization of soybean (*Glycine max*) urease on alginate and chitosan beads showing improved stability: Analytical applications. *Journal of Molecular Catalysis B: Enzymatic* 58(1), 138–145.
Mahdavi, S.A., Jafari, S.M., Ghorbani, M. and Assadpoor, E. (2014) Spray-drying microencapsulation of anthocyanins by natural biopolymers: a review. *Drying Technology* 32(5), 509–518.
Mariat, D., Firmesse, O., Levenez, F. *et al.* (2009) The firmicutes/bacteroidetes ratio of the human microbiota changes with age. *BMC Microbiology* 9, 123.
McGee, R.G., Bakens, A., Wiley, K., Riordan, S.M. and Webster, A.C. (2011) Probiotics for patients with hepatic encephalopathy. *Cochrane Database of Systematic Reviews* 9(11), CD008716.
Mooranian, A., Negrulj, R., Mikov, M., Golocorbin-Kon, S., Arfuso, F. and Al-salami, H. (2015) Novel chenodeoxycholic acid-sodium alginate matrix in the microencapsulation of the potential antidiabetic drug, Probucol. An *in vitro* study. *Journal of Microencapsulation* 32(6), 589–597.
Mortazavian, A.M., Mohammadi, R. and Sohrabvandi, S. (2012) Delivery of probiotic microorganisms into gastrointestinal tract by food products. In: Brzozwski, T. (ed.) *New Advances in the Basic and Clinical Gastroenterology.* InTech, Rijeka, Croatia, pp. 121–146.
Neish, A.S. (2009) Microbes in gastrointestinal health and disease. *Gastroenterology* 136(1), 65–80.
Prajapati, J.B. and Senan, S. (2014) Perspectives into probiotics-from phenotyping to pyrosequencing. *International Journal of Fermented Foods* 3(2), 115–126.
Reid, G. (2001) Probiotic agents to protect the urogenital tract against infection. *The American Journal of Clinical Nutrition* 73, 437S-443S.
Sarrela, M., Lateenmaki, L., Crittenden, R., Salminen, S., Mattila-Sandholm, T. (2002) Gut bacteria and health foods— the European Perspective. *International Journal of Food Microbiology* 15, 78 (1–2), 99–117.
Saier, M.H. and Mansour, N.M. (2005) Probiotics and prebiotics in human health. *Journal of Molecular Microbiology and Biotechnology* 10, 22–25.
Sanchez, S. and Demain, A.L. (2011) Enzymes and bioconversions of industrial, pharmaceutical, and biotechnological significance. *Organic Process Research & Development* 15, 224–230.
Stefan, D., Elena, B.G. and Gabriela, I. (2013) Sources, production and microencapsulation of probiotics. In: Ötleş, S. (ed.) *Probiotics and Prebiotics in Food, Nutrition and Health.* CRC Press, New York, pp. 25–49.
Tannis, A. (2008) *Probiotic Rescue: How you can use Probiotics to Fight Cholesterol, Cancer Superbugs, Digestive Complaints and More.* John Wiley & Sons, Ontario, Canada, p.269
Turino, L.N., Mariano, R.N. and Mengatto, L.N. (2015) In vitro evaluation of suspoemulsions for in situ forming polymeric microspheres and controlled release of progesterone. *Journal of Microencapsulation: Micro and Nano Carriers* 32(6), 538–546.
Urbanska, A.M., Bhathena, J. and Prakash, S. (2007) Live encapsulated *Lactobacillus acidophilus* cells in yoghurt for therapeutic oral delivery: preparation and in-vitro analysis of alginate-chitosan microcapsules. *Canadian Journal of Physiology Pharmacology* 85, 884–893.
Vyas, U. and Ranganathan, N. (2012) Probiotics, prebiotics, and synbiotics: gut and beyond. *Gastroenterology Research and Practice*, 2012, 872716.
Wandrey, C., Bartkowiak, A. and Harding, S.E. (2010) Materials for encapsulation. In: Zuidam, N.J. and Nedovic, V.A. (eds) *Encapsulation Technologies for Active Food Ingredients and Food Processing.* Springer, New York, pp. 31–100.
Wang, X., Jiang, Y., Wang, Y.W., Huang, M.T., Ho, C.T. and Huang, Q. (2008) Enhancing anti-inflammation activity of curcumin through o/w nanoemulsions. *Food Chemistry* 108, 419–424.
Xiao, Z., Liu, W., Zhu, G., Zhou, R. and Niu, Y. (2014) A review of the preparation and application of flavour and essential oils microcapsules based on complex coacervation technology. *Journal of the Science of Food and Agriculture* 94(8), 1482–1494.
You, H.J., Oh, D.K. and Ji, G.E. (2004) Anticancerogenic effect of a novel hiroinositol containing polysaccharide from *Bifidobacterium bifidum* BGN4. *FEMS Microbiology Letters* 240, 131–136.
Zuidam, N.J. and Eyal, S. (2010) Overview of microencapsulates for use in food products or processes and methods to make them. In: Zuidam, N.J. and Nedovic, V. (eds) *Encapsulation Technologies for Active Food Ingredients and Food Processing.* Springer, New York, pp. 3–29.

11 Chitosan in Drug Delivery and Targeting for Cancer Treatment

Anirbandeep Bose[1,2] **and Tin Wui Wong**[1,2*]
[1]*Particle Design Research Group, Faculty of Pharmacy, Universiti Teknologi MARA, Selangor, Malaysia;* [2]*Non-Destructive Biomedical and Pharmaceutical Research Centre, iPROMISE, Universiti Teknologi MARA, Selangor, Malaysia*

Abstract

Chitosan is a linear heteropolysaccharide consisting of β(1-4) linked 2-acetamido-2-deoxy-β-D-glucopyranose and 2-amino-2-deoxy-β-D-glycopyranose units. It is derived from chitin, the second most abundant polymer after cellulose, and which is available in the epidermis or exoskeletons of crustaceans such as crabs and shrimps; in insects such as grasshoppers and dragonflies; in fungal cell walls, including those of the enoki mushroom (*Flammulina velutipes*) and shiitake mushroom (*Lentinus edodes*); and in bacteria. Chitosan has been employed as the primary matrix former in pharmaceutical dosages with drugs, peptides, proteins and genes to treat gastric, duodenal, liver, breast, ovarian, lung, colorectal, pancreatic, leukaemia, nasal and kidney cancers. The chitosan-based dosage forms have been decorated with both passive (enhanced permeability and retention effect) and active targeting (receptor-mediated endocytosis) elements. Depending on the route of administration, these forms could be enteric coated, designed by means of chitosan–drug conjugation or equipped with superparamagnetic components for drug targeting and/or providing an alternative treatment strategy (e.g. hyperthermia) to kill the cancer cells. The drug targeting effectiveness and specificity of dosage forms can be enhanced through chemical modification of chitosan with ligands such as folate and galactose. Subjecting chitosan to chemical modification also leads to an increase in the transfection efficiency of therapeutics in the cancer cells. This chapter provides an introspective outlook on cancer-targeted carriers made of chitosan and its derivatives. It emphasizes the physicochemical aspects of chitosan and derivatives in relation to cancer targeting mechanisms of carriers.

Introduction

Chitosan is a linear, randomly distributed and binary heteropolysaccharide consisting of β(1-4) linked 2-acetamido-2-deoxy-β-D-glucopyranose and 2-amino-2-deoxy-β-D-glycopyranose units (Fig. 11.1). It is prepared by deacetylation of chitin, the second most abundant natural polymer after cellulose, which is a linear polymer of β(1-4) linked N-acetyl-D-glucosamine units composed of mucopolysaccharides and amino sugars (George and Abraham, 2006). Chitin is found in the epidermis or exoskeletons of crustaceans such as crabs and shrimps; in insects such as grasshoppers and dragonflies; in the cell walls of fungi, such as the enoki mushroom (*Flammulina velutipes*) and shiitake mushroom (*Lentinus edodes*); and in bacteria (Wong, 2009). The chitosan is prepared from chitin via chemical and heat processing (Kumar, 2000; Yong and Wong, 2013).

*Corresponding author. E-mail: wongtinwui@salam.uitm.edu.my

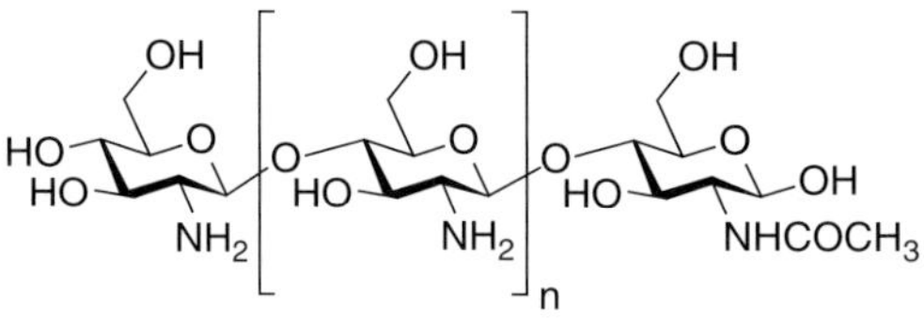

Fig. 11.1. Chemical structure of chitosan.

Physicochemical Properties

Chitosan is available in different molecular weights (700–1000 kDa) and degrees of deacetylation (66–95%) (Agnihotri *et al.*, 2004; Sinha *et al.*, 2004; Jon *et al.*, 2007). The amine functional group of chitosan is responsible for its solubility in aqueous acidic solutions (Yong and Wong, 2013). The solubilization occurs through protonation of $-NH_2$ functional groups on the C-2 position of D-glucosamine residues. The solubility is relatively low at a physiological pH of 7.4 or higher, as it is a weak base with pKa of 6.2–7.0 (Hejazi and Amiji, 2003; Issa *et al.*, 2005). In acidic media, the solubility increases for deacetylation 50% or more (Rinaudo, 2006). In solution it exists in an extended conformation due to the repelling effect of positively charged amine moieties on the neighbouring glucosamine units (Wong, 2009). Electrolytes such as sodium salts reduce the inter-chain repulsion and promote the interactions. These interactions induce a more random coil-like conformation in chitosan molecules and lead to salting out and precipitation of chitosan (Singla and Chawla, 2001).

The $-NH_2$ functional group on chitosan can exert specific interactions with metals. The intensity of chelation is dependent on the physical state, $-NH_2$ content and distribution, degree of polymerization, pH, and cation amount. The chitosan is able to form electrostatic complexes under acidic conditions with oppositely charged surfactants, polysaccharides, polymers and fats (Wong, 2009; Wong and Nurulaini, 2012; Yong and Wong, 2013). Examples include polyacrylic acid, sodium salt of polyacrylic acid, carboxymethylcellulose, xanthan, carrageenan, alginate, pectin, heparin, hyaluronan, sulfated cellulose, dextran sulfate and chondroitin sulfate. The polyelectrolyte complexation often takes place as an interfacial reaction, and could be conducted in single or multiple cycles. The latter produces layer-by-layer coacervates. The chitosan may also be crosslinked by reagents, namely epichlorohydrin, genipin, diisocyanate, 1,4-butanediol diglycidyl ether, glycerol phosphate, oxalic acid, pyrophosphate, tripolyphosphate, tetrapolyphosphate, octapolyphosphate, hexametaphosphate and $Fe(CN_6)^{-4}/Fe(CN_6)^{-3}$ (Singla and Chawla, 2001; George and Abraham, 2006; Rinaudo, 2006). The starch–chitosan blend can be crosslinked though oxidizing the starch to produce polyaldehyde that in turn reacts with the $-NH_2$ groups in the presence of reducing agents (Rinaudo, 2006). Alternatively, crosslinking can be accomplished using glyoxal and glutaraldehyde (George and Abraham, 2006). The aldehyde groups form covalent imine bonds with the amino groups of chitosan, due to the resonance established with adjacent double ethylinic bonds via a Schiff reaction. The use of dialdehyde promotes direct crosslinking without the need for a reducing agent.

One interesting feature of chitosan lies in its ability to be modified chemically to produce new covalent entities with altered physicochemical properties. The chemical reaction primarily involves the C2 position in quarternization of amino groups (Rinaudo, 2006). Regular and reproducible derivatives are largely obtained from highly deacetylated chitosan substrates in a chemical modification process. A few examples are O- and N-carboxymethylchitosan, chitosan 6-O-sulfate, trimethylchitosan ammonium, N-(2-hydroxyl)propyl-3-trimethyl ammonium chitosan chloride, carbohydrate branched chitosan, polyethylene grafted chitosan, cyclodextrin-linked chitosan, thiolated chitosan and N-alkylated chitosan (Wong, 2009; Yong and Wong, 2013). The chitosan is not only modified to tailor for specific physicochemical properties such as aqueous solubility (Kubota *et al.*, 2000); it can also be tagged with receptor-specific ligands covalently for biorecognition and targeted drug delivery (Wong and Bose, 2015).

Biological Properties

Chitosan is a biodegradable and biocompatible polymer (Wong, 2009). It is degraded *in vivo* by lysozymes and has a low oral toxicity with LD_{50} value in excess of 16 g/kg body weight of mouse (Yong and Wong, 2013). Its degradation propensity increases with a decrease in its degree of deacetylation, and highly deacetylated chitosan shows a low level of toxicity (Lubben *et al.*, 2001). It lacks irritant and allergic effects (Dodane and Vilivalam, 1998). Nonetheless, debatable data about its immunostimulating property are reported, which are dependent on the molecular weight, degree of deacetylation, purity and route of administration (Issa *et al.*, 2005).

Chitosan demonstrates a high capacity to adhere to the mucosa owing to ionic interaction between its positively charged amino groups and negatively charged mucus gel layers (George and Abraham, 2006). The interactions are strong at acidic and slightly acidic pH levels, at which the positive charge density of chitosan is high. However, the chitosan is less adhesive with amino groups participating in the crosslinking process or which have a low degree of deacetylation. High molecular weight chitosan, ~1400 kDa, demonstrates a stronger level of mucoadhesion than that of low molecular weight, ~500–800 kDa, mainly due to changes in viscosity (Dodane and Vilivalam, 1998).

Chitosan is also a bioactive polymer. This intrinsic property is due to its antioxidant nature, which scavenges hydroxyl and superoxide radicals as well as chelating metal catalysts that are responsible for free radical production. It is a haemostatic agent and possesses wound-healing, anti-ulcer, antacid, anti-microbial, anti-hyperlipidaemic, anti-inflammatory and anti-carcinogenic properties (Wong, 2009; Yong and Wong, 2013).

Drug Delivery Applications

Chitosan is widely used in the pharmaceutical industry and in research as a matrix former, binder, disintegrant and coating material due to its low toxicity, stable physicochemical profiles, biodegradability and biocompatibility (Wong, 2009). The mucoadhesive, absorption-enhancing and sustained release characteristics of chitosan have led its utility as an excipient in various dosage forms delivered via oral, nasal, ocular, transdermal and parenteral routes for small molecule drugs, peptides, proteins and vaccines. For the purpose of drug delivery, chitosan has been formulated as microparticles, directly compressible tablets, nanoparticles, beads, liposome, buccal discs, solutions, vesicles, film coats, spray dried particles and capsule delivery systems (Wong, 2009). It has been popularly used to prepare polyelectrolyte coacervates with natural polyanions such as carboxymethylcellulose, heparin, pectin, alginate, carrageenan and hyaluronate. Chitosan–polyion complexes have been investigated for drug and protein release applications as well as for cell transplants and enzyme immobilization (Wong, 2009).

Chitosan or its derivatives are used in dosage form design (Wong, 2009). Further, they can also be conjugated with drugs to increase physicochemical stability and release retardation capacity or with a targeting ligand to promote site-specific therapeutic action with reduced adverse effects (e.g. in cancer therapy). The drug delivery profile of the chitosan-based dosage form can be tailored to the release needs and targeting purposes by introducing chemical excipients or applying electromagnetic radiation subsequent to the assembly of drug and excipient (Nurjaya and Wong, 2005; Wong, 2008; Wong and Sumiran, 2008; Zakaria and Wong, 2009; Wong, 2010, 2011; Wong and Nurulaini, 2012).

Cancer

Cancer is a disease that involves unregulated cell growth of single or multiple organs. It is medically defined as a malignant neoplasm that denotes abnormal growth or division of cells (Bhadauria *et al.*, 2012). The root cause is related to DNA damage (Bernstein *et al.*, 2013). The central role of DNA damage and its progression to cancer is illustrated in Fig. 11.2.

Naturally, cellular metabolism is accompanied by the DNA damage, occurring at an average rate of 10,000 new instances of damage/human cell/day (Lindahl, 1993). Repairs are brought by exogenous factors that in-turn elicit cancer. Between 90% and 95% of cancers are attributed to environmental factors, and the rest to genetic defects. The common environmental contributors include tobacco (25–30%), diet and obesity (30–35%), infections (15–20%), radiation (ionizing and non-ionizing, up to 10%),

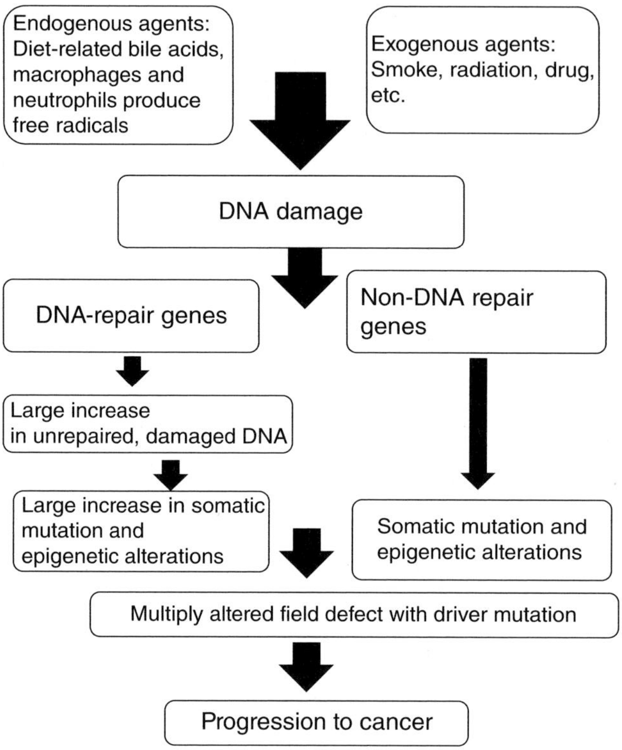

Fig. 11.2. Progression of DNA damage to cancer (adapted from a graphic describing the central role of DNA damage and epigenetic defects in DNA repair genes in carcinogenesis, created by Bernstein0275, licensed under the Creative Commons Attribution-Share Alike 3.0 Unported, and available at en.wikipedia.org/wiki/Cancer#/media/File:Diagram_Damage_to_Cancer_Wiki_300dpi.svg).

stress, lack of physical activity and environmental pollutants (Anand *et al.*, 2008; Wong *et al.*, 2011).

Cancer site-specific drug delivery

The major barrier to conventional cancer chemotherapeutics lies in their high drug toxicity due to non-selective delivery and inadequate drug accumulation at the diseased tissue following systemic administration, thereby leading to slow recovery of health (Moorthi *et al.*, 2011). In addition, most anti-cancer drugs are poorly water soluble (Liang *et al.*, 2010), and organic solvents may have been used to solubilize them in the process of formulation. The residual solvents affect the health of cancer patients, and cause undesirable adverse effects such as venous irritation and respiratory distress.

Pharmaceutically, a successful anti-cancer therapy could be created primarily through judicious design of a carrier with a targeting element, and having the drug dissolve, release and absorb at the specific cancer sites without drug solubility becoming an issue. Chitosan is widely used in dosage form design for cancer therapeutics to meet the stipulated demands (Park *et al.*, 2010), and could be chemically modified with ligands to equip the drug carrier with cancer site specificity (Park *et al.*, 2010). Further, the chitosan or its derivatives could be processed into nanoparticles or microparticles for efficient drug dissolution, release and absorption at cancer sites through matrix size, specific surface area, surface charges, hydrophobicity and mucoadhesiveness modulation (Wang *et al.*, 2011).

Ideally, targeting reduces drug concentration in normal tissue, ensures minimum release during transit and maximum release and absorption at the target site. Generally, it can be attained via passive or active delivery.

Passive targeting

Passive targeting refers to the accumulation of substrates at the target site due to enhanced permeability and retention effects brought about by the leaky vasculature and incomplete lymphatic system surrounding cancers of soft tissue and epithelial cell origin (McNeil, 2011) (Fig. 11.3). The enhanced permeability and retention effect is first discussed by Maeda (2001), who discovered selective accumulation of macromolecular drugs in the tumour tissues. This effect is envisaged to have derived from rapid tumour growth and angiogenesis mediated by the vascular endothelial growth factors and other growth factors to meet the nutrition and oxygen demands of tumour aggregates (Arias *et al.*, 2012). The newly formed tumour vessels are abnormal in form and architecture, constituting poorly aligned endothelial cells with wide fenestrations. The leaky vasculature is formed as a consequence of rapid and defective angiogenesis (Wang and Thanou, 2010). The size of endothelial pores varies from 10 to 1000 nm (Danhier *et al.*, 2010). For efficient extravasation from the fenestrations of the endothelial tissue, a substrate needs to be less than 400 nm in size. Practically, this has been possible with drugs delivered using polymer–drug conjugate systems, polymeric nanoparticles, micelles or liposomes (Ishida *et al.*, 2002).

Active targeting

The success of an anti-cancer therapy is not only dependent on selective drug accumulation in the tumour tissue. It is also crucial for drug molecules to reach the ultimate target site of the tumour successfully, namely the cell membrane, cytosol or nucleus (Park *et al.*, 2010). This can be facilitated through specific recognition of the drug substrate (via its surface decorated ligand) by the diseased site (which expresses biomarkers that distinguish such a site from the surrounding healthy tissues) (Danhier *et al.*, 2010; Wang and Thanou, 2010; McNeil, 2011). Active targeting requires the substrate being decorated with a biochemical moiety, namely a monoclonal antibody, leptin, transferrin, folic acid or others. Table 11.1 lists the ligands used for tumour-specific drug delivery system design. Of these, nanoparticles in particular have been recognized as vehicles that can be passively accumulated in tumour tissue as a result of enhanced permeation and retention effects (Maeda *et al.*, 2013). In addition, they can be internalized, with drug delivery being greatly improved by having a high-affinity targeting ligand attached on the matrix and ligand with the innate ability

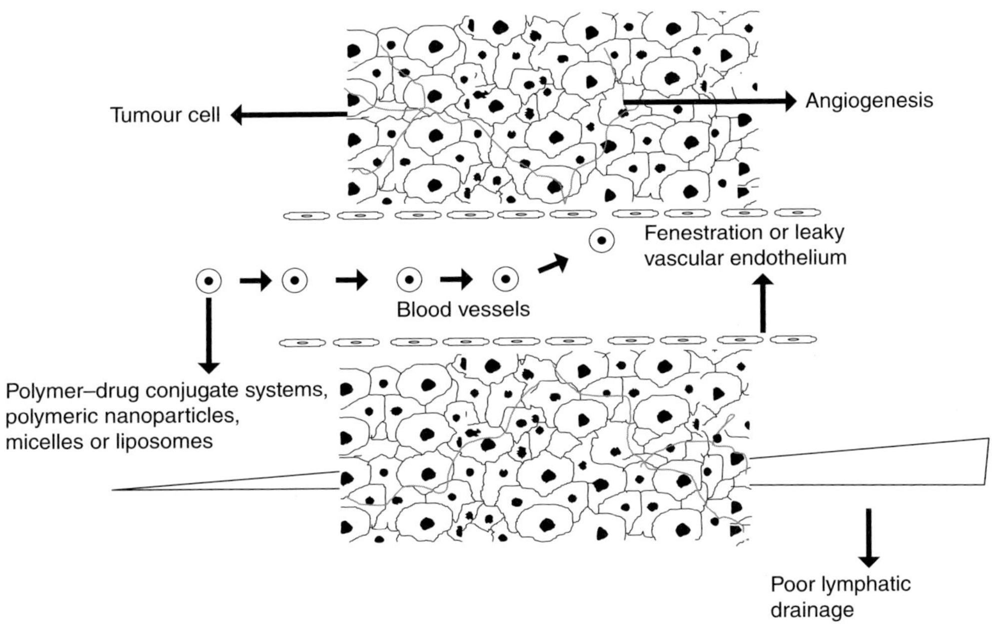

Fig. 11.3. Schematic diagram of passive targeting: enhanced permeability and retention effects.

Table 11.1. Targeting ligands for tumour-specific drug delivery (Danhier *et al.*, 2010; Kularatne and Low, 2010; Rothdiener *et al.*, 2010; Torchilin, 2010; Wang and Thanou, 2010).

Targeting ligand	Receptor site	Organ/tissue site
Folate	Folate receptor	Breast, lung, kidney, ovary, colon, brain, myelogenous leukaemia cancer
Galactosamine	Asialoglycoprotein receptor	Liver, melanoma
Hyaluronan	CD44 receptor	
Transferrin	Transferrin receptor	Ovarian cancer cell, glioma
Anti-HER-2	HER-2 receptor	Breast cancer, glioma
Anti-HER-2 monoclonal antibody	HER-2 receptor	
Anti-EGRF monoclonal antibody	EGRF receptor	
Anti-Flk1 monoclonal antibody	VEGFR-2	K1735-M2 and CT-26 tumour,
Anti-VEGF monoclonal antibody	VEGF	liver cancer
RGD peptide	Integrins ($\alpha_v\beta_3$)	Melanoma, breast cancer, pancreatic/renal orthopic tumour
Anti-VCAM-1	VCAM-1	Colon cancer
GPLPLR	MT1-MMP	Colon cancer, neuroblastoma
Anti-MT1-MMP F_{ab}	MT1-MMP	
NGR	Aminopeptidase N	
Urokinase plasminogen activator	Urokinase plasminogen activator receptor	Colon cancer, breast cancer

to activate receptor-mediated endocytosis (Park *et al.*, 2010). Through receptor-mediated endocytosis processes, the tumour cells could be killed directly, while cells at their periphery remain unharmed.

Figure 11.4 illustrates the major approaches to modify chitosan chemically into nanoparticles for passive targeting and its decoration with ligands to form an active targeting carrier with site-specific drug delivery characteristics. Extensive reviews and reports have been presented by Wang *et al.* (2013), Park *et al.* (2010) and Duncan (2006) on passive targeting strategies applied on chitosan nanomatrices. This chapter intends to address the enteric coating, superparamagnetic activated, polymer–drug conjugate and polymer–ligand conjugate aspects of cancer site-specific drug delivery by using chitosan as the primary excipient of interest.

Enteric-coated chitosan nanodelivery systems

Prior to enabling passive or active drug targeting at the tumour site, early drug release at unintended sites and its susceptibility to biodegradation are the primary concerns. Chitosan is acid-soluble and prone to solvation by an acidic medium. In the case of the oral chemotherapeutics used in the treatment of gastrointestinal (GI) cancer, chitosan will not be able to prevent premature drug release in the gastric region. Chemical crosslinking of chitosan with aldehyde is not entirely effective in encapsulating the drug along its GI transit (Lorenzo-Lamosa *et al.*, 1998). Instead, enteric coating by Eudragit S 100 is found to prevent drug release in an acidic gastric medium (pH 1.2) or in the small intestinal medium of pH between 5 and 6.8 (Oosegi *et al.*, 2008). The drug begins to release from the matrix only in a milieu characterized by a pH of 7.4. This implies that the enteric coat may be applicable to a chitosan nanodelivery system that could be used in the treatment of lower intestinal tumours such as colorectal cancer. The combination of enteric coat and chitosan core provides site-specific drug release through pH sensitivity and microbial digestibility of the dosage form (Omwancha *et al.*, 2011). The latter relies on enzymes secreted by colonic microflora that could degrade the glycosidic linkages of the

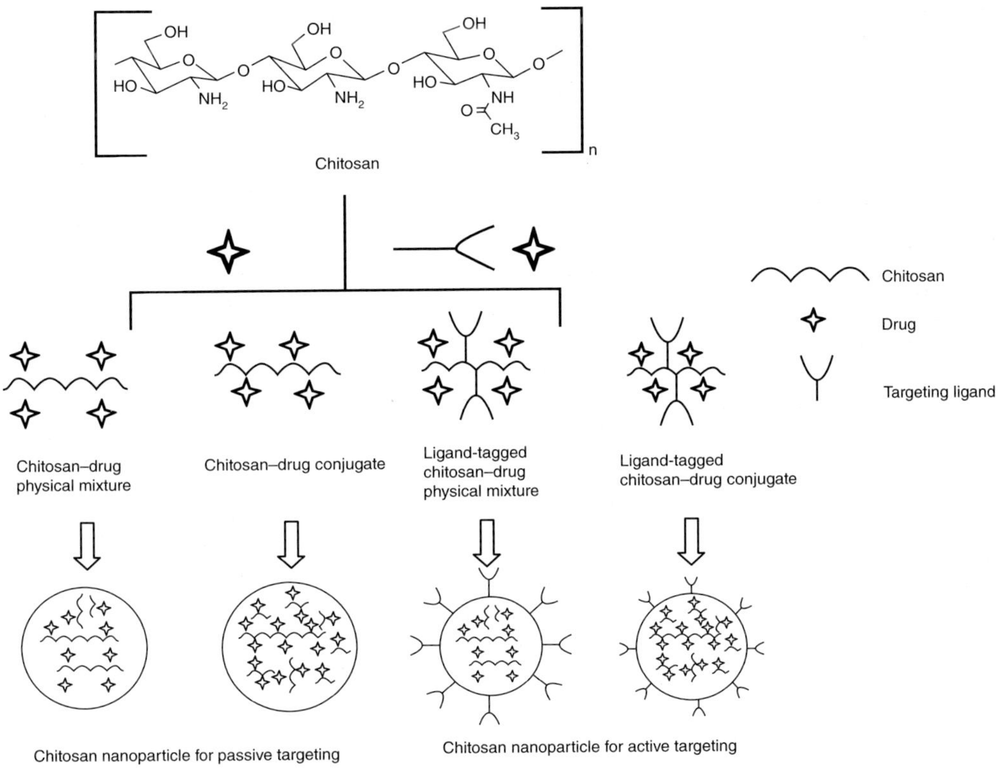

Fig. 11.4. Chitosan nanoparticles for passive and active targeting.

polymer and disaggregate the matrix to initiate drug release.

Chitosan and alginate are cationic and anionic polyelectrolytes, respectively (Sanna *et al.*, 2012). One common processing strategy of their blends is coacervation, which is deemed to have drug encapsulation and sustained release properties (Zhang *et al.*, 2011). Recent research by Liu *et al.* (2013) constructs alginic acid-coated chitosan nanoparticles to deliver legumain DNA vaccine orally. The coating is able to protect DNA from acidic degradation and in a way serves as an enteric barrier. Oral vaccination has been performed in the murine orthotopic 4T1 breast cancer model. The tumour volume becomes significantly small in mice receiving legumain DNA vaccine from alginic acid-coated chitosan nanoparticles. In spite of being protected by a coat, they exhibit an almost fivefold increase in cytotoxic T lymphocytes CD3(+)/CD8(+)/CD25(+), a critically improved autoimmune response to protect against breast cancer in mice.

Superparamagnetic activated chitosan nanodelivery system

The specificity of passive and active targeting of chitosan nanoparticles to a cancer site can be promoted by loading magnetic particles into these matrices, in addition to the drug cargo (Wadajkar *et al.*, 2013). Figure 11.5 shows that encircling the cancer site with a high gradient external magnetic field can concentrate the intravenously administered chitosan magnetic nanoparticles to the intended region. The efficiency of particulate accumulation is a function of magnetic field, rate of blood flow and surface characteristics of particles such as size, charge and hydrophobicity (Park *et al.*, 2010). Accumulation of particles at the cancer tissue magnetically aids drug release and absorption at specific regions of disease. The propensity of drug absorption and/or particle endocytosis, via passive or active targeting mechanisms, is expected to be enhanced to a greater extent thereby increasing

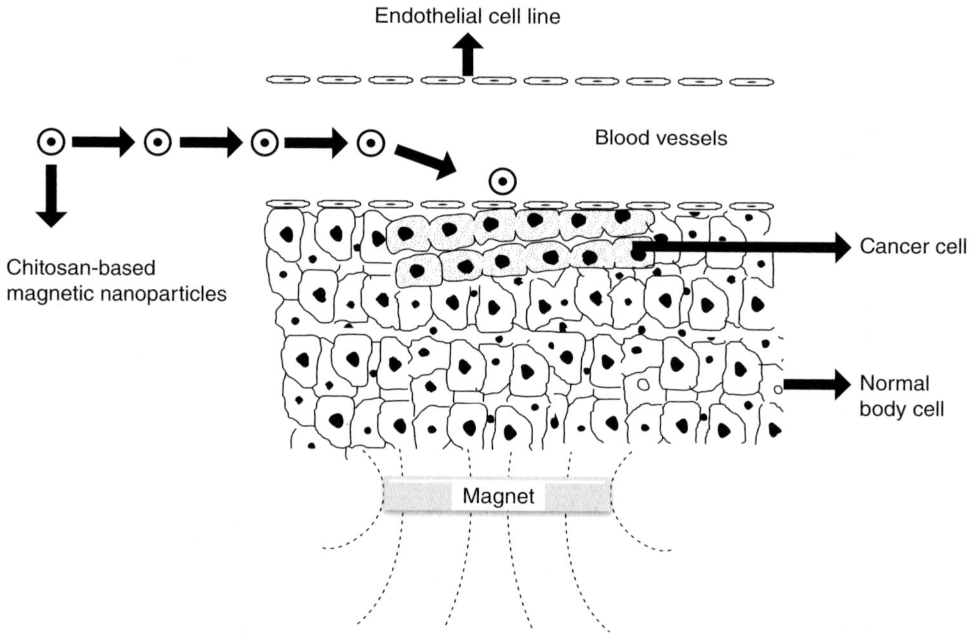

Fig. 11.5. Schematic diagram of chitosan magnetic nanoparticles in bloodstream subjected to a high gradient external magnetic field.

therapeutic efficiency and reducing unnecessary side effects of drugs. Table 11.2 summarizes various chitosan magnetic nanoparticles that have been designed for the treatment of breast and liver cancer.

The chitosan magnetic nanoparticles are not a mere representation of a targeting device. Magnetic hyperthermia can be induced in the cancer cell using such matrices through subjecting them to an alternating magnetic field at a specific amplitude and frequency, and the released heat can be employed to kill the cancer cells (Kobayashi, 2011). In the hyperthermia treatment, malignant tumours are typically heated to 43°C and maintained at this temperature for about 45 min (Lin *et al.*, 1998; Sterzer, 2002). Biologically the cancerous tissue is not able to survive above 41°C, while the healthy tissue survives up to 45°C, thereby allowing selective killing of cancerous cells (Vrba *et al.*, 1999). The hyperthermia treatment could increase the efficacy of conventional chemotherapy since it is often lethal to weak malignant cells, including those sublethally injured or resistant to conventional therapies owing to poor state of oxygenation. Several groups around the world have been pursuing this concept with the aim of developing the technology. Hyperthermia treatment has been regarded as adjuvant therapy in phase II human trials in Europe (Park *et al.*, 2010).

Chitosan–drug conjugate

The chitosan–drug conjugate comprises a polymer that is chemically conjugated to a drug via a biodegradable spacer (Park *et al.*, 2010). The spacer maintains physicochemical stability in the bloodstream or GI milieu, but is cleaved at the target site via enzymatic degradation or hydrolysis. It can be selectively accumulated at the tumour site by enhanced permeability and retention effects (Moghimi *et al.*, 2001). The parent drug then releases, either extracellularly or intracellularly (Dosio *et al.*, 2011). The conjugate can exert active targeting through tagging the same polymer with a ligand, which is recognizable by the target cancer site (Park *et al.*, 2010). Several polymer-drug conjugates such as N-(2-Hydroxypropyl) methacrylamide

Table 11.2. Profiles of chitosan magnetic nanoparticles for use in cancer treatment.

Cancer type	Drug	Formulation and design	Remarks	Reference
Breast cancer	Artesunate	The artesunate magnetic nanoparticles are formulated using two forms of chitosan: acetic acid- and water-soluble chitosans. The nanoparticles are evaluated for *in vitro* release, particle size distribution, zeta potential, drug encapsulation efficiency, drug loading capacity and magnetic susceptibility	Acetic acid-soluble chitosan, at low concentrations, exhibits the highest drug encapsulation efficiency whereas an increase in water-soluble chitosan concentration increases the drug encapsulation efficiency of the nanoparticles. The developed chitosan magnetic nanoparticles of artesunate shows controlled drug release characteristics and thus may have a higher *in vivo* anti-breast cancer activity than chitosan formulation developed without any magnetic particles	Subramanian *et al.* (2010)
Animal tumour model	Heat	Chitosanoligosaccharide-stabilized ferrimagnetic iron oxide nanocubes are developed as an effective heat nanomediator for cancer hyperthermia. Dynamic light scattering and transmission electron microscopic analyses indicate that the nanocubes are composed of multiple 30-nm ferrimagnetic iron oxide encapsulated by chitosan shell	An increase in the ferrimagnetic iron oxide content in the interior of the matrix increases the total magnetic moments, which leads to localized particulate accumulation under an applied magnetic field. The magnetically guided nanocubes eradicate target cancer cells through caspase-mediated apoptosis under the influence of hyperthermia. The nanocubes demonstrate excellent anti-tumour efficacy on animal tumour model without any severe toxicity	Bae *et al.* (2012)
Liver cancer	Carboplatin and heat	Carboplatin-Fe@C-loaded chitosan nanoparticles are prepared by reverse microemulsion method with the nanocages as the magnetic cores and chitosan as the matrix	Under an alternating magnetic field, the carboplatin-Fe@ C-loaded chitosan nanoparticles induce tumour temperature to 42.6 ± 0.2°C within 10 min of contact. The combination of drug and hyperthermia effect inhibits tumour growth significantly	Li *et al.* (2009)
SPCA-1 cancer cells	5-fluorouracil	Chitosan-coated magnetic nanoparticles containing 5-fluorouracil are prepared using reverse microemulsification method	The resultant nanoparticles release drug in a sustained manner *in vitro*. The chitosan-coated magnetic nanoparticles can effectively permeate into the SPCA-1 cancer cells to induce cellular apoptosis under the influence of external magnetic field	Zhu *et al.* (2009)
General	Doxorubicin	Doxorubicin-conjugated magnetite nanoparticles encapsulated in a thermosensitive polymer, chitosan-g-poly(N-isopropylacrylamide-co- N,Ndimethylacrylamide)	Doxorubicin is conjugated to the magnetite via acid-labile hydroazone bond. When the drug–magnetite is embedded in thermosensitive chitosan derivative, the nanoparticles release the drugs in response to changes in external temperature or pH	Yuan *et al.* (2008)

Continued

Table 11.2. Continued.

Cancer type	Drug	Formulation and design	Remarks	Reference
General	Epirubicin	Chitosan-bound magnetic nanoparticles loaded with epirubicin	The magnetic nanoparticles are stable at pH 3–7, and approximately 80% of the drug is released after 150–300 min The *in vitro* anti-cancer efficacy of the drug-loaded magnetic nanoparticles is comparable to that of the free drug. The external magnetic field induces hyperthermia and promotes cancer cell growth inhibition	Chang *et al.* (2005)
SK-BR-3 cell lines	Metho-trexate	Methotrexate is loaded in chitosan-coated superparamagnetic iron oxide nanoparticles (SPION)	The nanoparticles exhibit slow drug release behaviour with comparable SK-BR-3 cell line toxicity as the chitosan formulation developed without superparamagnetic iron oxide. The coating of SPION, though diminishing the paramagnetic behaviour of the matrix, has sufficiently strong superparamagnetic properties to allow particle separation from the immediate medium and thus is possibly being used to target drug delivery.	Mohammadi *et al.* (2013)

copolymer–doxorubicin conjugate containing galactosamine as a targeting moiety have entered into phase I/II clinical trials for the purpose of primary or secondary liver cancer treatment (Seymour *et al.*, 1991; Park *et al.*, 2010). Table 11.3 presents various chitosan−drug conjugates that have been designed and used in passive and/or active drug targeting. The conjugation is deemed to be advantageous due to reduced drug degradation during transit to the target site. Enhanced drug permeation into the cancer tissue can be attributed to chitosan's permeation enhancing property and reduced drug efflux in the extracellular region (which leads to loss of chemotherapeutic action due to drug–P-glycoprotein recognition) (Raub, 2006). In addition, the conjugate can self-aggregate into nanoparticles with geometry appropriate for passive and active targeting designs, depending on its hydrophile−lipophile balance in the immediate environment (Park *et al.*, 2010).

Chitosan-targeting ligand conjugate

Various ligands have been employed to decorate the chitosan or its derivatives for the purpose of active drug targeting in cancer treatment (Table 11.4). In some cases, they are introduced to the chitosan−drug conjugate or drug itself (Tables 11.3 and 11.4). The ligands can be tagged to the substrate via covalent or non-covalent bonding, namely coacervation and/or crosslinking. Examples of ligands include folic acid, iron-containing substances such as apo-transferrin, galactose, lactose, lactobionic acid, glycyrrhizin, glycyrrhetinic acid, herceptin, retinoic acid, cetuximab, trastuzumab and transactivating transcriptional (TAT) factor (De Benedetti and Fanelli, 2009). These ligands recognize receptors overexpressed on the surfaces of cancer cells. They increase the specificity and propensity of chitosan-based nanoparticles to interact with the cancer cells (Fig. 11.6).

A number of research studies have performed direct conjugation of ligand onto drug molecules (Park *et al.*, 2010). Nonetheless, most clinical investigations reveal lower therapeutic effects than pure drug molecules in cancer treatment (Allen, 2002). This is probably due to negation of biological activity of drugs by ligand conjugation. To overcome this predicament, most recent attempts are focusing on conjugating the ligand to chitosan polymers with drug entrapped in the formed matrix (Table 11.4).

Gene therapy

Deoxyribonucleic acid (DNA) is utilized to supplement or alter genes within cells to treat diseases (Friedmann and Roblin, 1972). The gene therapy uses DNA that encodes a functional therapeutic gene to replace the mutated gene or encodes a therapeutic protein drug (rather than a natural human gene). DNA that encodes a therapeutic protein is packaged within a 'vector'. It is expressed by the cell machinery and nucleus activity to produce therapeutic protein, which acts as a drug for the intended disease.

Gene therapy is widely investigated in cancer treatment as cancer can derive from gene defects. The obstacles encountered are summarized below:

1. DNA can be digested by DNAse enzyme in the extracellular environment prior reaching the nucleus target.
2. DNA is a macromolecule and its passage into cell is restricted and non-specific.
3. DNA possesses high electronegative charges due to the phosphoric acid group (pKa =1.5). The cell surface is similarly characterized by negative charges as a function of anionic proteoglycans. This hinders association of DNA with the cell surface and negates the process of transfection.
4. DNA is degraded in intracellular compartments by lysozymal enzyme.
5. DNA has to diffuse through dense microfibril network within the cytosol in order to reach the nucleus target. This is hindered by its macromolecular property.

Chitosan, its derivatives and ligand conjugates, could be used to circumvent issues of physicochemical stability, site-specific cellular association and entry (Table 11.5). The use of chitosan is regarded as a safe approach due to its biocompatibility, biodegradability, low toxicity and immunogenicity (Alves and Mano, 2008). Comparatively, cell transfection by viral vectors

Table 11.3. Chitosan–drug conjugates.

Cancer type	Drug type	Synthesis and formulation	Remarks	Reference
Epithelial colorectal adenocarcinoma cells (Caco-2)	Paclitaxel	Low molecular weight chitosan is conjugated with paclitaxel through a succinate linker	Low molecular weight chitosan–paclitaxel conjugate is absorbed in the small intestine after oral administration. It remains as intact conjugate until reaches the bloodstream. Advantages of such conjugation include: Upon oral delivery it bypasses the P-glycoprotein-mediated efflux in the gastrointestinal tract and CYP450-dependent metabolism in the intestine and liver. Low molecular weight chitosan is able to rapidly open the tight junctions between the human cancerous colorectal epithelial cells. This suggests that such conjugate could penetrate deep into the tumour and exert chemotherapeutic action.	Lee *et al*. (2008)
Breast and ovarian cancers	Doxorubicin	Doxorubicin is conjugated to chitosan using succinic anhydride as the spacer. Trastuzumab is conjugated to the self-assembled chitosan–doxorubin conjugate nanoparticles (particle size = 200 nm) via thiolation of its lysine residues and subsequent linking of the resultant thiols to chitosan	The chitosan–doxorubicin–trastuzumad conjugate is selectively taken up by $Her2^+$ cancer cells in comparison to the chitosan–doxorubicin nanoparticles and free drug. Antibody-conjugated nanoparticles are able to differentiate $Her2^+$ from $Her2^-$ cells. These are potentially useful in active targeting and reduction of drug adverse effects in association with treatment of $Her2^+$ breast and ovarian cancers	Yousefpour *et al*. (2011)

Hepatocarcinoma	Doxorubicin	Doxorubicin is conjugated to chitosan using transactivating transcriptional factor (TAT) as the spacer. A cell-penetrating peptide, the transactivating transcriptional factor domain from HIV, is linked to chitosan–doxorubicin conjugate to form a hybrid via peptide linkages	The chitosan–doxorubicin–TAT conjugate decreases the tumour volume by 77.4% and 57.5% with reference to free drug and chitosan–doxorubicin conjugate, respectively, in the tumour-bearing mice. Chitosan–doxorubicin conjugate is more active as an anti-cancer agent than the free drug. Its tagging by TAT further enhances the effectiveness in retarding the tumour growth	Lee *et al*. (2011a)
Murine leukaemias (L1210 and P388), B16 melanoma, sarcoma 180 solid tumour, murine liver metastatic tumour (M5076) and murine hepatic cell carcinoma (MH134)	Mitomycin C	N-succinyl-derivative of chitosan is conjugated with mitomycin C by means of carbodiimide chemistry	N-succinyl chitosan is hydrophilic and carries permanent positive charges. The polymer–drug conjugate formed is water-soluble when the mitomycin C content is less than 12% of the total weight. The N-succinylchitosan–mitomycin C conjugate exhibits good anti-tumour activity against various cancers due to drug action and possibly the permanent cationic character of the chitosan derivative	Song *et al*. (1992)

Table 11.4. Chitosan–ligand and drug–ligand conjugates for active drug targeting.

Cancer type	Drug type	Synthesis and formulation	Remarks	Reference
Colon cancer cells (HT-29)	5-fluorouracil	Hyaluronic acid is coupled with chitosan by means of coacervation to form nanoparticles bearing 5-fluorouracil	The hyaluronic acid-decorated nanoparticles show enhanced cellular uptake by HT-29 colon cancer cells compared to the uncoupled nanoparticles. The cytotoxicity of 5-fluorouracil incorporated in nanoparticles is higher than the free drug	Jain and Jain (2008)
Hepatoma	Future drug for liver cancer	Lactosaminated N-succinyl–chitosan is designed as the drug carrier	The lactose-succinate moiety exhibits specific binding to the asialoglycoprotein receptors in liver parenchymal cells. It localizes in the liver cells for prolonged periods, suggesting its potential utility as a ligand to target liver cancer cells	Kato *et al.* (2001)
Human hepatoma	5-fluorouracil, aminopentyl-carbomyl-5-fluorouracil, aminopentyl-ester-methylene-5-fluorouracil	Chitosan nanospheres containing 5-fluorouracil or immobilizing with aminopentyl-carbamoyl-5-fluorouracil or aminopentyl-ester-methylene-5-fluorouracil are prepared by the glutaraldehyde crosslinking and emulsification method. They are coated with anionic polysaccharides such as 6-O-carboxymethyl-N-acetyl-α-l,4-polygalactosamine sodium, 6-O-carboxymethyl-chitin sodium or sodium hyaluronate through polyelectrolyte complexation	The chitosan nanoparticles coated with 6-O-carboxymethyl-N-acetyl-α-l,4-polygalactosamine sodium exhibit a lectin-mediated aggregation phenomenon. They have the highest cytotoxic activity against HLE human hepatoma cell culture system *in vitro*	Ohya *et al.* (1994)
Orthotropic liver cancer	5-fluorouracil	The chitosan is conjugated with galactose with 5-fluorouracil loaded into nanoparticles made from such derivative	*In vivo* experiments show that galactosylated chitosan-5-fluorouracil nanoparticles could significantly inhibit tumour growth in an orthotropic liver cancer mouse model. The nanoparticles could significantly lower the tumour weight and increase the survival time of mice compared to free drug. Flow cytometry and the TUNEL assay indicate that the nanoparticles give higher rates of G_0-G_1 arrest and apoptosis in hepatic cancer cells than drug alone	Cheng *et al.* (2012)

Human hepatic carcinoma cells (QGY-7703 cells)	Doxorubicin	Chitosan-poly(ethylene glycol)-glycyrrhetinic acid nanoparticles containing doxorubicin are prepared by ionic gelation process with glycyrrhetinic acid decorated on the surfaces of chitosan-poly(ethylene glycol) matrix	The glycyrrhetinic acid-tagged nanoparticles are remarkably targeted to the liver and effectively inhibit tumour growth in H22 cell-bearing mice *in vivo* and QGY-7703 cells *in vitro*	Tian *et al.* (2010)
HeLa cells	Metho-trexate	The chitosan derivative N-[(2-hydroxy-3-trimethylammonium) propyl] chitosan chloride is conjugated with apo-transferrin and ionically crosslinked by sodium tripolyphosphate to form 200-nm microgels with methotrexate disodium as cytotoxic drug load	Killing of immortalized HeLa cells is promoted by receptor-mediated endocytosis of microgel via transferrin receptor and ability of chitosan in microgel to swell and release drug in low pH intracellular environment	Zhang *et al.* (2006)
Hepatocellular carcinoma	Doxorubicin	Retinoic acid-grafted chitosan copolymers are synthesized. Retinoic acid-tagged chitosan–albumin nanoparticles containing doxorubicin are prepared by coacervation method	IC_{50} of retinoic acid-decorated nanoparticles on HepG2 cell line is reduced to half and one-third of non-liganded nanoparticles and free drug, respectively	Varshosaz *et al.* (2013)
Pancreatic cancer	Gemcitabine	Herceptin is conjugated to drug gemcitabine. The ligand–drug conjugate is loaded in chitosan nanoparticles prepared by ionic gelation method	The nanoparticles display superior anti-proliferative activity along with an enhanced S-phase arrest leading to apoptosis, in comparison with unconjugated gemcitabine-loaded nanoparticles and free gemcitabine due to higher cellular binding with eventual uptake and prolonged intracellular retention	Arya *et al.* (2011)
Epidermal growth factor receptor positive cancer cells (A549, A431 and SKBR3)	Paclitaxel	O-carboxymethyl–chitosan nanoparticles with surface conjugation with cetuximab are prepared by simple ionic gelation method to encapsulate paclitaxel	The cetuximab conjugated nanoparticles are selectively taken up by epidermal growth factor receptor positive cancer cells The nanoparticles display a superior anti-cancer activity compared to non-liganded nanoparticles. Cancer cell death is enhanced due to higher cellular uptake	Maya *et al.* (2013)
Liver cancer cells (HepG2 cells)	Doxorubicin	Glycyrrhetinic acid-modified sulfated chitosan is synthesized and prepared in the form of nanoparticulate micelle to encapsulate doxorubicin	The micelles display rapid and significant ability to target the liver *in vivo*. The IC_{50} against HepG2 cells for doxorubicin loaded micelles is 54.7 ng/mL, which much lower than that of glycyrrhetinic acid-free micelles	Tian *et al.* (2012)

Continued

Table 11.4. Continued.

Cancer type	Drug type	Synthesis and formulation	Remarks	Reference
Squamous cell carcinoma (SCC-7)	Protoporphyrin IX	Protoporphyrin IX, a photosensitizer, is conjugated to glycol chitosan and processed in the form of nanoparticles	In tumour-bearing mice, protoporphyrin IX conjugated nanoparticles demonstrate prolonged blood circulation, enhanced tumour targeting ability and therapeutic efficiency compared to protoporphyrin-free nanoparticles. The protoporphyrin IX appears to act as a targeting ligand as well as being active for photodynamic imaging and therapy	Lee *et al.* (2011b)
Breast cancer cells (T47D)	Antisense oligonucleotide	N-acetyl cysteine-chitosan and N-acetyl penicillamine-chitosan are used to load antisense oligonucleotide in the form of nanoparticles by means of the gelation method	The thiolated nanoparticles significantly suppress epidermal growth factor receptor gene expression and downregulate protein expression in T47D cells compared to chitosan particles. Thiolated chitosan has high transfection and mucoadhesive properties	Talaei *et al.* (2011)
Colon cancer cells (HT 29)	Indigo carmine	Folic acid is conjugated to chitosan and prepared as nanoparticles to encapsulate indigo carmine	Bioassay indicates that the quantity of particles adhered to HT-29 colon cancer cell increases with incubation time and the conjugation ratio of folic acid to matrix. It implies that folic acid conjugated chitosan nanoparticles could serve as an ideal vector of a colon-specific targeting system	Yang *et al.* (2010)

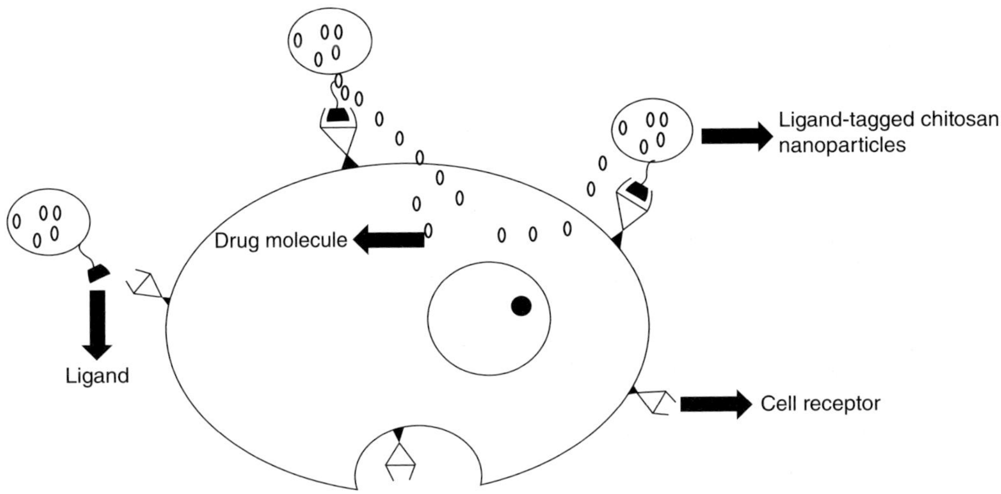

Fig. 11.6. Schematic diagram of receptor-mediated active drug targeting using ligand-tagged chitosan nanoparticles.

could be efficient. However, risks of undesirable effects brought by the immunological and carcinogenic responses are barriers to their clinical utility (Park *et al.*, 2010).

One main challenge of using chitosan as a DNA vector is inefficient intracellular targeting/transfection. This is partly attributed to difficulty in the release of the chitosan–DNA complex from the endosome. Endosomal escape could be promoted by the buffering capacity of chitosan polyplexes (Fig. 11.7; Kwon *et al.*, 2013). However, its capacity is not entirely satisfactory. The functional moieties such as histidine or imidazole equivalent could be incorporated into the chitosan matrix to improve its buffering capacity (Chang *et al.*, 2010). The buffering capacity of histidine is contributed by the imidazole ring that has a pKa around 6 and hence could be protonated in a slightly acidic pH medium (Chang *et al.*, 2010). It is evident that histidine could facilitate polyplexes escape from endosome through raising the buffering capacity of chitosan inside the acidic compartments. Figure 11.8 shows steps taken in DNA complexation with chitosan, surface decoration of the formed complex and its cellular association, endocytosis, endosomal entrapment and escape, DNA release and nucleus targeting. At each and every step, the chitosan could be chemically modified or formulated with functional excipients to meet the demand of drug delivery and targeting (Zhao *et al.*, 2013).

Over the years, various research studies have been conducted to investigate the physicochemical attributes of chitosan required for gene transfection; to modify chitosan to improve its aqueous solubility and gene binding affinity; to conjugate chitosan with a targeting ligand for cancer-specific uptake; and to equip chitosan with imidazole or equivalent for endosomal escape to realize gene therapy. Table 11.5 provides an overview of these studies and emphasizes the importance of all the steps in chitosan modification and formulation to succeed in drug targeting and delivery for cancer treatment.

Conclusions and Future Perspectives

Chitosan is a unique biopolymer characterized by free amine and hydroxyl moieties. The drug targeting effectiveness and specificity of dosage forms made of chitosan could be enhanced through chemical modification of these polymeric functional groups with ligand and/or drug conjugation. However, the associated experiments are largely conducted *in vitro* and at cell culture level. The translation of physicochemical modification of chitosan into pharmacodynamic expression requires more *in vivo* tests and challenges in the near future.

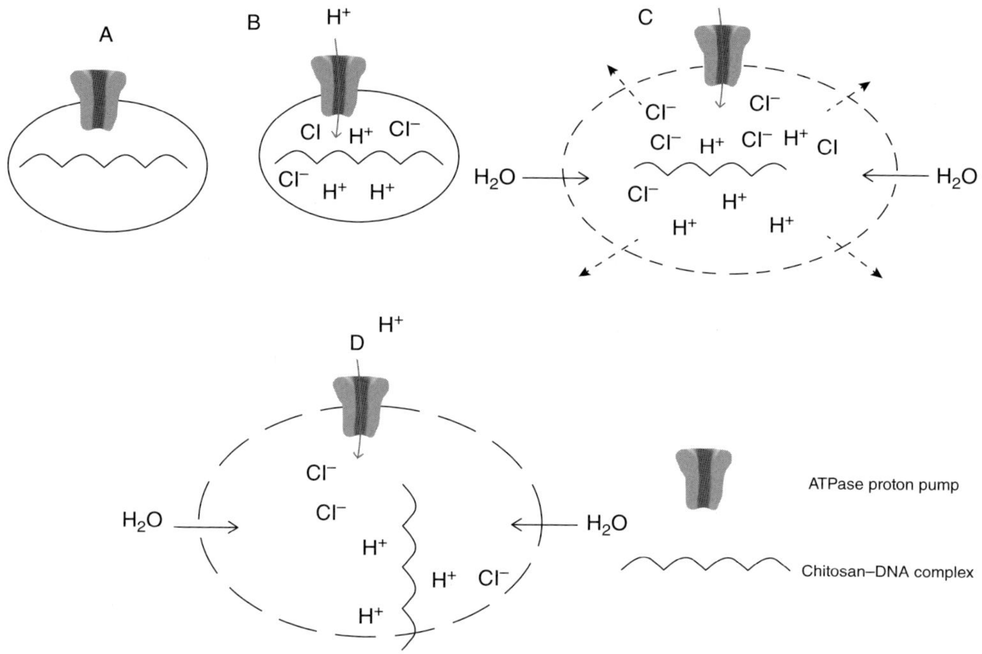

Fig. 11.7. pH-buffering effects of chitosan–DNA complex. (A) Chitosan–DNA complex enters the cell via endocytosis and is entrapped in endosome. (B) The ATPase proton pumps actively translocate protons (H^+) into endosomes. The chitosan–DNA complex is protonated and acidification of endosomes is prevented. (C) The proton pumping action is accompanied by entry of passive chloride ions. This increases ionic concentration and hence water influx. (D) High osmotic pressure induces the swelling and rupture of endosomes, giving rise to endosomal escape of polyplexes.

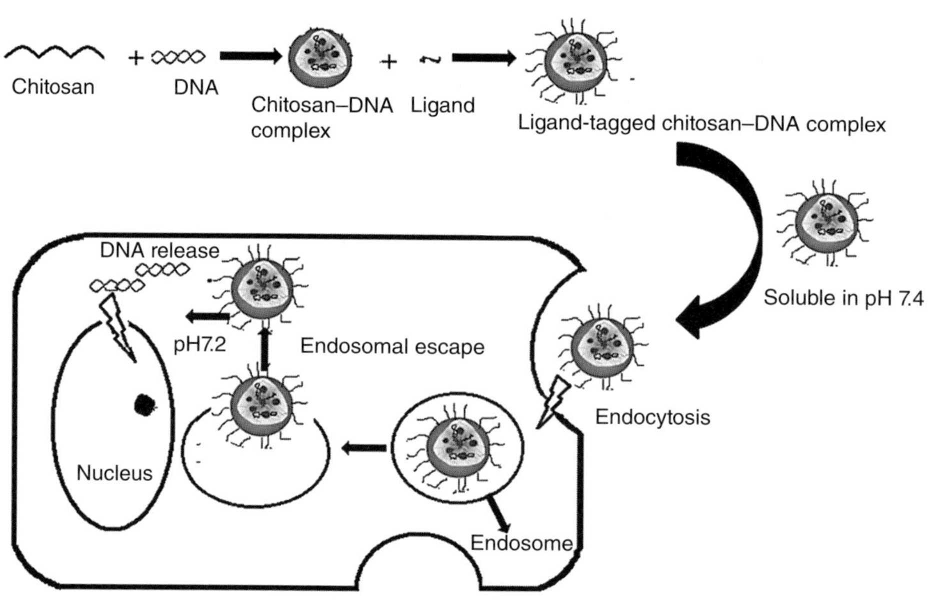

Fig. 11.8. Process of gene formulation and delivery using chitosan as carrier.

Table 11.5. Gene delivery systems of chitosan for cancer cell targeting.

Cancer type	Drug type	Synthesis and formulation	Remarks	Reference
Human lung carcinoma A549 cells, HeLa cells, and B16 melanoma cells	Luciferase plasmid	Chitosan and polygalactosamine are used as the vehicle of luciferase plasmid	Chitosan is more effective in luciferase plasmid transfection than polygalactosamine	Sato *et al.* (2001)
Human embryonic kidney 293 cells (HEK-293), human liver carcinoma cells (HepG2) and mouse myeloma cell line	Plasmid pEGFP	Magnetic Fe_3O_4 nanoparticles coated with chitosan and polyethylene glycol are synthesized by means of controlled chemical coprecipitation method to encapsulate plasmid pEGFP	Fe_3O_4 concentration of less than 2 mM has been verified as safe by means of a cytotoxicity test *in vitro*. A permanent magnetic field could be employed to increase the transfection efficiency of plasmid pEGFP	Kuang *et al.* (2012)
COS7 cells	Plasmid DNA	DNA-loaded chitosan nanobubbles are formed by means of homogenization. These nanobubbles are characterized with a mean diameter of less than 300 nm and a positive surface charge	The nanobubbles are stimulated to trigger transfection using ultrasound. The release of chitosan nanobubble-bound plasmid DNA occurs after 1 min of ultrasonication	Cavalli *et al.* (2012)
HeLa cell	Plasmid DNA VR1412 (pDNA) encoding β-galactosidase	Chitosan derivatives are synthesized with increasing amounts of diethylethylamine groups attached to the chitosan main chain. These derivatives are used to encapsulate plasmid DNA via titration method	Diethylethylamine–chitosan derivatives are able to condense DNA into small particles having a surface charge depending on the polymer/DNA ratio. Nanoparticles prepared with derivatives containing 15% and 25% of diethylethylamine groups exhibit transfection efficiencies ten times higher than those observed with underivatized chitosan	De Paula Pansani Oliveira *et al.* (2013)
HepG2 and HeLa cell lines	DNA	Galactosylated chitosan-hydroxypropyltrimethylammonium is synthesized and is used to entrap DNA by condensation method	The quarternized chitosan is water-soluble and more amorphous. It has a higher plasmid condensation capability than galactosylated chitosan. Cytotoxicity measurements indicate that galactosylated and quarternized chitosan has a significantly lower cytotoxicity in HepG2 and HeLa cell lines than positive control branched polyethylenimine (25 kDa)	Xiao *et al.* (2012)

Continued

Table 11.5. Continued.

Cancer type	Drug type	Synthesis and formulation	Remarks	Reference
HepG2, HeLa and primary hepatocyte cells	DNA	Polyethyleneimine-graft-chitosan is synthesized by performing cationic polymerization of aziridine in the presence of water-soluble oligochitosan. The polyethyleneimine-g-chitosan is used to encapsulate DNA via complexation method	Gel electrophoresis shows that DNA migration is retarded completely at a nitrogen/phosphorus ratio of 2.5/1, indicating good DNA condensation capability of polyethyleneimine-g-chitosan. Polyethyleneimine-g-chitosan shows a higher transfection efficiency than that of polyethyleneimine (25 K) both *in vitro* and *in vivo*	Wong *et al.* (2006)
Cancer cell line	DNA	The chitosan is conjugated to biotin through a functionalized poly(ethylene glycol) spacer. It is used to encapsulate DNA via complexation method. The complex has polyethyleneimine added and is further conjugated to TAT ligand via streptavidin–biotin linkage	The transfection efficiency of the chitosan–DNA complex is significantly correlated with the amount of polyethyleneimine incorporated in complex and is further enhanced when TAT is conjugated to the complex due to specificity in cellular interaction	Chung *et al.* (2010)
293T cells	DNA	Water-soluble chitosan is coupled with urocanic acid bearing imidazole ring which can play a crucial role in endosomal rupture through proton sponge mechanism	Urocanic acid–chitosan exhibits good DNA binding ability, high protection level of DNA from nuclease attack and low cytotoxicity. The transfection efficiency of DNA into 293T cells is promoted by the presence of urocanic acid and increases with an increase in urocanic acid in urocanic acid–chitosan matrix	Kim *et al.* (2003)
NIH3T3 fibroblasts and HEK293 kidney cells	Plasmid DNA	Polyelectrolyte multilayers of glycol–chitosan and hyaluronic acid are used to encapsulate lipoplexes of plasmid DNA. The film is designed to enable localized, surface-based transfection	Lipoplex-containing glycol–chitosan–hyaluronic acid films successfully transfect NIH3T3 fibroblasts and HEK293 kidney cells *in vitro*, maintaining transfection levels of approximately 20% for a period of at least 7 days	Holmes and Tabrizian (2013)

HEK 293 cells	Plasmid DNA	A chitosan-supported imidazole Schiff-base is developed as the vector of plasmid DNA	Introducing the imidazole Schiff-base to the branch of chitosan improves the water solubility of polymer. Chitosan-supported imidazole Schiff-base can thus effectively bind to plasmid DNA and protect it from DNase I digestion. The chitosan-supported imidazole Schiff-base does not induce remarkable cytotoxicity against HEK 293 cells and can enhance delivery of plasmid DNA into cytoplasm and nucleus efficiently. A transfection efficiency as high as 70% can be obtained. This may be associated with buffering capacity of imidazole and formation of Schiff-base to assure endosomal release of DNA	Shi *et al.* (2013)
HepG2 cells	Plasmid DNA	Chitosan–DNA complex is prepared with alpha-methoxy-omega-succinimidylpoly(ethylene glycol) conjugated to its surfaces using an active ester scheme	Plain chitosan nanoparticles can protect plasmid DNA from nuclease digestion and aid its entry into HepG2 cells. However, the transfection efficiency is very low. The transfection activity is improved using a chitosan–DNA–poly(ethylene glycol) system. *In vivo* experiments similarly indicate that poly(ethylene glycol) conjugated complexes mediate a higher gene expression in hepatoma tissues than plain nanoparticles	Zhang *et al.* (2007)
COS-7, HEp-2 and Vero cells	Syncytial virus F gene	Chitosan-coated poly(2-hydroxyethyl methacrylate) nanoparticles are used to encapsulate respiratory syncytial virus F gene construct, a model DNA vaccine	The positively charged chitosan-coated nanoparticles are able to bind to the negatively charged cell membrane and undergo endocytosis in COS-7 cells. Using quantitative polymerase chain reaction, the transfection efficiency of these nanoparticles in COS-7, HEp-2 and Vero cells is found to be improved. An *in vivo* transfection efficiency study using BALB/c mice indicates that intra-muscular injection of nanoparticles induces a high copy number of the transgene	Eroglu *et al.* (2013)

Continued

Table 11.5. Continued.

Cancer type	Drug type	Synthesis and formulation	Remarks	Reference
Hep-2 and HeLa cells	Plasmid DNA	Two stable ternary complexes are prepared by mixing chitosan with DOTAP–plasmid DNA lipoplex and DOTAP with chitosan–plasmid DNA polyplex via non-covalent conjugation	The transfection of lipopolyplexes is translated to a higher GFP gene expression than DOTAP–plasmid DNA and chitosan–plasmid DNA complexes in Hep-2 and HeLa cells, and luciferase gene expression is two- to threefold higher than DOTAP–plasmid DNA and 70–120-fold higher than chitosan–plasmid DNA in Hep-2 cells	Wang *et al.* (2012)
HEK 293 cells	Plasmid DNA	A series of hydrophobic amino acid-grafted chitosan derivatives are synthesized by carbodiimide-mediated coupling reaction	The resultant polyplex shows 3.4 to 5.4-fold greater cellular uptake and 13 to 30-fold higher transfection efficiency in HEK 293 cells than unmodified chitosan. The cellular uptake and transfection efficiency improve with a rise in hydrophobicity of amino acid. Hydrophobic amino acid substitution enhances plasmid DNA release at cytosolic pH	Layek and Singh (2013)
HEK293, MDCK and Hep-2 cells	Plasmid DNA	Thiolated chitosan is synthesized by reacting chitosan with thioglycolic acid. Nanocomplexes of thiolated chitosan with plasmid DNA encoding green fluorescent protein are prepared	Nanocomplexes of disulfide-crosslinked chitosan–plasmid DNA display sustained DNA release and continuous gene expression in cultured cells lasting up to 60 h post-transfection. Intranasal administration of nanocomplexes to mice yields gene expression that lasts for at least 14 days	Lee *et al.* (2007)
Human embryonic kidney HEK293T cells	Plasmid DNA	Novel ampholytical chitosan derivative, N-imidazolyl-O-carboxymethyl chitosan is used to encapsulate plasmid DNA	Chemical modification enhances the water solubility of chitosan over a wide pH range of 4–10, and increases its plasmid DNA binding ability with no remarkable cytotoxicity against human embryonic kidney HEK293T cells. The cell transfection efficiency of chitosan derivative is high. This is possibly due to imidazole moiety which buffers for endosomal release of DNA	Shi *et al.* (2012)

Human pancreatic cancer cells (AsPC-1)	Enhanced green fluorescent protein plasmid	N,N-diethyl N-methyl chitosan is used to encapsulate enhanced green fluorescent protein plasmid	Enhancement in the charge ratios of polyplexes from 5 to 40 results in 16.70-fold increase in transfection efficiency	Safari *et al.* (2011)
Gastric and duodenum mucosal cancer	Plasmid DNA	Quaternized chitosan is intercalated into the interlayers of rectorite to obtain a new polymer/layered silicate nanocomposite as plasmid DNA carrier	*In vitro* and *in vivo* toxicity studies reveal that the nanocomposites are biocompatible and non-toxic. At mass ratio of nanocomposite:plasmid DNA of 8:1, 100% plasmid DNA adsorption takes place	Wang *et al.* (2008)
Human kidney 293T and human nasopha ryngeal carcinoma CNE2 cells	Plasmid DNA	A dendronized chitosan derivative is prepared via copper-catalyzed azide alkyne cyclization reaction of propargyl focal point poly (amidoamine) dendron with 6-azido-6-deoxy-chitosan	With reference to polyethyleneimine (25 kDa), the dendronized chitosan exhibits enhanced transfection efficiency, in some cases with cellular toxicity reduced	Deng *et al.* (2011)
Nasal epithelial Calu-3 cells	Plasmid DNA	Plasmid DNA-loaded oligochitosan nanoparticles are prepared by ionic gelation using thiamine pyrophosphate as crosslinking agent	Oligochitosan nanoparticles are able to cause a significant and reversible decrease in transepithelial electrical resistance and promote efficient cellular uptake of oligochitosan nanoparticles via paracellular pathway	Ye *et al.* (2013)
BALB/c mice bearing CT-26 carcinoma cells	IL-12 gene	Mannosylated chitosan is used to encapsulate IL-12 gene via self-assembled complexation method	Mannosylated chitosan provides enhanced IL-12 gene transfer efficiency to dendritic cells rather than chitosan itself with respect to induction of murine IL-12 p70 and murine IFN-γ. This is attributed to mannosylated chitosan initiating active targeting via mannose receptor-mediated endocytosis. Intra-tumoural injection of mannosylated chitosan–plasmid encoding murine IL-12 complex into BALB/c mice bearing CT-26 carcinoma cells suppresses tumour growth and angiogenesis, and significantly induces cell cycle arrest and apoptosis.	Kim *et al.* (2006)

References

Agnihotri, S.A., Mallikarjuna, N.N. and Aminabhavi, T.M. (2004) Recent advances on chitosan-based micro- and nanoparticles in drug delivery. *Journal of Controlled Release* 100, 5–28.

Allen, T.M. (2002) Ligand-targeted therapeutics in anticancer therapy. *Nature Review Cancer* 2, 750–763.

Alves, N.M. and Mano, J.F. (2008) Chitosan derivatives obtained by chemical modifications for biomedical and environmental applications. *International Journal of Biological Macromolecules* 43, 401–414.

Anand, P., Kunnumakara, A.B., Sundaram, C., Harikumar, K.B., Tharakan, S.T., Lai, O.S., Sung, B. and Aggarwal, B.B. (2008) Cancer is a preventable disease that requires major lifestyle changes. *Pharmaceutical Research* 25(9), 2097–2116.

Arias, J.I., Aller, M.A., Prieto, I., Arias, A., Julian, Z.D., Yang, H. and Arias, J. (2012) The amazing power of cancer cells to recapitulate extraembryonic functions: the cuckoo's tricks. *Journal of Oncology* 26, 521284.

Arya, G., Vandana, M., Acharya, S. and Sahoo, S.K. (2011) Enhanced antiproliferative activity of Herceptin (HER2)-conjugated gemcitabine-loaded chitosan nanoparticle in pancreatic cancer therapy. *Nanomedicine* 7(6), 859–870.

Bae, K.H., Park, M., Do, M.J., Lee, N., Ryu, J.H., Kim, G.W., Kim, C., Park, T.G. and Hyeon, T. (2012) Chitosan oligosaccharide-stabilized ferrimagnetic iron oxide nanocubes for magnetically modulated cancer hyperthermia. *ACS Nano* 6(6), 5266–5273.

Bernstein, C., Prasad, A.R., Nfonsam, V. and Bernstein, H. (2013) DNA damage, DNA repair and cancer. In: Clark, C. (ed.), *New Research Directions in DNA Repair*. Intech, Rijeka, Croatia, pp. 413–465. Available at: www.intechopen.com/books/new-research-directions-in-dna-repair/dna-damage-dna-repair-and-cancer (accessed 16 September 2013).

Bhadauria, R.S., Joshi, K.S., Gunjan, J. and Diwaker, A.K. (2012) Introduction to neoplasm: a review article. *International Journal of Advanced Research in Pharmaceutical and Biosciences* 2(1), 65–71.

Cavalli, R., Bisazza, A., Trotta, M., Argenziano, M., Civra, A., Donalisio, M. and Lembo, D. (2012) New chitosan nanobubbles for ultrasound-mediated gene delivery: preparation and *in vitro* characterization. *International Journal of Nanomedicine* 7, 3309–3318.

Chang, K., Higuchi, Y., Kawakami, S., Yamashita, F. and Hashida, M. (2010) Efficient gene transfection by histidine-modified chitosan through enhancement of endosomal escape. *Bioconjugate Chemistry* 21, 1087–1095.

Chang, Y.C., Shieh, D.B., Chang, C.H. and Chen, D.H. (2005) Conjugation of monodisperse chitosan-bound magnetic nanocarrier with epirubicin for targeted cancer therapy. *Journal of Biomedical Nanotechnology* 1, 196–201.

Cheng, M., Liu, Z., Wan, T., He, B., Zha, B., Han, J., Chen, H., Yang, F., Li, Q., Wang, W., Xu, H. and Ye, T. (2012) Preliminary pharmacology of galactosylated chitosan/5-fluorouracil nanoparticles and its inhibition of hepatocellular carcinoma in mice. *Cancer Biology and Therapy* 13(14), 1407–1416.

Chung, Y.C., Chang, F.H., Wei, M.F. and Young, T.H. (2010) A variable gene delivery carrier-biotinylated chitosan/polyethyleneimine. *Biomedicals Materials* 5(6), 065012.

Danhier, F., Feron, O. and Préat, V. (2010) To exploit the tumor microenvironment: passive and active tumor targeting of nanocarriers for anti-cancer drug delivery. *Journal of Controlled Release* 148(2), 135–146.

De Benedetti, P.G. and Fanelli, F. (2009) Ligand-receptor communication and drug design. *Current Protein and Peptide Science* 10(2), 186–193.

De Paula Pansani Oliveira, F., Dalla Picola, I.P., Shi, Q., Barbosa, H.F., Tiera, V.A., Fernandes, J.C. and Tiera, M.J. (2013) Synthesis and evaluation of diethylethylamine–chitosan for gene delivery: composition effects on the *in vitro* transfection efficiency. *Nanotechnology* 24(5), 055101.

Deng, Z., Zhen, Z., Hu, X., Wu, S., Xu, Z. and Chu, P.K. (2011) Hollow chitosan-silica nanospheres as pH-sensitive targeted delivery carriers in breast cancer therapy. *Biomaterials* 32(21), 4976–4986.

Dodane, V. and Vilivalam, V.D. (1998) Pharmaceutical applications of chitosan. *Pharmaceutical Science & Technology Today* 1(6), 246–253.

Dosio, F., Brusa, P. and Cattel, L. (2011) Immunotoxins and anticancer drug conjugate assemblies: the role of the linkage between components. *Toxins* 3, 848–883.

Duncan, R. (2006) Polymer conjugates for drug targeting. From inspired to inspiration. *Journal of Drug Targeting* 14, 333–335.

Eroglu, E., Tiwari, P.M., Waffo, A.B., Miller, M.E., Vig, K., Dennis, V.A. and Singh, S.R. (2013) A nonviral pHEMA+chitosan nanosphere-mediated high-efficiency gene delivery system. *International Journal of Nanomedicine* 8, 1403–1415.

Friedmann, T. and Roblin, R. (1972) Gene therapy for human genetic disease? *Science* 175(4025), 949–955.

George, M. and Abraham, T.E. (2006) Polyionic hydrocolloids for the intestinal delivery of protein drugs: alginate and chitosan – a review. *Journal of Controlled Release* 114, 1–14.

Hejazi, R. and Amiji, M. (2003) Chitosan-based gastrointestinal delivery systems. *Journal of Controlled Release* 89, 151–165.

Holmes, C.A. and Tabrizian, M. (2013) Substrate-mediated gene delivery from glycol-chitosan/hyaluronic acid polyelectrolyte multilayer films. *ACS Applied Materials and Interfaces* 5(3), 524–531.

Ishida, T., Harashima, H. and Kiwada, H. (2002) Liposome clearance. *Bioscience Reports* 22(2), 197–224.

Issa, M.M., Köping-Höggård, M. and Artursson, P. (2005) Chitosan and the mucosal delivery of biotechnology drugs. *Drug Discovery Today* 2(1), 1–6.

Jain, A. and Jain, S.K. (2008) *In vitro* and cell uptake studies for targeting of ligand anchored nanoparticles for colon tumors. *European Journal of Pharmaceutical Sciences* 35, 404–416.

Jon, S., Lee, E., Lee, J.J. and Lee, I.-H. (2007) patent number US20070292387.

Kato, Y., Onishi, H. and Machida, Y. (2001) Biological characteristics of lactosaminated N succinylchitosan as a liver-specific drug carrier in mice. *Journal of Controlled Release* 70, 295–307.

Kim, T.H., Ihm, J.E., Choi, Y.J., Woon, J.N. and Su, C.C. (2003) Efficient gene delivery by urocanic acid-modified chitosan. *Journal of Controlled Release* 93(3), 389–402.

Kim, T.H., Jin, H., Kim, H.W., Cho, M.H. and Cho, C.S. (2006) Mannosylated chitosan nanoparticle-based cytokine gene therapy suppressed cancer growth in BALB/c mice bearing CT-26 carcinoma cells. *Molecular Cancer Therapy* 5, 1723–1732.

Kobayashi, T. (2011) Cancer hyperthermia using magnetic nanoparticles. *Biotechnology Journal* 6(11), 1342–1347.

Kuang, Y., Yuan, T., Zhang, Z., Mingyuan, L. and Yang, Y. (2012) Application of ferriferous oxide modified by chitosan in gene delivery. *Journal of Drug Delivery* Article ID 920764.

Kubota, N., Tatsumoto, N., Sano, T. and Toya, K. (2000) A simple preparation of half N-acetylated chitosan highly soluble in water and aqueous organic solvents. *Carbohydrate Research* 324, 268–274.

Kularatne, S.A. and Low, P.S. (2010) Targeting of nanoparticles: folate receptor. In: Grobmyer, S.R. and Moudgil, B.M. (eds) *Cancer Nanotechnology, Methods in Molecular Biology*. Vol. 624. Springer, New York, pp. 249–265.

Kumar, M.N.V.R. (2000) A review of chitin and chitosan applications. *Reactive Functional Polymers* 46, 1–27.

Kwon, O.J., Kang, E., Choi, J.W., Kim, S.W. and Yun, C.O. (2013) Therapeutic targeting of chitosan-PEG-folate-complexed oncolytic adenovirus for active and systemic cancer gene therapy. *Journal of Controlled Release* 169(3), 257–265.

Layek, B. and Singh, J. (2013) Amino acid grafted chitosan for high performance gene delivery: comparison of amino acid hydrophobicity on vector and polyplex characteristics. *Biomacromolecules* 14(2), 485–494.

Lee, D., Zhang, W., Shawna, A.S., Kong, X., Hellermann, G.R., Lockey, R.F. and Mohapatra, S.S. (2007) Thiolated chitosan/DNA nanocomplexes exhibit enhanced and sustained gene delivery. *Pharmaceutical Research* 24(1), 157–167.

Lee, E., Lee, J., Lee, I.H., Yu, M., Kim, H., Chae, S.Y. and Jon, S. (2008) Conjugated chitosan as a novel platform for oral delivery of paclitaxel. *Journal of Medicinal Chemistry* 51, 6442–6449.

Lee, J.Y., Choi, Y.S., Suh, J.S., Kwon, Y.M., Yang, V.C., Lee, S.J., Chung, C.P. and Park, Y.J. (2011a) Cell-penetrating chitosan/doxorubicin/TAT conjugates for efficient cancer therapy. *International Journal of Cancer* 128(10), 2470–2480.

Lee, S.J., Koo, H., Lee, D.E., Min, S., Lee, S., Chen, X., Choi, Y., Leary, J.F., Park, K., Jeong, S.Y., Kwon, I.C., Kim, K. and Choi, K. (2011b) Tumor-homing photosensitizer-conjugated glycol chitosan nanoparticles for synchronous photodynamic imaging and therapy based on cellular on/off system. *Biomaterials* 32, 4021–4029.

Li, F.R., Yan, W.H., Guo, Y.H., Qi, H. and Zhou, H.X. (2009) Preparation of carboplatin-Fe@C-loaded chitosan nanoparticles and study on hyperthermia combined with pharmacotherapy for liver cancer. *International Journal of Hyperthermia* 25(5), 383–391.

Liang, X.J., Chen, C., Zhao, Y. and Wang, P.C. (2010) Circumventing tumor resistance to chemotherapy by nanotechnology. *Methods in Molecular Biology* 596, 467–488.

Lin, J.C., Yuan, P.M.K. and Jung, D.T. (1998) Enhancement of anticancer drug delivery to the brain by microwave induced hyperthermia. *Bioelectrochemistry and Bioenergetics* 47, 259–264.

Lindahl, T. (1993) Instability and decay of the primary structure of DNA. *Nature* 362, 709–715.

Liu, Z., Lv, D., Liu, S., Gong, J., Wang, D., Xiong, M., Chen, X., Xiang, R. and Tan, X. (2013) Alginic acid-coated chitosan nanoparticles loaded with legumain DNA vaccine: effect against breast cancer in mice. *PLoS One* 8(4), e60190.

Lorenzo-Lamosa, M.L., Remunan-Lopez, C., Vila-Jato, J.L. and Alonso, M.J. (1998) Design of microencapsulated chitosan microspheres for colonic drug delivery. *Journal of Controlled Release* 52, 109–118.

Lubben, I.M.V.D., Verhoef, J.C., Borchard, G. and Junginger, H.E. (2001) Chitosan and its derivatives in mucosal drug and vaccine delivery. *European Journal of Pharmaceutical Sciences* 14, 201–207.

Maeda, H. (2001) The enhanced permeability and retention (EPR) effect in tumor vasculature: the key role of tumor-selective macromolecular drug targeting. *Advances in Enzyme Regulation* 41, 189–207.

Maeda, H., Nakamura, H. and Fang, J. (2013) The EPR effect for macromolecular drug delivery to solid tumors: improvement of tumor uptake, lowering of systemic toxicity, and distinct tumor imaging *in vivo*. *Advanced Drug Delivery Reviews* 65(1), 71–79.

Maya, S., Kumar, L.G., Sarmento, B., Rejinold, N.S., Menon, D., Nair, S.V. and Jayakumar, R. (2013) Cetuximab conjugated O-carboxymethyl chitosan nanoparticles for targeting EGFR overexpressing cancer cells. *Carbohydrate Polymers* 93(2), 661–669.

McNeil, S.E. (2011) Unique benefits of nanotechnology to drug delivery and diagnostics. In: McNeil, S.E. (ed.) *Characterization of Nanoparticles Intended for Drug Delivery, Methods in Molecular Biology*. Vol. 697. Springer, New York, pp. 3–8.

Moghimi, S.M., Hunter, A.C. and Murray, J.C. (2001) Long circulating and target-specific nanoparticles: theory to practice. *Pharmacological Reviews* 53, 283–318.

Mohammadi, S.S., Miri, R., Salmanpour, M., Khalighian, N., Sotoudeh, S. and Erfani, N. (2013) Preparation and assessment of chitosan-coated superparamagnetic Fe_3O_4 nanoparticles for controlled delivery of methotrexate. *Research in Pharmaceutical Sciences* 8(1), 25–33.

Moorthi, C., Manavalan, R. and Kathiresan, K. (2011) Nanotherapeutics to overcome conventional cancer chemotherapy limitations. *Journal of Pharmacy and Pharmaceutical Sciences* 14(1), 67–77.

Nurjaya, S. and Wong, T.W. (2005) Effects of microwave on drug release properties of matrices of pectin. *Carbohydrate Polymers* 62(3), 245–257.

Ohya, Y., Shiratani, M., Kobayashi, H. and Ouchi, T. (1994) Release behavior of 5-fluorouracil from chitosan-gel nanospheres immobilizing 5-fluorouracil coated with polysaccharides and their cell specific cytotoxicity. *Journal of Macromolecular Science Part A – Pure and Applied Chemistry* 31(5), 629–642.

Omwancha, W., Kouba, C., Yelamanchili, S. and Neau, S.H. (2011) Colon-specific drug delivery using ethylcellulose and chitosan in the coat of compression-coated tablets. *Drug Development and Industrial Pharmacy* 37, 945–953.

Oosegi, T., Onishi, H. and Machida, Y. (2008) Novel preparation of enteric-coated chitosan–prednisolone conjugate microspheres and *in vitro* evaluation of their potential as a colonic delivery system. *European Journal of Pharmaceutics and Biopharmaceutics* 68, 260–266.

Park, H., Saravanakumar, G., Kim, K. and Kwon, I.C. (2010) Targeted delivery of low molecular drugs using chitosan and its derivatives. *Advanced Drug Delivery Reviews* 62, 28–41.

Raub, T.J.(2006) P-glycoprotein recognition of substrates and circumvention through rational drug design. *Molecular Pharmaceutics* 3(1), 3–25.

Rinaudo, M. (2006) Chitin and chitosan: properties and applications. *Progress in Polymer Science* 31, 603–632.

Rothdiener, M., Beuttler, J., Messerschmidt, S.K.E. and Kontermann, R.E. (2010) Antibody targeting of nanoparticles to tumor-specific receptors: immunoliposomes. In: Grobmyer, S.R. and Moudgil, B.M. (eds) *Cancer Nanotechnology, Methods in Molecular Biology*. Vol. 624. Springer, New York, pp. 295–308.

Safari, S., Dorkoosh, F.A., Soleimani, M., Zarrintan, M.H., Akbari, H., Larijani, B. and Tehrani, M.R. (2011) N-diethylmethyl chitosan for gene delivery to pancreatic cancer cells and the relation between charge ratio and biologic properties of polyplexes via interpolations. *International Journal of Pharmaceutics* 420(2), 350–357.

Sanna, V., Roggio, A.M., Siliani, S., Piccinini, M., Marceddu, S., Mariani, A. and Sechi, M. (2012) Development of novel cationic chitosan-and anionic alginate-coated poly(D,L-lactide-co-glycolide) nanoparticles for controlled release and light protection of resveratrol. *International Journal of Nanomedicine* 7, 5501–5516.

Sato, T., Ishii, T. and Okahata, Y. (2001) *In vitro* gene delivery mediated by chitosan. Effect of pH, serum, and molecular mass of chitosan on the transfection efficiency. *Biomaterials* 22, 2075–2080.

Seymour, L.W., Ulbrich, K., Wedge, R., Hume, I.C., Stohalm, J. and Duncan, R. (1991) N-(2-hydroxypropyl) methacrylamide copolymers targeted to the hepatocyte galactose-receptor: pharmacokinetics in DBA2 mice. *British Journal of Cancer* 63, 859–866.

Shi, B., Shen, Z., Zhang, H., Bi, J. and Dai, S. (2012) Novel ampholytical chitosan derivative, N-imidazolyl-O-carboxymethyl chitosan is used to encapsulate plasmid DNA. *Biomacromolecules* 13(1), 146–153.

Shi, B., Zhang, H., Shen, Z., Bi, J. and Dai, S. (2013) Developing a chitosan supported imidazole Schiff-base for high-efficiency gene delivery. *Polymer Chemistry* 4, 840–850.

Singla, A.K. and Chawla, M. (2001) Chitosan: some pharmaceutical and biological aspects – an update. *Journal of Pharmacy and Pharmacology* 53, 1047–1067.

Sinha, V.R., Singla, A.K., Wadhawan, S., Kaushik, R., Kumria, R., Bansal, K. and Dhawan, S. (2004) Chitosan microspheres as a potential carrier for drugs. *International Journal of Pharmaceutics* 274, 1–33.

Song, Y., Onishi, H. and Nagai, T. (1992) Synthesis and drug release characteristics of the conjugates of mitomycin C with N-succinyl-chitosan and carboxymethyl-chitin. *Chemical Pharmaceutical Bulletin* 40, 2822–2825.

Sterzer, F. (2002) Microwave medical devices. *IEEE Microwave Magazine* 3(1), 65–70.

Subramanian, N., Abimanyu, S., Vinoth, J. and Sekar, P.C. (2010) Biodegradable chitosan magnetic nanoparticle carriers for sub-cellular targeting delivery of artesunate for efficient treatment of breast cancer. *AIP Conference Proceedings*, 1311, 416. Rostock, Germany, pp. 416–424.

Talaei, F., Azizi, E., Dinarvand, R. and Atyabi, F. (2011) Thiolated chitosan nanoparticles as a delivery system for antisense therapy: evaluation against EGFR in T47D breast cancer cells. *International Journal of Nanomedicine* 6, 1963–1975.

Tian, Q., Zhang, C.N., Wang, X.H., Wang, W., Huang, W., Cha, R.T., Wang, C.H., Yuan, Z., Liu, M., Wan, H.Y. and Tang, H. (2010) Glycyrrhetinic acid-modified chitosan/poly(ethylene glycol) nanoparticles for liver-targeted delivery. *Biomaterials* 31, 4748–4756.

Tian, Q., Wang, X., Wang, W., Zhang, C., Wang, P. and Yuan, Z. (2012) Self-assembly and liver targeting of sulfated chitosan nanoparticles functionalized with glycyrrhetinic acid. *Nanomedicine* 8(6), 870–879.

Torchilin, V.P. (2010) Passive and active drug targeting: drug delivery to tumors as an example. In: Schafer-Korting, M. (ed.) *Drug Delivery, Handbook of Experimental Pharmacology*. Vol. 197. Springer, New York, pp. 3–53.

Varshosaz, J., Hassanzadeh, F., Sadeghi, H., Khan, Z.G. and Rostami, M. (2013) Retinoic acid decorated albumin-chitosan nanoparticles for targeted delivery of doxorubicin hydrochloride in hepatocellular carcinoma. *Journal of Nanomaterials* 2013, Article ID 254127.

Vrba, J., Lapeš, M. and Oppl, L. (1999) Technical aspects of microwave thermotherapy. *Bioelectrochemistry and Bioenergetics* 48, 305–309.

Wadajkar, A.S., Menon, J.U., Kadapure, T., Tran, R.T., Yang, J. and Nguyen, K.T. (2013) Design and application of magnetic-based theranostic nanoparticle systems. *Recent Patents on Biomedical Engineering* 6(1), 47–57.

Wang, B., Zhang, S., Cui, S., Yang, B., Zhao, Y., Chen, H., Hao, X., Shen, Q. and Zhou, J. (2012) Chitosan enhanced gene delivery of cationic liposome via non-covalent conjugation. *Biotechnology Letters* (1), 19–28.

Wang, J.J., Zeng, Z.W. and Xiao, R.Z. (2011) Recent advances of chitosan nanoparticles as drug carriers. *International Journal of Nanomedicine* 6, 765–774.

Wang, M. and Thanou, M. (2010) Targeting nanoparticles to cancer. *Pharmacological Research* 62(2), 90–99.

Wang, X., Pei, X., Du, Y. and Li, Y. (2008) Quaternized chitosan/rectorite intercalative materials or a gene delivery system. *Nanotechnology* 19(37), 375102.

Wang, Y., Li, P. and Kong, L. (2013) Chitosan-modified PLGA nanoparticles with versatile surface for improved drug delivery. *AAPS PharmSciTech* 14(2), 585–592.

Wong, T.W. (2008) Basis of use of microwave in processing of drug delivery systems. *Current Drug Delivery* 5(2), 77–84.

Wong, T.W. (2009) Chitosan and its use in design of insulin delivery system. *Recent Patents on Drug Delivery and Formulation* 3(1), 8–25.

Wong, T.W. (2010) Design of oral insulin delivery systems. *Journal of Drug Targeting* 18(2), 79–92.

Wong, T.W. (2011) Alginate graft copolymers and alginate-co-excipient physical mixture in oral drug delivery. *Journal of Pharmacy and Pharmacology* 63(12), 1497–1512.

Wong, T.W. and Bose, A. (2015) Glyoxalated chitosan-5-fluorouracil/chitosan-folate as colon-specific and colon cancer cell-targeted device. *Journal of Controlled Release* 213, e105.

Wong, T.W. and Nurulaini, H. (2012) Sustained-release alginate-chitosan pellets prepared by melt pelletization technique. *Drug Development and Industrial Pharmacy* 38(12), 1417–1427.

Wong, T.W. and Sumiran, N. (2008) Drug release property of chitosan-pectinate beads and its changes under the influence of microwave. *European Journal of Pharmaceutics and Biopharmaceutics* 69, 176–188.

Wong, K., Sun, G., Zhang, X., Dai, H., Liu, Y., He, C. and Leong, K.W. (2006) PEI-g-chitosan, a novel gene delivery system with transfection efficiency comparable to polyethylenimine *in vitro* and after liver administration *in vivo*. *Bioconjugate Chemistry* 17(1), 152–158.

Wong, T.W., Colombo, G. and Sonvico, F. (2011) Pectin matrix as oral drug delivery vehicle for colon cancer treatment. *AAPS PharmSciTech* 12(1), 201–214.

Xiao, B., Wang, X., Qiu, Z., Ma, J., Zhou, L., Wan, Y. and Zhang, S. (2012) A dual-functionally modified chitosan derivative for efficient liver-targeted gene delivery. *Journal of Biomedical Material Research Part A* 101(7), 1888–1897.

Yang, S.J., Chen, J.W., Lin, F.H., Young, T.H., Lou, P.J. and Shieh, M.J. (2010) Colorectal cancer cell detection by folic acid-conjugated chitosan nanoparticles. *Biomedical Engineering – Applications Basis Communications* 22(1), 9–17.

Ye, Y., Xu, Y., Liang, W., Leung, G.P.H., Cheung, K., Zheng, C., Chen, F. and Lam, J.K.W. (2013) DNA-loaded chitosan oligosaccharide nanoparticles with enhanced permeability across Calu-3 cells. *Journal of Drug Targeting* 21(5), 474–486.

Yong, S.K. and Wong, T.W. (2013) Chitosan for body weight management: current issues and future directions. In: Se-Kwon, K. (ed.) *Marine Nutraceuticals: Prospects and Perspectives*. CRC Press, Boca Raton, Florida, pp. 151–168.

Yousefpour, P., Atyabi, F., Vasheghani-Farahani, E., Mousavi Movahedi, A.A. and Dinarvand, R. (2011) Targeted delivery of doxorubicin-utilizing chitosan nanoparticles surface-functionalized with anti-Her2 trastuzumab. *International Journal of Nanomedicine* 6, 1977–1990.

Yuan, Q., Venkatasubramanian, R., Hein, S. and Misra, R.D. (2008) A stimulus-responsive magnetic nanoparticle drug carrier: magnetite encapsulated by chitosan grafted-copolymer. *Acta Biomaterialia* 4, 1024–1037.

Zakaria, Z. and Wong, T.W. (2009) Chitosan spheroids with microwave modulated drug release. *Progress in Electromagnetics Research* 99, 355–382.

Zhang, H., Mardyani, S., Chan, W.C. and Kumacheva, E. (2006) Design of biocompatible chitosan microgels for targeted pH-mediated intracellular release of cancer therapeutics. *Biomacromolecules* 7, 1568–1572.

Zhang, Y., Chen, J., Zhang, Y., Pan, Y., Zhao, J., Ren, L., Liao, M., Hu, Z., Kong, L. and Wang, J.A. (2007) Novel PEGylation of chitosan nanoparticles for gene delivery. *Biotechnology and Applied Biochemistry* 46, 197–204.

Zhang, Y., Wei, W., Lva, P., Wanga, L. and Maa, G. (2011) Preparation and evaluation of alginate–chitosan microspheres for oral delivery of insulin. *European Journal of Pharmaceutics and Biopharmaceutics* 77, 11–19.

Zhao, X., Li, Z., Pan, H., Liu, W., Lv, M., Leung, F. and Lu, W.W. (2013) Enhanced gene delivery by chitosan-disulfide-conjugated LMW-PEI for facilitating osteogenic differentiation. *Acta Biomaterialia* 9(5), 6694–6703.

Zhu, L., Ma, J., Jia, N., Zhao, Y. and Shen, H. (2009) Chitosan coated magnetic nanoparticles as carriers of 5-fluorouracil: preparation, characterization andcytotoxicity studies. *Colloids and Surfaces B: Biointerfaces* 68, 1–6.

12 Polymers as Biodegradable Matrices in Transdermal Drug Delivery Systems

Bhanu Malhotra[1], Harsha Kharkwal[2,*] and Anuradha Srivastava[3]

[1]Amity Institute of Biotechnology and Amity Center for Carbohydrate Research, Amity University, Noida, India; [2]Amity Center for Carbohydrate Research and Amity Institute of Phytomedicine and Phytochemistry, Amity University Uttar Pradesh, Noida, India; [3]Biological Sciences and Geology, Queensborough Community College, Bayside, New York, USA

Abstract

The conventional forms of oral dosage have significant disadvantages including poor bioavailability in hepatic metabolism and drug degradation in the gastrointestinal (GI) tract due to enzymes and different pH ranges in these tracts. One effective route for drug absorption into the body and then into the systematic circulation to circumvent such issues is the skin. Transdermal drug delivery systems (TDDS) have emerged, combining high therapeutic efficacy with safety, reducing the number and size of dose administration significantly. TDDS are being pioneered in medical practices as alternatives to hypodermic injections and oral drug delivery systems. The therapeutic agents are introduced through the skin into the systemic circulation through the use of transdermal patches. This chapter presents an overview of TDDS practices, and the use of various biopolymers for drug delivery, and discusses the potential advantages and issues related to them.

Introduction

Oral drug delivery is the conventional form of drug delivery and faces many challenges of drug degradation before its absorption in the body. There has been considerable research in recent decades into transdermal drug delivery systems (TDDS), which deliver the therapeutic agents in the form of transdermal patches into the skin, from where they can be absorbed directly into the systemic circulation. Skin possesses 10–80 hair follicles and 200–250 sweat ducts per cm^2 and so provides the basic route for drug delivery.

The skin is also targeted for local delivery of therapeutic agents, when the drug acts only on various skin layers and does not penetrate it. This is referred to as 'topical delivery' and occurs in the case of some microbial skin infections, neoplastic formations and skin inflammation. The systemic circulation is considered an alternative route to oral and conventional drug delivery systems, and in this case the drug enters into the blood circulation directly, overcoming all the barriers to drug degradation before exerting its effects (Brown *et al.*, 2006). TDDS is a non-invasive method of drug administration in which sustained drug release is achieved by transdermal formulations over the skin and patient compliance is increased (Guy, 1996; Tanner and Marks, 2008).

*Corresponding author. E-mail: hkharkwal@amity.edu

Advantages and Challenges of Transdermal Drug Delivery Systems

The various advantages and challenges of transdermal drug delivery are:

Advantages

- It can substitute for oral drug administration in cases of vomiting and diarrhoea (Finnin and Morgan, 1999).
- It avoids the normal route of drug travel through the gastrointestinal (GI) tract and so avoids absorption difficulties caused by GI tract pH and enzymatic activity.
- It avoids the first-pass effect (the initial passage of the drug through the systemic, portal circulation followed by GI tract absorption) by avoiding the digestive and liver enzymes (Barry, 2002; Allen *et al.*, 2005).
- It avoids the issues of parenteral administration and is non-invasive (Barry, 2002; Walters and Brain, 2002; Allen *et al.*, 2005).
- It can be terminated rapidly by removal of its application from the skin surface (Barry, 2002).
- Single application provides extended therapy enhancing patient compliance over other dosage administration forms as it can be repeatedly administered (Allen *et al.*, 2005).
- It is easily identified in cases of unresponsiveness and unconsciousness because of its features and identifying marks.

At the same time, this system possesses a few challenges and disadvantages that limit its use.

Disadvantages

- It cannot be used for drugs which need significantly higher blood levels (Allen *et al.*, 2005).
- Its use may be uneconomical (Barry, 2002).
- It may not be suitable to deliver some drugs. Only relatively potent drugs are suitable candidates, because of the natural limits to drug delivery imposed by the impermeability of the skin (Barry, 2002; Allen *et al.*, 2005).
- It may cause contact dermatitis in some patients, a side reaction at the site of application which requires the drug to be immediately discontinued.
- Long term adherence is a potential disadvantage.

The structure of the skin plays a very important role in allowing penetration, and researchers should take this into into account if our understanding of transdermal drug delivery procedures is to be improved (Barry, 2002).

Transdermal Patches

TDDS employ the use of patches which are dosage forms required to deliver a therapeutic drug into the skin through controlled release in a specified amount of time (Vasil'ev *et al.*, 2001; Aulton, 2007; Tiwary *et al.*, 2007). When attached to the skin, transdermal patches remain for a specified amount of time (hours, days or weeks) in order to deliver the drug. The route of absorption could be intracellular, intercellular or transappendageal. The drug has to cross the barrier of various skin layers after release from the patch. The main barrier is posed by the stratum comenum, which lowers drug permeation, as shown in Fig. 12.1.

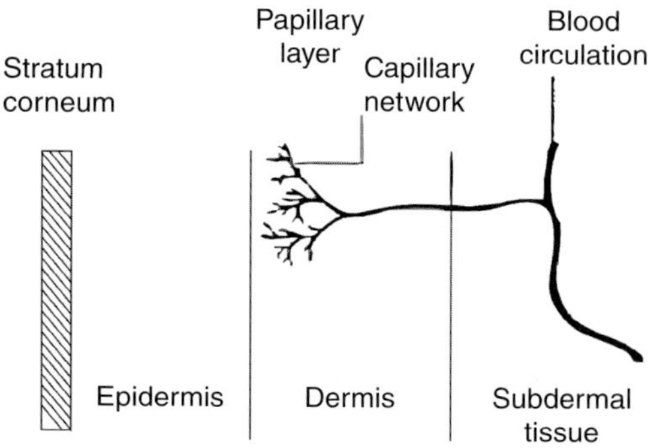

Fig. 12.1. Barriers between drug and systemic circulation through transdermal routes. Source: Gaikwad, 2013.

A TDDS comprises an adhesive layer, drug reservoir, backing membrane and release control liner. The drug can be applied through a transdermal patch in relatively higher doses worn on the skin for a period of time. The drug enters into the bloodstream through diffusion through the skin pores. The difference in concentration of drug in the patch and the blood will drive the movement of the drug into the bloodstream until a constant concentration of the drug is maintained in the bloodstream. Transdermal patches come in varying sizes and usually contain more than one drug. A patch can be characterized for toxicological administration by checking its size, quality, time of onset, adhesiveness, thickness and other qualities. Drug systems approved for transdermal delivery include nitroglycerin, scopolamine, fentanyl, nicotine, isosorbide dinitrate, selegeline, estradiol, clonidine, testosterone, oxybutynin, buprenorphine, methylphenidate and rivastigmin (Guy, 2010).

Transdermal patches are prepared (Meathrel, 2010; Quan, 2011) by considering:

- Drug daily dose and size.
- Compatibility of drug with various polymers, plasticizers, adhesives and other excipients of the formulation.
- Size of the patch.
- Stability of the patch.
- A proper balance between adhesion and patch removal over the duration for which the patch is applied.

Types of Transdermal Drug Delivery Systems

Polymers are the backbone of transdermal drug delivery procedures as they are biodegradable, biocompatible, non-toxic and non-antigenic in nature. Transdermal skin formulations are made as multilayered polymeric laminates contacting a fabricated drug reservoir or possessing a drug polymer matrix system sandwiched between two homogenous polymeric layers. Both layers then hold the drug in position in a transdermal patch. The outer layer acts as a backing layer, preventing loss of the drug from the patch during administration, and the inner layer acts as a rate-controlling drug release membrane and as an adhesive. There are three broad types of transdermal drug delivery systems, and these are described in Fig. 12.2.

Reservoir systems

Such systems mainly contain the drug/therapeutic agent in a reservoir of liquid. Drugs can be contained as a suspension of one or more agents in a viscous liquid which can be dissolved in a solvent in the storage part. In another type, a membrane of a polymer with a different structure separates the drug reservoir from the adhesive layer. Basically, the drug reservoir is embedded between an impervious backing layer which has a rate-controlling membrane. The membrane can either be microporous or non-porous, through which drugs can be released in a sustained manner. The drug can either be in the form of a suspension, a gel or as a solution dispersed throughout a polymer matrix. Membrane thickness and adhesiveness control drug release through these membranes (Delgado-Charro and Guy, 2001; Williams, 2003; Padula *et al.*, 2007). Transderm-Scop (scopolamine), Transderm-Nitro (nitroglycerin), Estraderm (estradiol) and Catapress-TTS (clonidine) are some of the commercially available diffusion reservoir systems.

Matrix systems

These systems include a homogenously dispersed and distributed drug within a suitable polymer matrix carrier which is either hydrophilic or lipophilic in nature. One side (possibly the outer side) has a backing layer. The transdermal patch is held on the skin with an adhesive polymeric strip. Such formulations can be prepared using the polymeric matrix and dispersing the drug in an adhesive polymer that is sensitive and subjected to direct pressure which is covered later with a backing layer. In such systems drug release from the semi-solid matrix is not controlled by any membrane; rather, it is a factor of surface area of the applied patch (Delgado-Charro and Guy, 2001; Williams, 2003; Padula *et al.*, 2007). Commercially available systems of this type include Exelon (rivastigmine), Minitran (nitroglycerin), Sancuso (granisetron), Emsam (selegeline) and Oxytrol (oxybutyne).

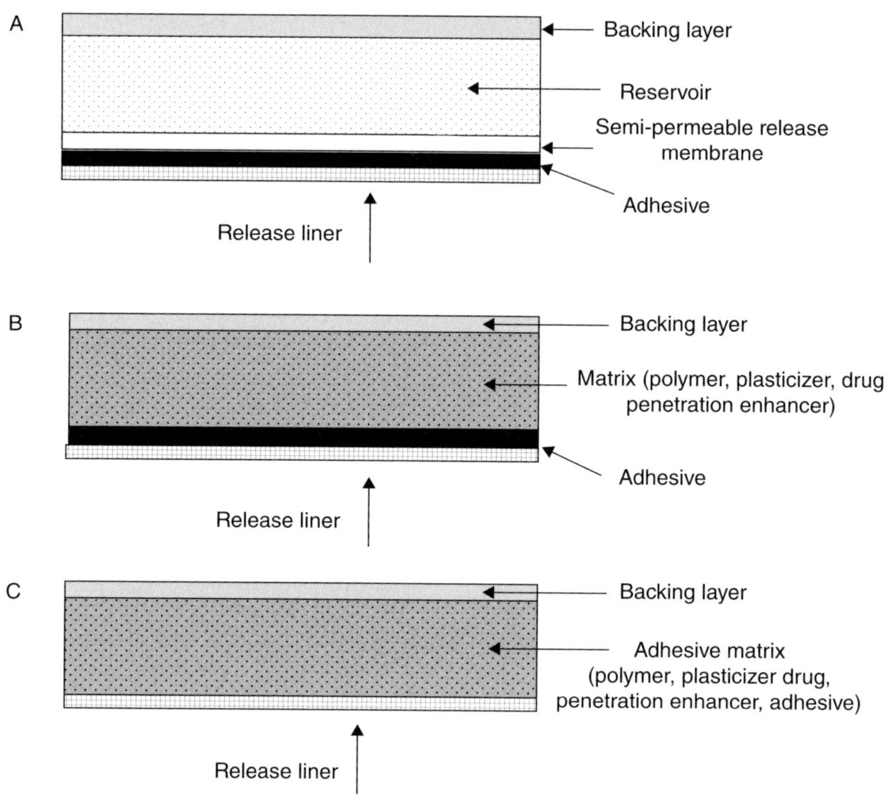

Fig. 12.2. Schematic representation of transdermal patch types: A. Reservoir, B. Matrix, C. Drug-in-Adhesive transdermal systems. (Adapted, with permission, from Güngör *et al.*, 2012.)

Adhesive systems

Drug dispersed is adhesive polymer in such systems. An adhesive membrane made up of polymer lies beneath the drug reservoir layer, controlling its release rate. The release of drugs in such systems is controlled both by the membrane and by drug dispersion. Such systems can be single layered or multilayered (Delgado-Charro and Guy, 2001; Williams, 2003; Padula *et al.*, 2007). Commercially available systems of this type include Nitrodur (nitroglycerin), Duragesic (fentanyl) and Daytarana (methyl phenidate).

Composition of Transdermal Drug Delivery Systems

Various different types of transdermal drug delivery systems are prepared but the basic composition of such systems remains almost the same and includes the following components (Williams, 2003):

- Matrix
- Reservoir
- Drug
- Semi-permeable/release membrane
- Backing layer
- Adhesive
- Release liner
- Plasticizers
- Solvents, penetration enhancers

The drug is dispersed and dissolved evenly in a polymer matrix carrier type system (Delgado-Charro and Guy, 2001; Williams, 2003). The matrix and polymer will control the drug release rate. Natural polymers such as pectin, chitosan and sodium alginate; synthetic polymers such as Polyvinyl Alcohol (PVA) and Eudragit; and semi-synthetic derivatives of cellulose and starch are used (Lin *et al.*, 1991; Amnuaikit *et al.*, 2005;

Nicoli *et al.*, 2006; Schroeder *et al.*, 2007a; Güngör *et al.*, 2008).

Reservoir

This system uses a semi-permeable membrane which controls drug release. The drug is present in the reservoir of liquid or solid (Delgado-Charro and Guy, 2001; Williams, 2003).

Drug

Drugs need to be designed appropriately for transdermal drug delivery procedures. They should have some specific physicochemical characteristics; for example, they should be of low molecular weight, be water soluble and have a moderate amount of lipophilic carrier. The drug molecule should be as potent as possible to provide an effective response at low concentrations (Guy, 1996; Quan, 2011).

Semi-permeable membrane

Multilayer adhesive and reservoir systems employ the use of a semi-permeable membrane made of either high-density polyethylene, silicones, an ethylene–vinyl acetate copolymer, cellulose nitrate or cellulose acetate, which controls the drug release rate (Williams, 2003).

Backing layer

The backing layer enables protection from external effects during transdermal patch application. It also maintains the integrity of the transdermal delivery system during the storage period. Materials impermeable to drugs and therapeutic agents are used as backing layers. The layer must be chemically inert and non-toxic, providing absolutely no side reactions with the drug to be administered through the patch. Some of the agents used for a backing layer include polypropylene, ethylene–vinyl acetate, polyethylene and polyvinylidene chloride (Williams, 2003).

Adhesive

The adhesive attaches the TDDS easily to the skin surface. Its purpose is to facilitate adherence by being non-toxic to the skin; it should not be a skin irritant or an allergen. Some pressure-sensitive adhesives have commonly been used in transdermal drug delivery procedures. Generally used agents include acylates, polyisobutilene adhesives and polysiloxan agents (Williams, 2003).

Release liner

This is the lining that protects the transdermal patch before it is adhered to the skin from the external environment. Aluminium foil, ethylene–vinyl acetate and paper are commonly used. A release liner in the form of a thin layer is peeled from the transdermal patch, which is to be applied to the skin without damaging the skin layer. Perfluorocarbons and silicone can be used (Williams, 2003).

Plasticizers

Plasticizers in transdermal patches improve the brittleness of the polymer matrix and provide the patch with some flexibility (Williams, 2003). Flexibility improves flow and also enhances the tear strength of the polymer used (Bergo and Sobral, 2007). Table 12.1 indicates the percentage of plasticizers used in transdermal patch preparations.

Solvents and penetration enhancers

Different solvents can be used to disperse the polymer, adhesive and the drug used in preparation of the TDDS. Common solvents include methanol, chloroform, acetone and dichloromethane. Penetration enhancers increase the permeation of the drug into the skin; agents generally used include terpenes, water, dimethyl sulfoxide (DMSO) and azone (Williams, 2003).

Table 12.1. The plasticizers used in transdermal drug delivery formulations. (Adapted, with permission, from: Güngör *et al.*, 2012)

Plasticizer	Percentage	Type of transdermal formulations made	Polymer used	Reference
Sorbitol Sucrose	20 20	Drug-free film	Polyvinyl alcohol: chitosan	Arvanitoyannis *et al.*, 1997
Glycerine	4	Matrix	Polyvinyl alcohol 72000	Padula *et al.*, 2003
Sorbitol solution	2	Matrix	Polyvinyl alcohol	Nicoli *et al.*, 2005
Dibutyl phthalate propylene glycol	20	Matrix	Polyvinyl alcohol Xanthan gum	Kumar *et al.*, 2007
Polyethylene glycol 600	10–50	Membrane	Cellulose acetate	Wang *et al.*, 2002
Triethyl citrate Triacetin Dibutyl phthalate	1–6 2.1 4	Film forming polymeric solution	NE 40D Eudragit RL PO, E100, S100 Acrylate copolymer Polyvinylpyrrolidone Acrylate/octylacrylamide copolymer Hydroxypropylcellulose Polyvinyl alcohol Silicon gum Polyisobutylene	Schroeder *et al.*, 2007b

Factors Affecting Drug Bioavailabilities in Transdermal Systems

Two important factors determine the bioavailability of the drug delivered through TDDS:

Formulation factors

- device used;
- penetration enhancers;
- vehicles and membrane preferred;
- method of application; and
- physical chemistry of transport.

Physiological factors

- skin metabolism;
- skin condition and disease;
- stratum corneum of the skin;
- anatomic site of application; and
- skin irritation and sensitization.

Polymers in Transdermal Drug Delivery Systems

Natural polymers and synthetic polymers, along with their derivatives, are the backbone of TDDS. As we have seen, there are three different types of TDDS: the reservoir, matrix and adhesive systems. These contain multilayered polymeric laminates which create and surround the drug reservoir. The layers of polymers provide integrity to the transdermal patch and release the drug at a specified, controlled rate. A number of drugs marketed as TDDS make use of various polymeric membranes, as described in Table 12.2.

A stable and monolithic solid state design is the basic preference for transdermal delivery patches, and provides them an appropriate cosmetic appeal. A number of polymers are used for rate-controlled delivery of the drug; the others just provide integrity to the transdermal patch. Therefore, polymers are used for

Table 12.2. List of transdermal products currently on the US market. (Reproduced, with permission, from Paudel *et al*., 2010).

Active ingredients	Type of delivery system	Name	Company	Type of patch	Dose and application	Regulatory status	Uses
Clonidine	Transdermal patch extended release	Catapres TTS	Boehringer Ingelheim	Drug in reservoir and in adhesive formulation	3.5–10.5-cm^2 patches deliver 0.1–0.3 mg/day for 7 days Applied to hairless skin on the upper outer arm or chest	Rx	Essential hypertension
		Clonidine	Par Pharm	Reservoir type	0.1–0.3 mg/24 h for 7 days		
Estradiol	Transdermal patch extended release	Alora	Watson Laboratories	Adhesive matrix drug reservoir	9–36-cm^2 patches deliver 0.025–0.1 mg/day and continuous delivery for twice weekly dosing Applied to lower abdomen	Rx	Menopause, postmenopausal and osteoporosis, in case of lowered estrogen levels
		Climara	Bayer Healthcare	Adhesive matrix containing drug	6.5–25-cm^2 patches deliver 0.025–0.1 mg/day for 7 days Applied to lower abdomen or upper quadrant of buttock		
		Estraderm	Novartis	Reservoir type	0.05 or 0.1-mg/day and continuous delivery for twice weekly application		
		Estradiol	Mylan Technologies	Adhesive matrix containing drug	0.025–0.1 mg/day continuous delivery once weekly patch		
		Menostar	Bayer Healthcare	Adhesive matrix containing drug	3.25-cm^2 delivers 14 μg/day for 7 days Applied near lower abdomen		
		Vivelle/ Vivelle-Dot	Novartis/ Novogyne	Adhesive formulation contains drug	Patches having active surface area of 2.5–10 cm^2 deliver 0.025–0.1 mg/day and twice weekly application Applied to the abdomen		
	Transdermal gel	Divigel	Upsher-Smith Laboratories	0.1% gel	0.25–1 g dose available Applied to a small area (200 cm^2) of the thigh in a thin, quick-drying layer		

Continued

Table 12.2. Continued.

Active ingredients	Type of delivery system	Name	Company	Type of patch	Dose and application	Regulatory status	Uses
		Elestrin	Azur Pharma	0.06% gel supplied in a non-aerosol, metered-dose pump container	Applied once daily to the upper arm using a metered-dose pump that delivers 0.87 g of Elestrin® gel per actuation		
		Estrogel	Ascend Therapeutics	0.06% estradiol in an absorptive hydroalcoholic gel	1.25 g in a single dose and applied to 750-cm^2 area Applied to arm between wrist and shoulder		
	Transdermal spray	Evamist	KV Pharm/ Ther-Rx	Topical application to the skin of a rapidly drying homogeneous solution of 1.7% drug from a metered-dose pump	One, two or three sprays/day (90 μl/spray) to adjacent nonoverlapping 20-cm^2 areas on the inner surface of the arm between the elbow and the wrist and allowed to dry	Rx	Menopause, postmenopausal and osteoporosis, in case of lowered estrogen levels
Estradiol and levonorgestrel	Transdermal patch extended release	Climara Pro	Bayer Healthcare Pharmaceuticals	Drug in adhesive layer	22-cm^2 Climara Pro™ system contains 4.4 mg estradiol and 1.39 mg levonorgestrel and delivers 0.045 mg estradiol and 0.015 mg levonorgestrel/day for 7 days applied to lower abdomen	Rx	Menopausal symptoms
Estradiol and norethindrone acetate	Transdermal patch extended release	Combipatch	Novartis	Adhesive layer contains both drugs	9–16-cm^2 patches deliver 0.05/0.14 or 0.05/0.25 mg estradiol/norethindrone acetate per day and applied twice weekly to lower abdomen	Rx	Menopausal symptoms

Ethinyl estradiol and norelgestromin	Transdermal patch extended release	Ortho Evra	Ortho McNeil Janssen	Adhesive matrix containing drug	6.00 mg norelgestromin and 0.75 mg ethinyl estradiol in each 20-cm^2 patch and delivers for 7 days Applied to buttock, abdomen, upper outer arm or upper torso	Rx	Contraception
Fentanyl	Transdermal patch extended release	Fentanyl transdermal system	Actavis, Mylan Technologies, Lavipharm Labs, Noven, Watson Laboratories and Teva Pharms	Matrix type (Mylan technologies and Teva Pharms) and reservoir (Actavis and Watson laboratories)	10–40-cm^2 patches deliver 25–100 µg/h	Schedule II	Chronic pain (opioid tolerant) that cannot be managed by any other means
		Duragesic	Ortho McNeil Janssen	Drug in reservoir and in adhesive formulation	5–40-cm^2 patches deliver 12.5–100 µg/h continuous systemic delivery for 72 h applied to flat surface such as the chest, back, flank or upper arm		
Granisetron	Transdermal patch extended release	Sancuso	Prostraken	Adhesive matrix containing drug	52-cm^2 patch containing 34.3 mg of granisetron. The patch releases 3.1 mg of granisetron per 24 h for up to 7 days Applied to upper outer arm	Rx	Chemotherapy-induced nausea and vomiting
Methylphenidate	Transdermal patch extended release	Daytrana	Shire	Adhesive-based matrix type patch	12.5–37.5-cm^2 patches deliver 10–30 mg/9 h per patch applied to the hip area 2 h before an effect is needed and should be removed 9 h after application	Schedule II	Attention-deficit hyperactivity disorder
Nicotine	Transdermal patch extended release	Nicoderm CQ	Aventis	Matrix type patch	7–21 mg over 24 h at different stages of the treatment	OTC	Smoking cessation

Continued

Table 12.2. Continued.

Active ingredients	Type of delivery system	Name	Company	Type of patch	Dose and application	Regulatory status	Uses
		Nicotine transdermal system	Novartis Consumer, Watson Laboratories, Cardinal Health and Aveva	Matrix type patch	7–21 mg/24 h at different stages of treatment		
		Habitrol	Novartis and Novartis Consumer	Reservoir type	17.5–52.5 mg that delivers 7–21 mg/day for the duration of treatment		
Nitroglycerin	Transdermal patch extended release	Nitro Dur	Key Pharmaceuticals	Drug in adhesive	5–40-cm^2 patch delivers 0.1–0.8 mg/h for 12–14 h	Rx	Angina prophylaxis
		Nitroglycerin	Noven, Hercon Laboratories, Kremers Urban and Mylan Technologies	Drug in adhesive	Delivers nitroglycerin at 0.2 mg/h		
		Minitran Transdermal System	Graceway Pharmaceuticals	Drug in adhesive	Delivers 0.1–0.6 mg/h		
	Transdermal ointment	Nitroglycerin	Fougera	2%	7.5–30 mg applied in the morning and again 6 h later to a 36-$inch^2$ area of truncal skin		
Oxybutynin	Transdermal patch extended release	Oxytrol	Watson Laboratories	Adhesive matrix containing drug	39 cm^2 system containing 36 mg and has a nominal *in vivo* delivery rate of 3.9 mg oxybutynin per day consistently for 3–4 days Applied to abdomen, hip or buttock	Rx	Bladder muscle dysfunction

Oxybutynin chloride	Transdermal gel	Gelnique	Watson Labs	10% gel	100 mg applied once daily to dry, intact skin on the abdomen, upper arms/ shoulders or thighs (area of application rotated)	Rx	Bladder muscle dysfunction
Rivastigmine	Transdermal patch extended release	Exelon	Novartis	Matrix reservoir containing drug	4.6–9.5 mg/24 h from 5–10-cm^2 patches Preferable application to upper or lower back	Rx	Dementia associated with Alzheimer's disease and Parkinson's disease
Scopolamine	Transdermal patch extended release	Transderm Scop	Novartis	Matrix reservoir containing drug	2.5-cm^2 patch delivers 1.0 mg for 3 days Applied to the hairless area behind one ear	Rx	Motion sickness, Postoperative nausea and vomiting (prophylaxis).
Selegiline	Transdermal patch extended release	Emsam	Somerset	Drug in adhesive	6–12 mg/24 h from 20–40-cm^2 patch Applied to the upper torso, upper thigh or the outer surface of the upper arm	Rx	Major depressive disorder
Testosterone	Transdermal patch extended release	Androderm	Watson Laboratories and Watson Pharma	Reservoir type	2.5 or 5 mg/day from 37–44-cm^2 patch	Schedule III Rx	Hypogonadism (testosterone deficiency)
	Transdermal gel	Androgel	Unimed Pharma and Solvay/ Abbott	1% gel	5–10 g contains 50–100 mg, 10% of the applied testosterone dose is absorbed across skin of average permeability during a 24-h period Applied 5 g once daily to shoulders and upper arms and/or abdomen		

OTC, Over-the-counter drug; Rx, Prescription drug.

both purposes. Poly ethylene glycol (PEG) polymers are crosslinked networks with enhanced biocompatibility. They are, therefore, a good choice for various biomedical applications. A large number of proteins are delivered by PEGs which can be crosslinked by the use of a urethane–allophanate bond using tris(6-isocyanatohexyl) isocyanurate. This forms polymeric networks which swell in phosphate buffer and ethanol, and form good gels; they release the solutes in a biphasic technique (Bromberg, 1996).Transdermal delivery systems have also employed the use of acrylic-acid matrices and plasticizers to make drug–polymeric matrix films. Reported polymers include Eudragit RS PM, Eudragit RL PM, Eudragit E-100 and Eudragit S-100. A square root time model describes the release of drugs by such polymeric systems (Minghetti *et al.*, 1999).

Ethyl cellulose and polyvinylpyrrolidone matrix films with plasticizers such as 30% dibutyl phthalate deliver indomethacin and diltiazem hydrochloride in transdermal delivery systems. The release rates of these polymers are enhanced by hydrophilic components such as ethyl cellulose. Higher dissolution rates are attributed to the leaching of the soluble component, decreasing the path length of diffusion of drugs (Rao and Diwan, 1998). Hydroxypropyl methylcellulose (HPMC) is a hydrophilic derivative of cellulose which swells and is used widely for oral drug delivery systems. It is used to form a matrix in transdermal patches made from propranolol hydrochloride. This polymer yields absolutely clear films because of the increased solubility of the drug in it. Fast release of drugs through HPMC-based transdermal matrices is due to the fact that it is hydrated and swells easily (Guyot and Fawaz, 2000). Table 12.3 shows the various diseases which are targeted through TDDS worldwide.

The use of rate-controlling membranes is an important aspect of transdermal patches. Reservoir-type transdermal systems possess an inert membrane which encloses an active agent that diffuses through the membrane at a particular finite and controllable rate. These release rate membranes are essentially non-porous and release the drug directly. The membrane thickness and solubility of the drug in the membrane are two important factors that determine the release passage of drugs. Therefore the choice of membrane is an important parameter and it should conform with the drug being delivered. Membranes are made of various materials, including EVA, silicone rubber, polyurethane, polyisobutylene and polyacrylate.

Future Considerations

Transdermal drug systems are potentially one of the fastest growing sectors in the pharmaceutical industry. They are effective in bringing about precise, controlled release of a drug at the site of delivery, and further research into these systems may be expected. A number of drugs are incompatible with transdermal patch delivery, and alternative systems are needed for these, which is unfortunate since TDDS is non-toxic and one of the safest means of administering drugs. One of the challenges to transdermal drug delivery is the nature of the skin, which poses various barriers to drug transport through these patches. Despite the use of skin permeation techniques, and a variety of topical delivery systems, delivery remains a challenge. More research and focus on such deliveries will enhance the application of TDDS; for example, it has been found that various nanoparticles have enhanced potential to deliver the drug at the site but some do not penetrate the entire Stratum corneum (SC) layer of the skin. Understanding the exact mechanism by which the carriers act as perfect skin drug delivery agents still needs to be explored.

Acknowledgements

The authors are thankful to Dr Ashok K. Chauhan, Founder President, Ritanand Balved Educational Foundation, for support. They are also thankful to Shri. Atul Chauhan, Chancellor, Amity University Uttar Pradesh and Prof. Dr (Mrs) Balvinder Shukla, Vice Chancellor, Amity University Uttar Pradesh, for their constant support and guidance. The authors acknowledge the immense help received from the scholars whose articles are cited and included in references of this manuscript.

Table 12.3. Transdermal drugs approved by the US FDA. (Reproduced, with permission, from Prausnitz and Langer, 2008.)

Approval year	Drug	Indication	Product Name	Marketing company
1979	Scopolamine	Motion sickness	Transderm-Scop	Novartis Consumer Health (Parsippany, NJ)
1981	Nitroglycerin	Angina pectoris	Transderm-Nitro	Novartis (East Hannover, NJ)
1984	Clonidine	Hypertension	Catapres-TTS	Boehringer Ingelhim (Ridgefield, CT)
1986	Estradiol	Menopausal symptoms	Estraderm	Novartis (East Hannover, NJ)
1990	Fentanyl	Chonic pain	Duragesic	Janssen Phamaceutica (Titusville, NJ)
1991	nicotine	Smoking cessation	Nicoderm, Habitrol, ProStep	GlaxoSmithKline (Philadelphia, PA), Novartis Consumer Health (Parsippany, NJ) Elan (Gainesville, GA)
1993	Testosterone	Testosterone deficiency	Testoderm	Alza, Mountain View, CA
1995	Lidocaine/epinephrine (iontophoresis)	Local dermal analgesia	Iontocaine	Iomed (Salt Lake City, UT)
1998	Estradiol/norethidrone	Menopausal symptoms	Combipatch	Novartis (East Hannover, NJ)
1999	Lidocaine	Post-herpetic neuralgia pain	Lidoderm	Endo Phamaceuticals (Chadds Ford, PA)
2001	Ethinyl estradiol/ norelgestromin	Contraception	Ortho Evra	Ortho-McNeil Pharmaceutical (Raritan, NJ)
2003	Estradiol/levonorgestrel	Menopausal symptoms	Climara Pro	Bayer Healthcare Pharmaceuticals (Wayne, NJ)
2004	Lidocaine (ultrasound)	Local dermal anaesthesia	SonoPrep	Echo Therapeutics (Franklin, MA)
2005	Lidocaine/tetracaine	Local dermal analgesia	Synera	Endo Pharmaceuticals (Chadds Ford, PA)
2006	Fentanyl HCl (iontophoresis)	Acute postoperative pain	Ionsys	Alza, Mountain View, CA
2006	Methylphenidate	Attention deficit hyperactivity disorder	Daytrana	Shire (Wayne, PA)
2006	Selegiline	Major depressive disorder	Emsam	Bristol-Myers Squibb (Princeton, NJ)
2007	Rotigotine	Parkinson's disease	Neupro	Schwarz Pharma (Mequon, WI)
2007	Rivastigmine	Dementia	Exelon	Novartis (East Hannover, NJ)

References

Allen, L.V., Popovich, N.G. and Ansel, H.C. (2005) *Ansel's Pharmaceutical Dosage Forms and Drug Delivery Systems*, 8th edition. Lippincott Williams & Wilkins, Baltimore, pp. 298–315.

Amnuaikit, C., Ikeuchi, I., Ogawara, K., Higaki, K. and Kimura, T. (2005) Skin permeation of propranolol from polymeric film containing terpene enhancers for transdermal use. *International Journal of Pharmaceutics* 289, 167–178.

Arvanitoyannis, I., Kolokuris, I., Nakayama, A., Yamamoto, N. and Aiba, S. (1997) Physicochemical studies of chitosan-polyvinylalcohol blends plasticized with sorbitol and sucrose. *Carbohydrate Polymers* 34, 9–19.

Aulton, M.E. (2007) Plasticizers in transdermal drug delivery systems. In: Taylor, K. (ed.) *Aulton's Pharmaceutics: The Design and Manufacture of Medicines*. 3rd edn. Elsevier, London, pp. 565–597.

Barry, B. (2002) Transdermal drug delivery. In: Aulton, M.E. (ed.) *Pharmaceutics: The Science of Dosage Form Design*. Churchill Livingston, London, pp. 499–533.

Bergo, P.V.A. and Sobral, P.J.A. (2007) Effects of plasticizer on physical properties of pig skin gelatin films. *Food Hydrocolloids* 21(8), 1285–1289.

Bromberg, L. (1996) Crosslinked poly(ethylene glycol) networks as reservoirs for protein delivery. *Journal of Applied Polymer Science* 59(3), 459–466.

Brown, M., Martin, G., Jones, S. and Akomeah, F. (2006) Dermal and transdermal drug delivery systems: current and future prospects. *Drug Delivery* 13(3), 175–187.

Delgado-Charro, M.B. and Guy, R.H. (2001) Transdermal drug delivery. In: Hillery, A.M., Lloyd, A.W. and Swarbrick, J. (eds) *Drug Delivery and Targeting for Pharmacists and Pharmaceutical Scientists*. Taylor & Francis, London, pp. 189–214.

Finnin, B. and Morgan, T. (1999) Transdermal penetration enhancers: applications, limitations, and potential. *Journal of Pharmaceutical Sciences* 88(10), 955–958.

Gaikwad, A.K. (2013) Transdermal drug delivery system: Formulation aspects and evaluation. *Comprehensive Journal of Pharmaceutical Sciences* 1(1), 1–10.

Güngör, S., Bektas, A., Alp, F.I., Uydes-Dogan, B.S., Özdemir, O., Araman, A. and Özsoy, Y. (2008) Matrix type transdermal patches of verapamil hydrochloride: in vitro permeation studies through excised rat skin and pharmacodynamic evaluation in rats. *Pharmaceutical Development and Technology* 13(4), 283–289.

Güngör, S., Sedef Erdal, M. and Özsoy, Y. (2012) Plasticizers in transdermal drug delivery systems. In: Luqman, M. (ed.) *Recent Advances in Plasticizers.* Intech, Rijeka. Available at: http://www.intechopen.com/books/recent-advances-in-plasticizers/plasticizers-in-transdermal-drug-delivery-systems- (accessed 4 July 2016).

Guy, R.H. (1996) Current status and future prospects of transdermal drug delivery. *Pharmaceutical Research* 13(12), 1765–1769.

Guy, R.H. (2010) Transdermal drug delivery. In: Schäfer-Korting, M. (ed.) *Drug Delivery, Handbook of Experimental Pharmacology*. Springer, Berlin.

Guyot, M. and Fawaz, F. (2000) Design and in vitro evaluation of adhesive matrix for transdermal delivery of propranolol. *International Journal of Pharmaceutics* 204(1–2), 171–182.

Kumar, T.M.P., Umesh, H.M., Shivakumar, H.G., Ravi, V. and Siddaramaiah, V. (2007) Feasibility of polyvinyl alcohol as a transdermal drug delivery system for terbutaline sulphate. *Journal of Macromolecular Science, Part A* 44(6), 583–589.

Lin, S.Y., Lee, C.J. and Lin, Y.Y. (1991) The effect of plasticizers on compatibility, mechanical properties and adhesion strength of drug free Eudragit E films. *Pharmaceutical Research* 8(9), 1137–1143.

Meathrel, W.G. (2010) Challenges in formulating adhesives for transdermal drug delivery. *Transdermal Magazine* 9, 1–4.

Minghetti, P., Cilurzo, F., Casiraghi, A., Molla, F. and Montanari, L. (1999) Dermal patches for the controlled release of miconazole: influence of the drug concentration on the technological characteristics. *Drug Development and Industrial Pharmacy* 25(5), 679–684.

Nicoli, S., Colombo, P. and Santi, P. (2005) Release and permeation kinetics of caffeine from bioadhesive transdermal films. *American Association of Pharmaceutical Scientists Pharmaceutical Science and Technology* 7(1), 218–223.

Nicoli, S., Penna, E., Padula, C., Colombo, P. and Santi, P. (2006) New bioadhesive transdermal film containing oxybutynin: in vitro permeation across rabbit ear skin. *International Journal of Pharmaceutics* 325(1–2), 2–7.

Padula, C., Colombo, G., Nicoli, S., Catellani, P.L., Massimo, G. and Santi, P. (2003) Bioadhesive film for the transdermal delivery of lidocaine: in vitro and in vivo behavior. *Journal of Controlled Release* 3(88), 277–285.

Padula, C., Nicoli, S., Aversa, V., Colombo, P., Falson, F., Pirot, F. and Santi, P. (2007) Bioadhesive film for dermal drug delivery. *European Journal of Dermatology* 17(4), 309–312.

Paudel, K.S., Milewski, M., Swadley, C.L., Brogden, N.K., Ghosh, P. and Stinchcomb, A.L. (2010) Challenges and opportunities in dermal/transdermal delivery. *Therapeutic Delivery* 1(1), 109–131.

Prausnitz, N.R. and Langer, R. (2008) Transdermal drug delivery. *Nature Biotechnology* 26, 1261–1268.

Quan, D. (2011) Passive transdermal drug delivery systems (TDDS): challenges and potential. *Transdermal Magazine* 3, 6–12.

Rao, P. and Diwan, P. (1998) Formulation and in vitro evaluation of polymeric films of diltiazem hydrochloride and indomethacin for transdermal administration. *Drug Development and Industrial Pharmacy* 24(4), 327–336.

Schroeder, I.Z., Franke, P., Schaefer, U.F. and Lehr, C.M. (2007a) Delivery of Ethinylestradiol from film forming polymeric solutions across human epidermis in vitro and in vivo in pigs. *Journal of Controlled Release* 118, 196–203.

Schroeder, I.Z., Franke, P., Schaefer, U.F. and Lehr, C.M. (2007b) Development and characterization of film forming polymeric solutions for skin drug delivery. *European Journal of Pharmaceutics and Biopharmaceutics* 65, 111–121.

Tanner, T. and Marks, R. (2008) Delivering drugs by the transdermal route: review and comment. *Skin Research and Technology* 14(3), 249–260.

Tiwary, A.K., Sapra, B. and Jain, S. (2007) Innovations in transdermal drug delivery: formulations and techniques. *Recent Patents on Drug Delivery and Formulation* 1(1), 23–36.

Vasil'ev, A.E., Krasnyuk, I.I. and Tokhmakhchi, V.N. (2001) Transdermal therapeutic systems for controlled drug release. *Pharmaceutical Chemistry Journal* 35, 613–626.

Walters, K.A. and Brain, K.R. (2002) Dermatological formulation and transdermal systems. In: Walter, K.A. (ed.), *Dermatological and Transdermal Formulations*. Marcel Dekker, New York, pp. 319–400.

Wang, F.J., Yang, Y.Y., Zhang, X.Z., Zhu, X., Chung, T. and Moochala, S. (2002) Cellulose acetate membranes for transdermal delivery of scopolamine base. *Materials Science and Engineering* 20(2), 93–100.

Williams, A. (2003) *Transdermal and Topical Drug Delivery: From Theory to Clinical Practice*. Pharmaceutical Press, London, pp. 169–194.

13 Ocular Drug Delivery Systems

Bhanu Malhotra[1,*], Harsha Kharkwal[2] and Anupam Pradhan[3]
[1]Amity Institute of Biotechnology and Amity Center for Carbohydrate Research, Amity University Uttar Pradesh, Noida, India; [2]Amity Center for Carbohydrate Research and Amity Institute of Phytomedicine and Phytochemistry, Amity University Uttar Pradesh, Noida, India; [3]Global Health, College of Public Health University of South Florida, Tampa, Florida, USA and Queensborough Community College, City University of New York, Bayside, New York, USA

Abstract

Topical eye drugs are the most convenient and conventional ways of drug administration to the eyes, especially in the cases of anterior segment ailments. Drug delivery is restricted due to the presence of various static barriers such as the presence of the corneal layer, sclera, retina, blood retina barriers, and certain dynamic barriers including lymphatic clearance, conjunctival blood flow and tear dilution. A major challenge of the ocular drug systems is the delivery of drugs to the posterior segments of the eye. In recent years certain influx transporters to the ocular tissues have been researched and discovered. Liposome-, nanoparticle- and nanomicelle-mediated drug transport can overcome static and dynamic barriers to drug delivery in the eye. The use of biodegradable polymer materials as novel drug carriers for sustained release of the drug at the target site is nowadays a thoroughly researched field. Non-invasive biopolymer-based ocular drug delivery systems, which overcome all the limitations of topical delivery, are attracting considerable interest. This chapter presents a detailed description of various biopolymers used in ocular delivery strategies, and discusses their promising future.

Introduction

One of the major challenges posed by topical drug delivery or conventional methods of drug delivery is that there are various static and dynamic barriers to treating the anterior segment of the eye. In addition, drug levels are not maintained for long in the target tissues of the eye (Le Bourlais *et al.*, 1998). Strategies such as the use of viscosity enhancers, ointments, suspensions and emulsions for topical drug delivery are used to overcome such limitations, but the current era of ocular drug delivery is based on the use of novel drug carriers such as biopolymers and the sustained release of drugs, especially for the posterior segment of the eye. Certain novel nanoformulations are being used for treatment of chronic vitreoretinal diseases of the posterior segment. Such nanoformulations overcome not only the ocular barriers but also the limitations of topical delivery. The new strategies of using biopolymers involve high precorneal residence time and sustained drug release, consequently enhancing the bioavailability of the ocular drugs (Gaudana *et al.*, 2008, 2010).

*Corresponding author. E-mail: malhotrabhanu91@gmail.com

Anatomy of the Eye: Barriers to Ocular Drug Delivery

The eye exhibits a unique anatomy and physiology. The overall structure of the eye is divided into the anterior and the posterior segments, as shown in Fig. 13.1. Only one-third is referred to as the anterior segment; the rest of the space is occupied by the posterior segment. The anterior portion is made up of conjunctiva, cornea, aqueous humour, ciliary body, iris and the lens; while the posterior part constitutes choroid, sclera, retinal pigment epithelial lining, vitreous humour, neural retina and optic nerve. Both segments are subject to the occurrence of various vision-threatening diseases. Cataract, allergic conjunctivitis and anterior uveitis affect the anterior portion of the eye; and diseases such as age-related macular degeneration and retinopathy due to diabetes affect the posterior part.

Conventional Methods of Ocular Drug Delivery

Ophthalmic drug delivery is successfully achieved by overcoming the physiological and anatomical barriers. The major ocular drug delivery procedures include local, topical and systemic administration. Because of its ease of administration, topical administration is the most preferred route of drug delivery for anterior segment diseases. However, a constraint is that less than 5% of the drug is available at the target site due to the limited bioavailability of the topical drugs. This poor drug bioavailability is caused by blinking, nasolacrimal drainage and transient delay in the cul-de-sac (a pouch-like tubular cavity which is closed at one end). The physiochemical properties of the drug are also important parameters, which determine the impermeability and diffusion resistance of various ocular tissues (Alexander, 1965). There are various routes for drug administration including topical, injectable and systemic. Table 13.1 presents the summary of available routes of administration, the benefits they provide and the various challenges they pose.

For posterior segment diseases, local and systemic modes of drug delivery are generally preferred due to poor corneal drug permeation. Subconjunctival and intravitreal injections are local drug delivery procedures for vitreoretinal diseases. Higher dosage and repeated administration are required for systemic drug delivery. Sometimes these repeated doses for maintaining the concentration can cause clinical complications and discomfort (Urtti, 2006). The presence of breast cancer-resistant protein and efflux pumps for P-glycoprotein on the ocular tissues restrict the proper flow of drug to the eye (Weiner and Gilger, 2010).

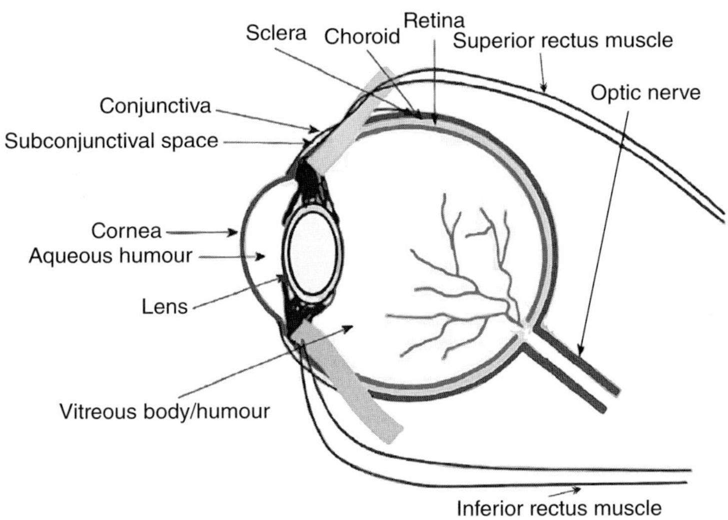

Fig. 13.1. Structure of the human eye.

Table 13.1. Summary of routes of administration, benefits, and challenges in ocular delivery. (Reproduced, with permission, from Gaudana *et al.*, 2010.)

Route	Benefits	Challenges	Application in the treatment of diseases
Topical	High patient compliance, self-administrable and noninvasive	Higher tear dilution and turnover rate, cornea acts as barrier, efflux pumps, BA <5%	Keratitis, uveitis, conjunctivitis, scleritis, episcleritis, blepharitis
Oral/Systemic	Patient compliant and noninvasive route of administration	BAB, BRB, high dosing causes toxicity, BA <2%	Scleritis, episcleritis, CMV retinitis, PU
Intravitreal	Direct delivery to vitreous and retina, sustains drug levels, evades BRB	Retinal detachment, hemorrhage, cataract, endophthalmitis, patient incompliance	AMD, PU, BRVO, CRVO, DME, CME, UME, CMV retinitis
Intracameral	Provides higher drug levels in the anterior chamber, eliminates usage of topical drops, reduces corneal and systemic side effects seen with topical steroid therapy	TASS, TECDS	Anaesthesia, prevention of endophthalmitis, inflammation and pupil dilation
Subconjunctival	Delivery to anterior and posterior segment, site for depot formulations	Conjunctival and choroidal circulation	Glaucoma, CMV retinitis, AMD, PU
Subtenon	High vitreal drug levels, relatively noninvasive, fewer complications unlike intravitreal delivery	RPE, chemosis, subconjunctival hemorrhage	DME, AMD, RVO, uveitis
Retrobulbar	Administer high local doses of anaesthetics, more effective than peribulbar, minimal influence on IOP	Retrobulbar hemorrhage, globe perforation, respiratory arrest	Anaesthesia
Posterior juxtascleral	Safe for delivery of depot formulations, sustain drug levels up to 6 months to the macula, avoids risk of endophthalmitis and intraocular damage	Requires surgery and RPE acts as barrier	AMD

BA, bioavailability; BAB blood–aqueous barrier; BRB, blood–retinal barrier; AMD, age-related macular degeneration; DME, diabetic macular edema;, BRVO, branched retinal vein occlusion; CRVO, central retinal vein occlusion; RVO, retinal vein occlusion; CME, cystoid macular edema; UME, uveitic macular edema; CMV, cytomegalovirus; IOP, intraocular pressure; TASS, toxic anterior segment syndrome; TECDS, toxic endothelial cell destruction syndrome; RPE, retinal pigmented epithelium; PU, posterior uveitis.

Controlled drug release is a potential approach, releasing drug slowly over a prolonged period of time at the target site, and it overcomes all the limitations of conventional drug delivery systems. It avoids repeated drug administration, which is problematic in many drug delivery systems and especially in chronic eye disorders. It can also be used with drugs with very short half-lives by protecting them in situ (Shell, 1984; Daugherty and Mrsny, 2003). The first controlled release device was developed in the 1960s and was based on erodible or non-erodible matrices. A number of synthetic polymers such as Polyglycolic acid (PGA) and Polylactic acid (PLA) have been used in pharmaceutical applications, but the focus has now shifted to the use of biodegradable polymers for ocular drug delivery. Liposomes, microspheres, hydrogels and nanoparticles have been used for controlled drug release strategies (Park *et al.*, 2005). The release of the drug from these agents is due to the change in the physicochemical property of the drug, degradation of the polymeric matrices for drug release, molecular weights of drugs, or the shape and sizes of the polymers carrying the drugs (Weiner and Gilger, 2010).

Biodegradable Polymers in Ocular Drug Delivery

Biodegradation is defined as an enzymatic or a non-enzymatic process of hydrolysis of the polymer backbone into simpler compounds and water-soluble products. Biodegradation occurs with the involvement of complementary processes such as degradation and erosion. Degradation involves cleavage of the polymer backbone into low molecular weight fractions, and erosion involves diffusion and dissolution of the low molecular weight fractions formed in degradation. These simpler products that result from biodegradation are excreted from the body via normal metabolic pathways (Katti *et al.*, 2002).

Types of Biodegradation

Polymers can be classified on the basis of three basic mechanisms of biodegradation (Fig. 13.2) (Heller, 1984).

Type I polymers

Polymers which follow type I erosion are crosslinked and water soluble. Examples include collagen, gelatin, poly vinyl pyrrolidone and poly vinyl alcohol; these form a water-insoluble crosslinking structure in the form of a hydrogel. Biodegradation of type I polymers generates: (i) type IA polymers, which are of higher molecular weight; and (ii) type IB polymers, which are of lower molecular weight. Research into the mechanisms of biodegradation has revealed that type IA polymers are good for topical administration of drugs while type IB polymers are suitable for the formation of implants. These type I polymers are water soluble and unsuitable for drugs with appreciable water solubility, but can be used to deliver antigens and enzymes which can be released from the hydrogels through simple diffusion processes, followed by degradation of the polymer (Heller, 1984).

Type II polymers

The polymers that follow type II erosion are initially water insoluble and are converted into water soluble polymers using simple hydrolysis, protonation and ionization of the chemical groups. The molecular weight of the polymers remains the

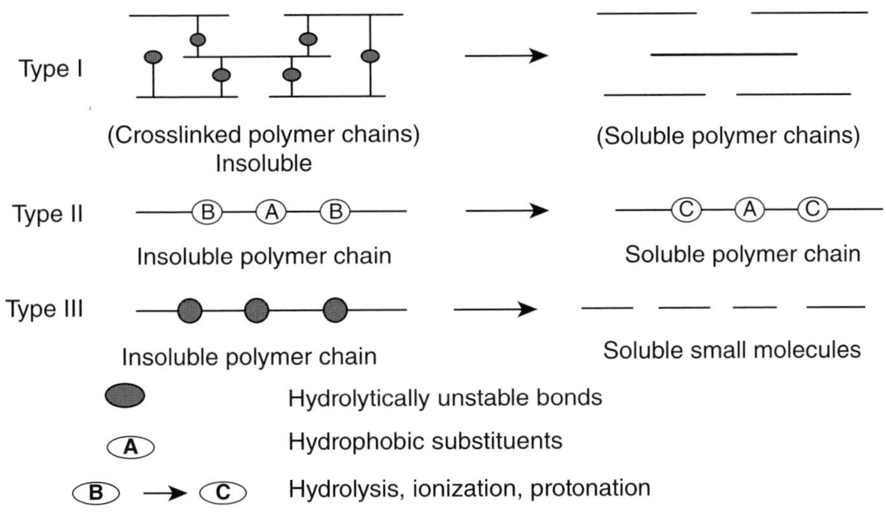

Fig. 13.2. Classification of polymers on the basis of biodegration mechanisms.

same since no backbone degradation is involved. They are also used for topical application. Type II erosion mechanism polymers are those of alkyl vinyl ethers (Woodruff *et al.*, 1972).

Type III polymers

This type of degradation generates low molecular weight molecules through cleavage of water-insoluble aggregates. The products produced by them are completely safe and non-toxic and are therefore used for systemic and topical administration. They are used to treat chronic ocular diseases and for post-surgery wound healing. They have a wide range of distributed physicochemical properties and are moulded easily for drug delivery. Examples include copolymers of PLA, poly caprolactones and poly ortho esters (Heller, 1984).

Transporters in the Eye

Drugs can be modified to obtain the appropriate solubility and lipophilicity to increase their bioavailability. Another rational approach is the exploitation of various transporters present in the eye. Transporters are present in the membrane which facilitates the active transport of substances across the biological membranes. A number of transporters are distributed across the ocular tissues, of which the conjunctiva, retina and cornea are particularly important in drug delivery. Transporters are classified into two types: (i) influx, belonging to the solute carrier superfamily (SLC); and (ii) efflux, belonging to the Adenosine Triphosphate-binding cassette family (ATP). Efflux transporters include multi-drug-resistant protein (MRP) and breast cancer-resistant protein (BCRP) (Dey *et al.*, 2003). The traditional approach to improve ocular bioavailability is to modify the drug chemically to achieve the desired solubility and lipophilicity. However, a more rational approach would be a transporter-targeted modification of the drug. Transporters are membrane-bound proteins that play an important role in active transport of nutrients across biological membranes. The presence of transporters has been reported in various ocular tissues (Yang *et al.*, 2007).

Influx transporters supply essential nutrients, substances, solutes and xenobiotics across biological membranes. The basic ones include peptide transporters and those for glucose, vitamins, lactate, nucleosides, etc. Valacyclovir was transported using the influx transporters for targeted prodrug delivery. The influx transporters facilitate the solubility of the drug by enhancing absorption of poorly permeating drugs. The influx transporters most used include peptide and amino acid transporters. Amino acid transporters are SLC1, 6, 7 gene families in ocular tissues (Katragadda *et al.*, 2005). Certain transporter targeting prodrugs have been made for ocular drug delivery procedures, exploiting the presence of influx and efflux transporters. These studies of ocular transporters have proved be a boon for the improvement of bioavailability of poorly permeating parent drug moieties. Prodrugs are recognized by the membrane transporters as substrates resulting in their translocation across the epithelia (Saha *et al.*, 1996). Figure 13.3 shows a schematic diagram of efflux evasion by a prodrug mediated by the influx transporter, and Table 13.2 lists various transporters along with their prodrugs.

Biodegradable Polymers in Ocular Drug Delivery

Biodegradable polymers are important interventions for controlled release of drugs in ocular tissues as they need not be surgically removed post-application, so they avoid all the problems caused by the use of non-biodegradable polymers in surgery. Some natural and other synthetic biopolymers show good biocompatibilities with ocular tissues, and mechanical strength and biodegradability. They produce non-toxic compounds as a result of biodegradation. The wide range of properties exhibited by biodegradable polymers means they can easily be tailored to carry drugs to ocular tissues. Polymers such as gelatin, albumin, chitosan and some synthetic polymers such as Polyvinylpyrrolidone (PVP), poly ($\in$-caprolactone) (PCL) and Poly(vinyl alcohol) (PVA) have shown good potential for ocular drug delivery systems. Some polymers approved by the US Food and Drug Administration (FDA) for ocular drug delivery are presented in Table 13.3.

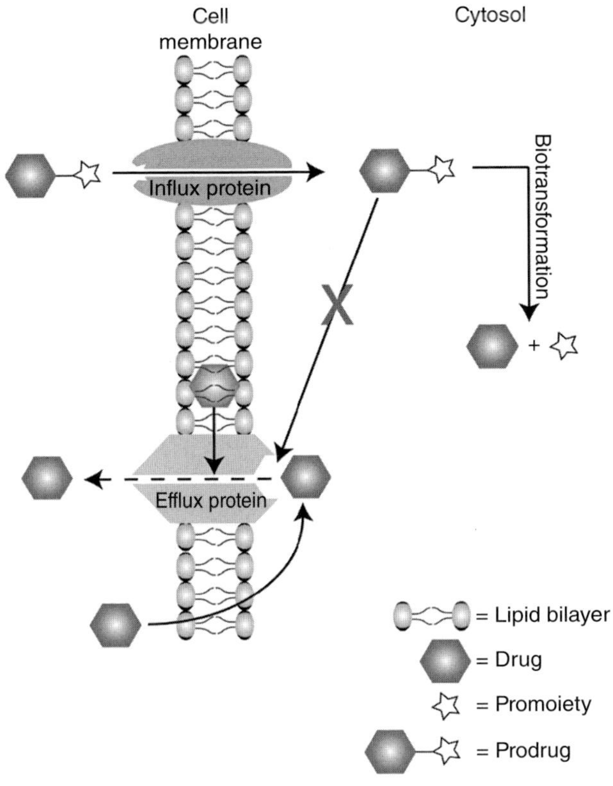

Fig. 13.3. Circumvention of efflux proteins by prodrug approach. (Reproduced with permission, from Gaudana *et al.*, 2010.)

Gelatin

Gelatin is made from collagen and shows good biocompatibility and biodegradability compared with the native protein, collagen. It is an FDA-approved polymer for drug delivery, widely used in pharmaceutical applications, and is present in two forms, type A and type B. Type A gelatin is an acid-processed form of collagen and type B is an alkaline-treated collagen. These have different isoelectric points in the range of 7–9 (type A) and 4–5 (type B). Hydrogels can be made from gelatin by chemical and thermal modification. Hydrogels are a unique class of macromolecular networks that may contain a large fraction of aqueous solvent within their structure. They are particularly suitable for biomedical applications because of their ability to simulate biological tissues (Park *et al.*, 1993). The electrical and physical properties of gelatin can be changed for ocular drug delivery procedures (Young *et al.*, 2005). A water solution of gelatin can be converted to thermoreversible hydrogels, and these hydrogels show good mechanical strength at 30°C, resulting in drug release. Crosslinking of gelatin chains provide thermal stability. Considerable work has been done to characterize the crosslinked carbodiimide gelatin hydrogel filled with pilocarpine hydrochloride for treating disorders such as glaucoma. Water uptake by the hydrogel was found to decrease considerably by crosslinking, resulting in the slow release of pilocarpine from the matrix of the gel. Lyophilisates of gelatin also showed similar results from crosslinking (Natu *et al.*, 2007). Another intervention with gelatin includes the preparation of nanoparticles encapsulating hydrocortisone steroid and pilocarpine hydrochloride for topical administration. Gelatin showed a prolonged residence time over the ocular tissues, and is used in the form of microspheres and microcapsules for ocular drug delivery (Leucuta *et al.*, 1997).

Table 13.2. Transporter-targeted prodrugs for ocular drug delivery. (Reproduced, with permission, from Gaudana *et al.*, 2010.)

Transporter system-targeted tissue/cell line	Drug/prodrugs employed	Observation
B(0,+) on the cornea	L-aspartate ACV	Fourfold higher transcorneal permeability of L-aspartate ACV compared to ACV
B(0,+) on the cornea	Gamma-glutamate-ACV (EACV)	Higher aqueous solubility of the prodrug along with the transporter recognition
B(0,+) on the cornea	Phenylalanine-ACV and gamma-glutamate-ACV	The prodrugs inhibited the transport of L-arginine[a] across the cornea implying that they are substrates of B(0,+)
OPT system on the cornea	L-valine ACV	Threefold higher transcorneal permeability of L-valine ACV compared to ACV
OPT system on the cornea	Gly-Val-GCV, Val-Val-GCV, and Tyr-Val-GCV	Significant transcellular passive diffusion and transporter recognition resulted in higher AUC and C_{max}
OPT system on rPCEC cells and the cornea	Val-quinidine and Val-Val-quinidine	Prodrugs were not recognized by P-gp efflux pump and further found to be substrates of peptide transporters
OPT system on the retina	Gly-Val-GCV, Val-Val-GCV, and Tyr-Val- GCV	Twofold higher RCS tissue permeability than that of GCV due to higher lipophilicity and translocation mediated by OPT across RPE
SMVT on the retina	Biotin-GCV	Higher biotin-GCV permeability into the retina–choroid and slower elimination from vitreous
GLUT1 on the HRPE cells	Glu-dopamine	Transporter recognizes prodrug, not the parent drug

OPT, oligopeptide transporter; SMVT, sodium-dependent multiple vitamin transporter; B(0,+), amino acid transporter; GLUT, glucose transporter; rPCEC, rabbit primary corneal epithelial cells; HRPE, human retinal pigment epithelium cells; RCS, retina–choroid–sclera; ACV, acyclovir; GCV, ganciclovir; RPE, retinal pigment epithelium.
[a]A substrate of B(0,+)

Collagen

Collagen, a structural biopolymer, is both biodegradable and biocompatible, and is distributed all over the ocular tissues. The sclera, cornea and stroma are made up of collagen only. It provides support and is an important structural biopolymer in animal tissues. Type I collagen is the most important protein in humans. Collagen can be used in drug delivery applications in humans and its source can be from animal origin (e.g. bovine and porcine) because these collagens are structurally similar to the human form. Collagen can be degraded *in vitro* by metalloproteases and collagenases, which form the simpler amino acids in the body. In ocular drug delivery collagen can be used in the form of nanofibres, powders, foam and dispersions, and in injectable solutions (Nair and Laurencin, 2007). Collagen has low immunogenicity, a good absorption capacity and excellent biocompatibility (Silbiger and Stern, 1992). It not only enhances the drug release time but improves biodegradability by behaving as a drug reservoir. Collagen-based ocular drugs also treat the dry eye syndrome. In a recent development efficacy of collagen is very high in releasing porcine sclera into the scleral part of the eye (Liu and Wang, 2013). Derivatization of collagen with succinic acid can easily be tolerated by dry-eyed patients. Tear film stability improved significantly, as did the symptomatic relief of decrease in tear production (Ros *et al.*, 1991). Some collagen shields interfere with the use of contact lenses and can cause some patient discomfort (Ros

Table 13.3. Some examples of ophthalmic drug-delivery systems approved by the US FDA. (Reproduced, with permission, from Kompella *et al*., 2010.)

Product category	Description	Drug	Year of approval	Indication
Topical delivery systems				
Ophthalmic solution (Acuvail)	Preservative free 0.45% ketorolac tromethamine solution containing carboxymethylcellulose as viscosity enhancer	Ketorolac tromethamine	2009	NSAID for the treatment of pain and inflammation after cataract surgery
Ophthalmic solution (Lumigan)	0.03% ophthalmic solution of bimatoprost with 0.005% benzalkonim chloride as preservative	Bimatoprost (prodrug)	2001	Reduction of elevated IOP
Ophthalmic solution (Travatan)	0.004% ophthalmic solution of travoprost with 0.015% benzalkonim chloride as preservative	Travaprost (prodrug)	2001	Reduction of IOP
Ophthalmic solution (Xalatan)	0.005% ophthalmic solution of latanoprost with 0.02% benzalkonim chloride as preservative	Latanoprost (prodrug)	1996	Reduction of IOP
Ophthalmic solution (Propine)	0.1% ophthalmic solution of dipivefrin-HCl with benzalkonim chloride as preservative	Dipivefrin (prodrug)	1980	Control of IOP in chronic open-angle glaucoma
In situ gel-forming solution (Timoptic-XE)	Sterile ophthalmic gel forming solution containing 0.25 or 0.5% of timolol equivalent timolol maleate, GELRITE gellan gum, tromethamine and mannitol, with 0.012% benzalkonim chloride as preservative	Timolol maleate	1993	Treatment of glaucoma
Ophthalmic solution (AzaSite)	1% sterile aqueous topical ophthalmic solution of azithromycin formulated in DuraSite (polycarbophil, disodium EDTA, sodium chloride) with 0.003% benzalkonim chloride as preservative	Azithromycin	2007	Bacterial conjunctivitis
Ophthalmic gel (Zirgan)	Sterile topical ophthalmic gel containing 0.15% of ganciclovir formulated with carbopol and 0.0075% benzalkonim chloride as preservative	Ganciclovir	2009	Herpetic keratatis
Ophthalmic emulsion (Durezol)	Sterile topical aqueous ophthalmic emulsion containing 0.05% of difluprednate formulated with boric acid, castor oil, glycerin, and polysorbate 80, with 0.1% sorbic acid as preservative	Difluprednate	2008	Steroid for treatment of inflammation and pain associated with ocular surgery
Ophthalmic emulsion (Restasis)	Sterile preservative free ophthalmic emulsion containing 0.05% of cyclosporine formulated with glycerin, castor oil, polysorbate-80 and carbomer1342.	Cyclosporine	2002	Increase tear production in dry eye diseases

Continued

Table 13.3. Continued.

Product category	Description	Drug	Year of approval	Indication
Ophthalmic suspension (Betoptic-S)	0.25% ophthalmic suspension of betaxolol formulated with poly(styrene-divinyl benzene) sulfonic acid (Amberlite IRP-69), Carbomer 934P, edetate disodium and 0.01% benzalkonim chloride as preservative	Betaxolol	1990	Treatment of glaucoma
Opthalmic inserts (Ocusert)	Pilocarpine and alginic acid enclosed in thin ethylene-vinyl acetate membrane to control release of drug for 7 days	Pilocarpine	1974	Treatment of glaucoma
Intravitreal products				
Intravitreal implant (Ozurdex)	Intravitreal implant containing 0.7 mg (700 μg) dexamethasone in the NOVADUR™ solid polymer drug delivery system	Dexamethasone	2009	Macular oedema
Intra-vitreal implant (Retisert)	Intravitreal implant containing 0.59 mg (590 μg) of flucinolone acetonide formulated with microcrystalline cellulose, polyvinyl alcohol and magnesium stearate	Flucinolone acetonide	2005	Chronic noninfectious uveitis
Intra-vitreal implant (Vitrasert)	Intravitreal implant containing 4.5 mg of ganciclovir formulated in the nonbiodegradable polymers, polyvinyl alcohol and ethylene-vinyl acetate	Ganciclovir	1996	Cytomegalovirus retinitis

EDTA, Disodium ethylene diamine tetra-acetic acid; IOP, Intraocular pressure; NSAID, non-steroidal anti-inflammatory drug.

et al., 1991). They also provide protection against bacterial, fungal and viral agents.

Chitosan

Deacetylation of chitin produces chitosan. Chemically, chitin is composed of D-glucosamine in 1-4 glycosidic linkage and is obtained from the exoskeletons of arthropods (Silbiger and Stern, 1992). *In vitro* degradation of chitosan occurs through enzymes such as papain, lysozyme and chitosanase, whereas *in vivo* degradation primarily occurs through the lysozyme enzyme (Başaran and Yazan, 2012). The rate of chitin degradation is inversely proportional to the degree of acetylation. Chitosan can be used for its ence of an amino group which can bind easily to the negatively charged sclera and corneal tissue. It also increases ocular bioavailability (Badawy, 2007). The viscoelastic and pseudoelastic properties of chitosan enhance controlled drug release across the corneal tissue (Karatas *et al.*, 2015). It enhances the permeability of corneal epithelial tissues in a reversible nature and can cause changes in the paracellular and transcellular pathways, maintaining cell integrity. Chitosan nanoparticles have also been prepared for ocular drug delivery (Paolicelli *et al.*, 2009). Chitosan nanoparticles were found to be tenfold higher in response than the suspension of the normal drug in a chitosan solution (Rajendran *et al.*, 2010).

Future Implications

Since targeted drug delivery to the ocular tissues has been a challenge for researchers for decades, tremendous efforts are being made to develop novel classes of polymer-based drug delivery systems to contribute towards ocular research. The introduction of polymer-based nanomedicines to the eyes is not only working hard to improve targeted drug delivery with controlled release but is also improving patience compliance by combaling conventional drug delivery problems to the eye. The encapsulation of the drug moiety into non-invasive nanosized carrier systems like liposomes and dendrimers are convincingly benefiting patient requirements, minimizing drug related toxicities and loss of vision. The future belongs to nanosized carrier systems and devices which increase patient compliance in drug delivery to the ocular tissues.

Acknowledgements

The authors are thankful to Dr Ashok K. Chauhan, Founder President, Ritanand Balved Educational Foundation, for support. They are also thankful to Shri. Atul Chauhan, Chancellor, Amity University Uttar Pradesh, and Prof. Dr (Mrs) Balvinder Shukla, Vice Chancellor, Amity University Uttar Pradesh, for their constant support and guidance. The authors also acknowledge the immense help received from the scholars whose articles are cited and included in references of this manuscript.

References

Alexander, J.A. (1965) Physiology of the Eye by Francis Heed Adler reviewed in *The Australian Journal of Optometry* 48(7), 216–219.

Badawy, M. (2007) Chemical modification of chitosan: synthesis and biological activity of new heterocyclic chitosan derivatives. *Polymer International* 57(2), 254–261.

Başaran, E. and Yazan, Y. (2012) Ocular application of chitosan. *Expert Opinion on Drug Delivery* 9(6), 701–712.

Daugherty, A. and Mrsny, R. (2003) Emerging technologies that overcome biological barriers for therapeutic protein delivery. *Expert Opinion on Biological Therapy* 3(7), 1071–1081.

Dey, S., Anand, B., Patel, J. and Mitra, A. (2003) Transporters/receptors in the anterior chamber: pathways to explore ocular drug delivery strategies. *Expert Opinion on Biological Therapy* 3(1), 23–44.

Gaudana, R., Jwala, J., Boddu, S. and Mitra, A. (2008) Recent perspectives in ocular drug delivery. *Pharmaceutical Research* 26(5), 1197–1216.

Gaudana, R., Ananthula, H., Parenky, A. and Mitra, A. (2010) Ocular drug delivery. *The AAPS Journal* 12(3), 348–360.

Heller, J. (1984) Biodegradable polymers in controlled drug delivery. *Critical Reviews in Therapeutic Drug Carrier Systems* 1(1), 39–90.
Karatas, A., Boluk, A. and Algan, A. (2015) Poloxamer/chitosan in situ gelling system for ocular delivery of ofloxacin. *Current Drug Therapy* 9(4), 219–225.
Katragadda, S., Talluri, R., Pal, D. and Mitra, A. (2005) Identification and characterization of a Na+-dependent neutral amino acid transporter, ASCT1, in rabbit corneal epithelial cell culture and rabbit cornea. *Current Eye Research* 30(11), 989–1002.
Katti, D., Lakshmi, S., Langer, R. and Laurencin, C. (2002) Toxicity, biodegradation and elimination of polyanhydrides. *Advanced Drug Delivery Reviews* 54(7), 933–961.
Kompella, U.B., Kadam, R.S. and Lee, V.H. (2010) Recent advances in ophthalmic drug delivery. *Therapeutic Delivery*, 1(3), 435–456.
Le Bourlais, C., Acar, L., Zia, H., Sado, P., Needham, T. and Leverge, R. (1998) Ophthalmic drug delivery systems—recent advances. *Progress in Retinal and Eye Research* 17(1), 33–58.
Leucuta, S., Ponchel, G. and Duchene, D. (1997) Oxprenolol release from bioadhesive gelatin/poly(acrylic acid) microspheres. *Journal of Microencapsulation* 14(4), 511–522.
Liu, T. and Wang, Z. (2013) Collagen crosslinking of porcine sclera using genipin. *Acta Ophthalmologica* 91(4), e253–e257.
Nair, L. and Laurencin, C. (2007) Biodegradable polymers as biomaterials. *Progress in Polymer Science* 32(8–9), 762–798.
Natu, M., Sardinha, J., Correia, I. and Gil, M. (2007) Controlled release gelatin hydrogels and lyophilisates with potential application as ocular inserts. *Biomedical Materials* 2(4), 241–249.
Paolicelli, P., de la Fuente, M., Sánchez, A., Seijo, B. and Alonso, M.J. (2009) Chitosan nanoparticles for drug delivery to the eye. *Expert Opinion on Drug Delivery* 6(3), 239–253.
Park, J., Ye, M. and Park, K. (2005) Biodegradable polymers for microencapsulation of drugs. *Molecules* 10(1), 146–161.
Park, K., Shalaby, W. and Park, H. (1993) *Biodegradable Hydrogels for Drug Delivery*. Technomic, Lancaster, Pennsylvania, USA.
Rajendran, N., Natrajan, R., Kumar, R. and Selvaraj, S. (2010) Acyclovir-loaded chitosan nanoparticles for ocular delivery. *Asian Journal of Pharmaceutical Sciences* 4(4), 220.
Ros, F.E., Tijl, J.W. and Faber, J.A. (1991) Bandage lenses: collagen shield vs. hydrogel lens. *CLAO Journal* 17(3), 187–190.
Saha, P., Uchiyama, T., Kim, K. and Lee, V. (1996) Permeability characteristics of primary cultured rabbit conjunctival epithelial cells to low molecular weight drugs. *Current Eye Research* 15(12), 1170–1174.
Shell, J. (1984) Ophthalmic drug delivery systems. *Survey of Ophthalmology* 29(2), 117–128.
Silbiger, J. and Stern, G. (1992) Evaluation of corneal collagen shields as a drug delivery device for the treatment of experimental *Pseudomonas keratitis*. *Ophthalmology* 99(6), 889–892.
Urtti, A. (2006) Ocular drug delivery. *Advanced Drug Delivery Reviews* 58(11), 1129–1130.
Weiner, A. and Gilger, B. (2010) Advancements in ocular drug delivery. *Veterinary Ophthalmology* 13(6), 395–406.
Woodruff, C., Peck, G. and Banker, G. (1972) Dissolution of alkyl vinyl ether–maleic anhydride copolymers and ester derivatives. *Journal of Pharmaceutical Sciences* 61(12), 1916–1921.
Yang, J., Ann, D., Kannan, R. and Lee, V. (2007) Multidrug resistance protein 1 (MRP1) in rabbit conjunctival epithelial cells: its effect on drug efflux and its regulation by adenoviral infection. *Pharmaceutical Research* 24(8), 1490–1500.
Young, S., Wong, M., Tabata, Y. and Mikos, A. (2005) Gelatin as a delivery vehicle for the controlled release of bioactive molecules. *Journal of Controlled Release* 109(1–3), 256–274.

14 Polymers Targeting Habitual Diseases

Bhanu Malhotra[1], Preeti Panthari[2], Harsha Kharkwal[2,*] and Madhav P. Yadav[3]
[1]Amity Institute of Biotechnology and Amity Center for Carbohydrate Research, Amity University Uttar Pradesh, Noida, India; [2]Amity Institute of Phytomedicine and Phytochemistry and Amity Center for Carbohydrate Research, Amity University Uttar Pradesh, Noida, India; [3]Sustainable Biofuels and Co-Products Research Unit, USDA, Wyndmoor, Pennyslvania, USA

Abstract

The use of polymeric drug conjugates mainly as a cancer therapy treatment has been addressed, but these polymers also find their way into the treatment of various lifestyle disorders such as diabetes, hypertension and cardiovascular diseases. Focus is on the development of biodegradable, polymer-based drug conjugates which can be administered easily and pose no side effects. This chapter illustrates the role and applications of polymer–drug conjugates for the treatment of diabetes, atherosclerosis and colon-specific diseases, and their future prospects. Although cutting-edge research is yet to emerge, polymeric drugs stand out as an exciting example of how their horizon is expanding beyond cancer therapy to other therapeutic applications.

Introduction

Present-day lifestyle diseases such as diabetes mellitus, acute hypertension, cardiovascular disease and obesity are affecting the well-being of millions of people across the globe. Great advances in polymer sciences have led to the development of a wide array of polymeric drug conjugates which not only treat different types of cancers in the body but are also applicable to the delivery of drugs to cure other diseases like rheumatoid arthritis, Crohn's disease, gout, etc. Table 14.1 represents approved polyethylene glycol (PEG)-based drugs for such diseases.

A large amount of work has been done on polymeric drug conjugates which deliver an anti-cancer agent to the target cell and increase the efficacy of cancer treatment; but the use of these polymer-based conjugates has also expanded to treat habitual diseases. One of these is diabetes mellitus (Wild *et al.*, 2004). This disorder threatens people in both developed as well as developing countries, and is responsible for 5% of deaths worldwide. It is a chronic disorder caused by a decrease in the production of insulin by pancreatic islet cells, and in consequence an increase in blood glucose level, a situation called hyperglycaemia. Diabetes mellitus is characterized by polyuria, polyphagia, polydipsia and excessive weight loss (Vftter, 1986; Alberti and Zimmet, 1998). Diabetes insipidus is characterized by the elimination of severely diluted urine in large amounts, which does not reduce even when fluid intake is stopped. The prime cause behind this is deficiency in secretion of vasopressin, known as the antidiuretic hormone (ADH), by the posterior pituitary. Diabetes mellitus is classified as either Type 1 diabetes (insulin-dependent diabetes mellitus) or

*Corresponding author. E-mail: hkharkwal@amity.edu

Table 14.1. Drug delivery systems stabilized with polyethylene glycol (PEG) that have received regulatory approval in the USA and EU. (Reproduced, with permission, from Kadajji and Betageri, 2011.)

PEG drug candidate	Company	Indication	Year of approval
Adagen (11-17 × 5 kDa mPEG per adenosine deaminase)	Erzon Inc. (USA & Europe)	Immunodeficiency	1990 (USA)
Oncospar (5 kDa mPEG-L-asparginase)	Erzon Inc. (USA/ Rhone-Poulenc (Europe)	Acute lymphoblastic leukaemia	1994 (USA)
Doxil/Cadyx (SSL formulation of doxorubicin)	Alza Corp. (USA)/ Schering Plough Corp. (Europe)	Ovarian and breast cancer, multiple myeloma	1995 (USA) 1999 (USA) 1996 (EU)
PEG-Intron (2 × 20 kDa mPEG-interferon-α-2a)	Schering Plough Corp. (USA & EU)	Chronic hepatitis C	2000 (EU) 2001 (USA)
(Pegasys) 12 kDa interferon mPEG-interferon-α-2b)	Hoffmann-Laa-Roche (USA & EU)	Chronic hepatitis C	2002 (USA & EU)
Neulasta (20 kDa mPEG-GCSF)	Amgen Inc. (USA & EU)	Febrile neutropenia	2002 (USA & EU)
Somavert (4–6 × 5 kDa mPEG per structurally modified HG receptor antagonist)	Pfizer (USA & EU)	Acromegaly	2002 (EU) 2003 (USA)
Macugen (2 × 20 kDa mPEG anti-VEGF-aptamer)	Pfizer (EU/OSI Pharm Inc.) and Pfizer (USA)	Age-related macular degeneration	2004 (USA) 2006 (EU)
Cimzia (2 × 40 kDa mPEG anti TNFα)	CB S.A (USA & EU)	Crohn's disease rheumatoid arthritis	2008 (USA) 2009 (USA) 2009 (EU)
Krystexxa (Pegloticase) PEGylated Uric acid	Savient Pharmaceutical Inc. (USA)	Chronic gout	2010 (USA)

mPEG, methoxypoly(ethylene glycol); SSL, sterically stabilized liposome; G-CSF, granulocyte-colony stimulating factor; HG, human growth; VEGF, vascular endothelial growth factor; TNF, Tumour necrosis factor.

Type 2 diabetes (non-insulin-dependent diabetes, or gestational diabetes) (Genuth *et al.*, 2003). Insulin-dependent diabetes mellitus is caused by the loss of beta Langerhans cells in the pancreas (following autoimmune attack mediated by T cell populations), leading to an insulin deficiency. In children this is termed juvenile diabetes. In Type 2 diabetes there is reduced insulin production and sensitivity. The response of cells to insulin is mediated by receptors present in the cell membranes (Engelgau *et al.*, 1997).

A number of natural polymers have been identified for enhancement of the therapeutic effects of pre-existing drugs. Their use has also been extended to the treatment of various colonic diseases, such as Crohn's disease and ulcerative colitis, which are mainly caused by lifestyle changes. The colon, a very dynamic and diverse environment, is considered to be the site of absorption of residual carbohydrates and water (Reddy, 1999). An increase in the prevalence of colonic diseases has interested researchers worldwide in the development of colon-targeted drug delivery systems. Polymers that can produce the same intestinal microflora can be used as potential carriers for colon drug delivery. Polymers along with biodegradable carriers, a gelling agent and micelle formation can be combined with specialized control release systems to treat colon problems. The main polymers used as drug carriers in colon disease treatment include guar gum, pectin, chondroitin sulfate and chitosan. In the colon, systemic and local drug absorption can take place (Kotla *et al.*, 2014). Colonic drugs can be administered using oral or rectal routes, the transit time being 20–30 h (Chourasia and Jain, 2004). Specialized drug delivery systems can prevent the premature release of the colonic drugs at other sites, such as the stomach or intestine, and release them only upon entry to the dynamic environment of the colon (Jain *et al.*, 2007).

Another important disease which causes global adult mortality is coronary artery disease, triggered by atherosclerosis, and polymer-based drug conjugates are being used in its treatment. Atherosclerosis is the deposition of lipoproteins within the walls of blood vessels, narrowing their diameter and restricting the flow of blood to the heart (Libby, 2002). Approximately 81 million people in the USA have one or more cardiovascular diseases. Cardiovascular diseases cause 50% of deaths in western countries (the 34 highest-income countries in the Organisation for Economic Co-operation and Development) and diabetics are more susceptible than non-diabetics (Moon, 2010). Atherosclerosis is defined as the building up of lipid plaques inside the walls of arteries, leading to thinning of their diameter and dominates the clinical manifestations of chronic stable angina, myocardial infarction and stroke (Bressler *et al.*, 2011). The American Heart Association has reported a significant 20% reduction in cardiovascular diseases (CVDs) mortality rates from the application of novel techniques of drug delivery (Lloyd-Jones *et al.*, 2010).

Polymers for Diabetes Treatment

The current treatment and therapy of diabetes mellitus via oral and subcutaneous insulin supply routes has serious disadvantages. As a result, many such patients develop hypoglycaemia. Work has, therefore, focused on delivering insulin through oral routes which can mimic the normal physiological process of its release from the pancreatic cells. The oral route is convenient and should be preferred as it has very few side effects. Nanopolymeric particles ('microparticles') using natural and biodegradable polymers as matrices are administered to deliver insulin orally. One widely used and well researched polymer for this is chitosan. It can be modified easily, is readily available and is considered to have promise in oral insulin delivery systems (Sonia and Sharma, 2012). Polymer nanoparticles are colloidal, solid particles which contain macromolecular substances from 10 nm to 1000 nm in size (Brigger *et al.*, 2002). Basically, they can be of two types: nanocapsules and nanospheres. Vesicles of nanocapsules possess the drug in a cavity surrounded by the membrane, while in the matrix system of nanospheres, the drug is widely and uniformly distributed. The nanospheres finally degrade into biological products and deliver the encapsulated drug to the target tissue. The end products of their degradation are lactic acids which can be converted to CO_2 via the Krebs cycle. Elimination occurs in bulk as the matrix degrades, and the polymer is released at a rate which is related to the surface area of the nanoparticle. The engineering and specificity of these polymer-based nanoparticles targeted for insulin delivery to cells can be improved further.

Basically, the nanoparticle consists of a biodegradable polymer. The insulin matrix is covered by a nanoporous membrane which releases a grafted enzyme (glucose oxidase — an oxidoreductase that catalyzes the oxidation of glucose to hydrogen peroxide and D-glucono-δ-lactone). When the blood glucose level increases it triggers a response in these polymer insulin-carrying drugs which results in their biodegradation and the release of insulin to the target cells. It has been reported from the past that a so called molecular gate system consists of a insulin reservoir surrounded by a membrane of poly (methacrylic acid) (polyethylene glycol) copolymer which controls the delivery rate of insulin to cells (Bathe *et al.*, 2013). At normal body pH the polymer swells up and gates for insulin are closed. When the pH decreases, the nanoparticle opens the gates to release insulin.

An interesting achievement using chitosan is the use of nanoparticles with targeting ligand FQSIYPpIK (FQS) as a modified ligand; this enables increased interaction between the nanocarrier and intestinal epithelial cells receptor integrin αvβ3. Insulin is attached to poly(lactide-co-glycolide)-monomethoxy-poly(polyethylene glycol) micelle cores using the ligand FQS and trimethyl chitosan chloride to prepare the targeting nanoparticles. Good results have been obtained *in vitro*, indicating enhanced enzymatic stability along with a good drug release profile. Enhanced uptake of insulin by the cells mediated by active ligand receptor interaction has also been shown. In diabetic rats it has been observed that these nanoparticles lead to increased serum insulin levels and a noticeable hypoglycaemic response. Results from both *in vitro* and *in vivo* analysis have shown that ligand FQS-modified nanoparticles are an efficient system for oral insulin delivery (Liu, 2015). One chitosan derivative, thiomalyl chitosan, has shown good potential for delivering proteins orally. This derivative of chitosan is pH sensitive, and demonstrated insulin release restriction at pH 1.2 for about 10% in 3 h. The diabetic rats have shown good results with a pharmacological availability of 1.5% (Chopra *et al.*, 2006).

The technology of cell encapsulation for the treatment of diabetes involves insulin-producing pancreatic cells which can be masked from clearance by the host immune system and rejection by the body. One of the polymers, cellulose sulfate, is a pioneer for cell encapsulation and has shown profound and safe results in rat models (Wang *et al.*, 2009).

Glucagon-like peptide-1 (GLP-1) hormone and exendin-4 (which is its receptor agonist with an increased half-life), are very important for glucose homeostasis as well as the insulinotropic effect. The enzyme DPP-IV degrades GLP-1 causing the elimination of exendin 4, and so clinical application is rather restricted. Frequent administration is still required for treating diabetes. The delivery of GLP-1 and exendin-4 genes for enhanced production through viral and non-viral systems for delivery is done. Non-viral methods make use of cationic polymers for plasmid condensation, including liposomes as carriers. Arginine-grafted polymers, poly(ethylenimine) and chitosan have been used and promising results have been obtained. Adenoviruses are used as vectors to deliver the genes of GLP-1 and exendin 4, maintaining controlled release through the use of Triblock Copolymer of PLGA-PEG-PLGA (PLGA = poly lactic-co-glycolic acid) which is called as Regel. This is a biodegradable, water-based gel and forms a sol–gel transition in different temperature regimes. GLP-1 and Regel have been mixed and introduced into animal models through the subcutaneous route. Immediately after injection into the body, Regel releases GLP-1 slowly and gradually. This has been found to be one of the efficient and successful carriers of GLP-1, and so it can help to treat diabetes and maintain glucose levels in the body (Gappa *et al.*, 2001; Jeong *et al.*, 2007; Panyam and Labhasetwar, 2012).

Another important strategy for oral insulin delivery is surface engineering of silica nanoparticles. This work has focused on insulin- and silica-based nanoparticles surfaced with natural polymers such as chitosan and sodium alginate, and on synthetic polymers such as PEG. Evaluation is done on two important cell lines, Caco-2 and HepG2. These cell lines are investigated for oral insulin delivery, and act as molecular mimics of insulin–nanoparticle-based targets. The silica particles are used in conjunction with these mucoadhesive polymers to increase the bioavailability of insulin to the intestinal mucosal gut. Interesting results have been obtained using Fourier transform infrared spectroscopy (FTIR) spectra, indicating low toxicities in the above cell lines (Andreani *et al.*, 2014).

The synthesis of recombinant insulin-like growth factor-I (IGF-I) improves insulin sensitivity in patients showing resistance. The production of recombinant protein growth factor is the alternative therapy if the patients become insensitive to external insulin treatment. Clinical trials show good convenience with a slow release of PLGA in recombinant IGF-I with 15–20% of protein loading with continuous release. Spray freeze-drying was used in PGLA formulation, and a 14-day slow and sustained release recombinant IGF-I formulation was developed with trials in diabetic rats (Lam *et al.*, 2000). A technique known as phase inversion nanoencapsulation uses zinc insulin in polyanhydride and polyester nanosphere formulations. It has been demonstrated that this insulin is released from the nanospheres over a 6-h time span and maintains its biological activity perfectly. Researchers have described the composition of a specific formulation containing 1.6% zinc insulin in PLGA with iron oxide, as additives, and fumaric anhydride as oligomer (Carino *et al.*, 2000). This formulation can be administered orally. An approximate efficacy of 11.4% intraperitoneally has been achieved to maintain plasma glucose levels. A number of factors, particularly size, bioadhesiveness, release kinetics and ability to penetrate the intestinal epithelium, are important for its oral delivery (Carino *et al.*, 2000). Work has also been done on chitosan TBA (chitosan-4-thiobutylamidine) enzyme inhibitor conjugates with reduced glutathione, which may represent a promising strategy for oral delivery of insulin (Krauland *et al.*, 2004). A thiolated polymer of chitosan (2-Iminothiolane) covalently linked to chitosan was prepared. The resulting Chitosan–TBA conjugated insulin tablets allow the controlled release of insulin for 8 h as compared to the control tablets of unmodified chitosan, as shown in Fig. 14.1 (Krauland *et al.*, 2004).

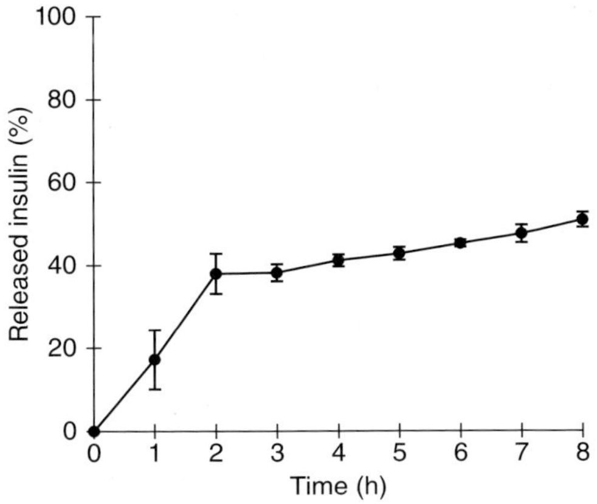

Fig. 14.1. Release profile of insulin from the chitosan–TBA–insulin tablets. The dosage form was incubated in 10 ml release medium (25 mM phosphate buffer pH 6.0 containing 30% dimethylsulfoxide) at 37 +/- 0.5 °C. Each point represents the mean +/-S.D. of three experiments. (Reproduced, with permission, from Krauland *et al.*, 2004.)

A number of dietary factors are responsible for the cure of Type II diabetes. It has been found that cinnamon extracts enhance insulin activity (Anderson, 2008). In the past insulin-enhancing complexes were isolated as well as characterized from cinnamon which helps in the alleviation of glucose intolerance (Anderson, 2008; Sartorius, 2014). Isolation and study of aqueous polyphenol polymers of cinnamon, using resonance and mass spectroscopic techniques, revealed that they not only possess antioxidant activity but also enhance insulin-dependent glucose metabolism 20-fold *in vitro* (Anderson *et al.*, 2004). The *ex vivo* technique of therapeutic gene transfers for prolonging the lifespan of cell transplants is a most important strategy. Non-viral vectors are generally safe but they have very low transfection efficiency. An insulin-secreting cell line (RINm5F, expressing sulfonylurea receptor) has been studied and PLL SU containing poly(l-lysine)-g-sulfonylurea was assessed and made for gene delivery to it (Remaury and Paris, 1992). It was reported that the polymer complexes with pCMV-Luc DNA secreted higher amounts of insulin, giving a 5% insulinotropic effect (Kang *et al.*, 2005). Therefore PLL Su can be used safely for the transferring of genes with therapeutic effect to the cells secreting insulin (Kang *et al.*, 2005).

Insulin receptors and IGF receptors are from a class of structurally same transmembrane proteins. They work when the ligand attaches to the receptor's extracellular domain and generates a tyrosine protein kinase, phosphorylating the further substrates. In a study by Sahal *et al.* (1988), it was found that the tyrosine kinases were mediating all activities of the insulin-like growth factor IGF-I. Receptor activity was studied in human placental membranes. The study also showed that (Y,A,E)n and (Y-A-E)n have shown the inhibition of the substrate phosphorylation and β subunit autophosphorylation while $(E^4Y^1)n$ and $(E^6,A^3,Y^1)n$ are best substrates for both the receptor kinases. It was found that IGF-I is ten times less sensitive than the insulin receptor to these inhibitors (Sahal *et al.*, 1988).

Polymers for Treatment of Colonic Diseases

The incidence of colon diseases such as ulcerative colitis, inflammatory bowel disease and Crohn's disease is rising, and these diseases are due to the generation of de-escalated immune responses to the commensal microflora in a genetically favourable host organism (Abraham and Cho, 2009). In genetically susceptible individuals, the extrinsic environment, intestinal microflora and immune system deregulation generates the disease. 163 alleles (alternating forms of the same gene present at the same loci in homologous chromosomes) which are responsible for Crohn's disease and ulcerative colitis have been identified. Inflammatory bowel disease shows a decrease in populations of firmicutes and a reduction in bacterial species (Gentschew and Ferguson, 2012), and diet has been hypothesized to play an important role in the pathogenesis of inflammatory bowel disease (Rajendran, 2010). Epidemiologic studies have shown the prevalence of the disease in locations where it used to be uncommon, and indicators suggest this is related to westernized lifestyles (Thia *et al.*, 2008). Disease in these westernized countries has increased due to an increase of polyunsaturated fatty acids (PUFAs) in diets. It has been found that the dietary intake of n-3 PUFAs inhibits formation of leukotrienes and pro-inflammatory prostaglandins competitively via the arachidonic acid pathway

(Marion-Letellier *et al.*, 2013). The result is the inhibition of angiogenesis, vascular adhesion/migration and adaptive immune responses via the PPARγ receptor through NF-κB-mediated pathways (Calder, 2012). The increase in the incidence of colonic diseases, and the westernized lifestyle, demand specialized site-specific drug delivery systems that are able to release drugs specifically in the colon in a reproducible manner. Usually the treatment strategy for ulcerative colitis requires aminosalicylates as well as daily administration of corticosteroids such as Prednisolone (Klotz and Schwab, 2005). Currently the therapeutic formulations available for treatment of colon diseases include slow-release oral preparations known as orenemas, and foams for rectal administration. The approximate length of the colon is 150 cm and it is divided into five segments: caecum; ascending colon; hepatic flexure (with the right half of the transverse colon on the right side of it and the left half of the transverse colon); splenic flexure; and descending colon (with the sigmoid on the left side of it). The colon not only stores the faeces but also selectively reabsorbs all essential electrolytes and water from it (Van den Mooter and Kinget, 1995).

The pH of the gastrointestinal tract (GIT) is regulated by inter- and intra-subject variations influenced by food intake, diet and the diseased state. The change in pH across the GIT has been used as a means of targeting colon drug delivery applications. The dynamic environment of the colon contains 400 distinct bacterial species with populations of 1011–1012 CFU/ml, including the genera *Bifidobacterium*, *Lactobacillus*, *Bacteroides* and *Eubacterium*. The colon epithelial junctions across the mucosal lining are very tight, therefore poor paracellular absorption of drugs in the colon is a disadvantage; there is also a low surface area which limits the efficiency of drug delivery. Another fact that makes drug targeting to the colon very complicated is its location at the distal end of the alimentary canal. The varied microbial flora (which may cause the degradation of various drugs) and different pH regimes in the colon contribute to the further complexity of delivering the drug to the target site.

A number of advances have been made for improving the bioavailability of the drugs for treating inflammatory bowel syndrome. Enhanced oral delivery of drugs has enabled localized treatment and improved drug efficacy for treatment of colon-specific diseases, and we have already mentioned that colonic drug delivery relies on different physiological factors, the most important being pH. The microbial and physiological changes of the GIT in inflammatory bowel disease (IBD) are shown in Fig. 14.2. Patients suffering from IBD show increased orocaecal transit times due to the complete absence of small intestinal bacterial overgrowth (SIBO) (Rana *et al.*, 2013). They have also shown a small increase in the pH of the small intestine, but there is a huge decrease in pH in cases of Crohn's disease and ulcerative colitis (Lucas *et al.*, 1978).

Polymers containing the drugs for colon drug delivery are synthesized and mainly formulated as tablets, capsules or pellets. A number of pH sensitive polymers have been intended for use. The usage of pH dependent polymers utilizes the variations of the pH of the GIT. The polymeric drug conjugate is stimulus sensitive (i.e. it is a pH-sensitive polymer protecting the drug inside the stomach as well as in the proximal part of the small intestine), and can be released slowly in the lower part of the small intestine. Examples includes Eudragit S-100, Eudragit FS 30D, Hydroxy Propyl Methyl Cellulose Phthalate HPMC50 and Cellulose Acetate Phthalate (Philip and Philip, 2010). Delivery of polymeric drugs to the colon also exploits the different luminal pressure conditions resulting from peristaltic movement inside the colon; the pressure delivery systems deliver the drug to the target site using the muscular contractions of the visceral organs (Elkhodairy *et al.*, 2014).

Natural polymers have the advantage of remaining intact in the physiological environment of the stomach as well as in the intestine, and offer selective controlled release of the drug by polymer degradation into the colonic environment (Rajpurohit *et al.*, 2010). The degradation pattern of polymers is well understood. The polymer matrix can swell with hydration throughout its journey in the GIT, preventing drug diffusion. The colonic microflora degrades these matrices, releasing the drug. Polymers from plant sources for delivery to the colon include starch, amylase, cellulose, pectin and guar; from animal sources include chitosan and its derivatives, chondroitin sulfate; from microbial sources include dextrins and xanthan gum; and from fungal sources include pullulan. The deacetylation of chitin produces a polycationic

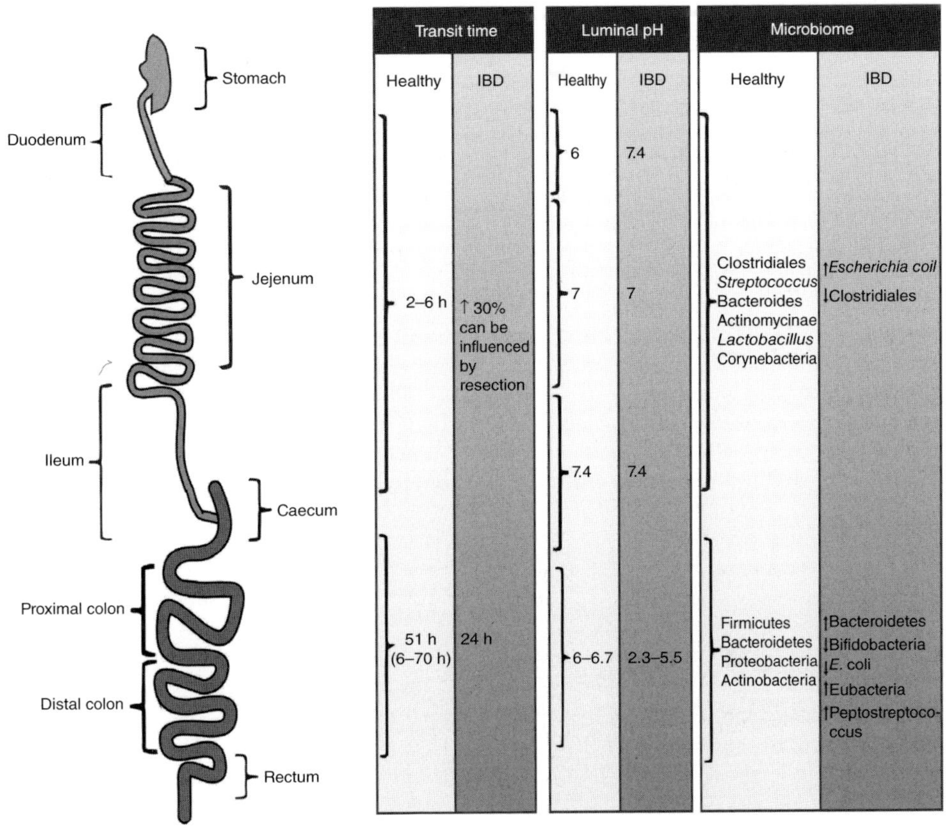

Fig. 14.2. Comparison of microbial and physiological changes in the gastrointestinal tract under normal conditions and incidence of inflammatory bowel disease (from Hua *et al.*, 2015).

polymer known as chitosan which can be used for colonic drug delivery. Chitosan is non-toxic, readily available, shows good biocompatibility, is soluble in dilute acids and precipitates at a pH above 7. It can be used over different pH regimes in the colon for targeted drug delivery. The cores of the chitosan-based drug conjugates are acted upon by microflora including the bacteroides of the colon, degrading the polymeric matrix of chitosan and thereby releasing the drug. One such formulation includes microspheres of sodium diclofenac using the spray drying technique to encode Eudragit L-100. It is possible that chitosan glutamate may be able to control the slow and controlled release of sodium diclofenac from the core (Lorenzo-Lamosa *et al.*, 1998).

Chondroitin sulfate is a glycosaminoglycan and a heteropolysaccharide, and is used as a substrate by the bacteroid species inhabiting the colon. It is readily soluble in water, adding to the disadvantages of its ability to carry the drug in the GIT. Crosslinked forms of chondroitin sulfate, however, are less hydrophilic and can shield the drug core, effectively controlling drug release in the colon. Crosslinking was achieved using the catalyst 1, 12-diaminododecane and indomethacin as the drug. It was found that release of the drug indomethacin was dependent on biodegradation by the caecal microflora (Rubinstein, 1995).

Dextran ester-based prodrugs have been formed in the past for treating chorionic colitis by selective release of the drug to the colon (Larsen, 1989). Dextran consists of linear chains of α-D glucose molecules linked via 1:6 linkage and side chains of 1:3. These are obtained from microorganisms of the genus *Lactobacillus* and are colloidal, water soluble, hydrophilic and inert.

Pullulan comes from fungi and the composition includes maltotriose in α-(1, 4) linkage with branching at α-1, 6 linkage. There has been research into the delivery of drugs from pullulan-based polymers (Witczak, 1997).

Pectins, which are linear polysaccharides, consist of 1, 2 D-rhamnose and α-1, 4 D-galacturonic acid with D-galactose and D-arabinose as the side chains. They can be completely reduced to soluble oligalactorunates by colonic microbes (Englyst and Macfarlane, 1986). The use of pectin, being highly water soluble, is limited in a colonic drug delivery system. The underlining principle is that it swells in contact with the watery fluids of the GIT and so releases the entrapped drug by diffusion. Since pectin is non-toxic, considerable research is going on to develop drug delivery systems based on it. Tablets of calcium pectinate have shown good results *in vitro*. When indomethacin was formulated in pectin-based matrices, targeted drug release was seen to be triggered by the rat caecal content (Rubinstein, 1995).

With the use of nanotechnology, reduction in the size of drug delivery carriers has considerably improved the colonic residence time of the drugs in inflamed regions of the intestine. A diagrammatic comparison of a healthy mucosal lining and changes caused by IBD are shown in Fig. 14.3. Reduction in size enhances selective drug delivery of active molecules into the colitis tissue as with the previously discussed enhanced permeability and retention (EPR) effect. Carriers are eliminated rapidly due to recurrent diarrhoea in IBD patients, and so reduction in their diameter is an important nano-based idea that prevents it (Beloqui *et al.*, 2013).

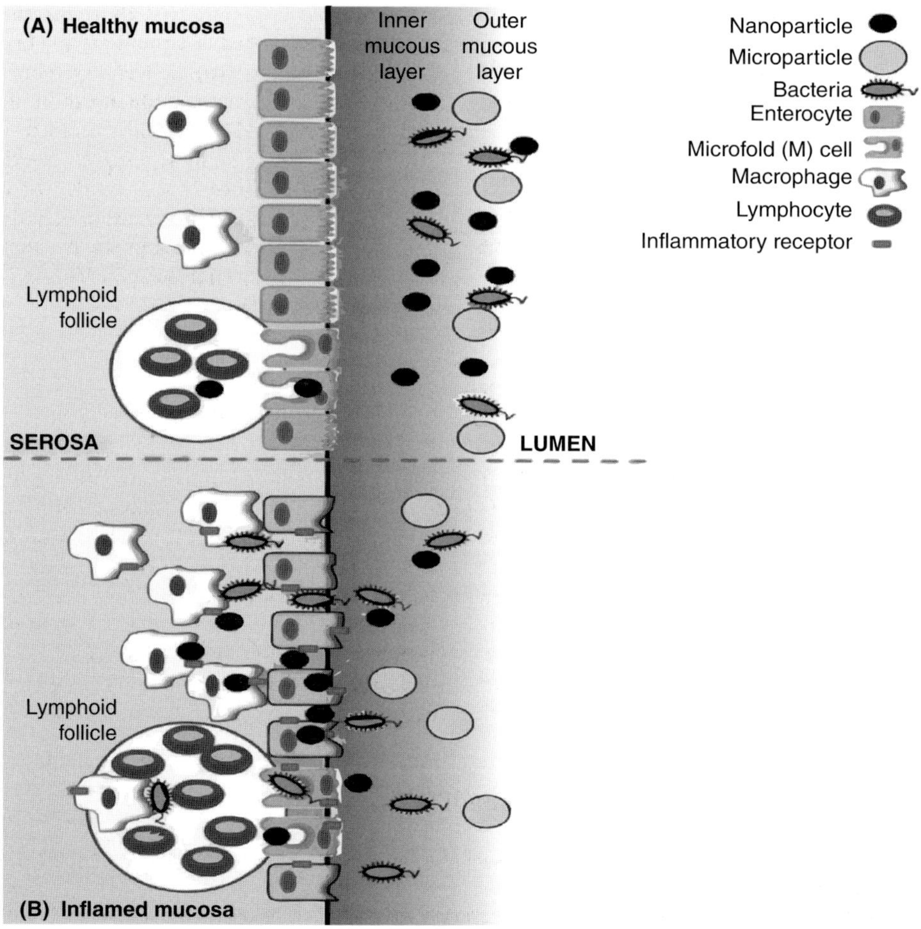

Fig. 14.3. Physiology of (A) healthy and (B) inflamed intestinal mucosa (from Hua *et al.*, 2015).

Polymers for Treating Cardiovascular Diseases

The use of drug-eluting stents in the treatment of cardiovascular diseases has a broadened application. The use of endovascular devices to treat the stenotic artery has its own disadvantages of causing vascular inflammation and stent thrombosis due to the presence of a foreign body. Bioresorbable scaffolds are used to release vessels from the cage and are currently undergoing clinical trials. They are made using a tyrosine-derived polycarbonate polymer known as the REVA scaffold which uses a slide lock mechanism; this provides considerable radial strength to devices after implantation (Pollman, 2009). One of the stents used is the Ideal BioStent, which has a core backbone and is a bioresorbable balloon which gets expanded. The backbone consists of a trimer of two salicylic acid molecules attached by sebacic acid and polylactide anhydride. The preclinical trials showed the salicylate had anti-platelet properties, as well as anti-inflammatory activity (not only promoting retenosis, but healing the vessels after the polymer degraded) (Jabara *et al.*, 2008). Another scaffold is the Amaranth BRS, comprising Poly-L-Lactide (PLLA) and including proprietary tube fabrication. This can remain intact for 3–6 months and shows enhanced mechanical stability and improved radial strength, showing less recoil than other bioresorbable scaffolds (Han *et al.*, 2012).

Nanoparticles, including organic and inorganic nanostructures, are very attractive for therapeutic treatment of CVD. These nanosized carriers of 10–200 nm are good enough to interact with receptors and are also able to transport drugs easily, protecting them from metabolic deactivations because of renal clearance (Mailänder and Landfester, 2009). Most atherosclerosis therapies use polymeric micelles to basically mimic the amphilicity as well as polyanionic charge distribution of oxLDL (Bijsterbosch and van Berkel, 1990). The design includes lauroyl-modified mucic acid and a chain of PEG, giving a good micelle concentration of 7–10 M and being 15–20 nm in size. These drugs could sequester the LDL ('bad' cholesterol) but shows little affinity towards oxidized forms of LDL (Zinn *et al.*, 2008).

Future Implications

As highlighted in the other chapters, polymers have a 'tuneable' role in biomedical applications in their ability to be tailor-made to release drugs at targeted locations, thereby improving their residence times in the body. There is certainly considerable promise for the future; however, further research leading to a thorough understanding is warranted, if completely biodegradable and non-toxic drug delivery systems that can be excreted from the body are to be developed. As lifestyle-related health disorders continue to increase, today's research needs to prepare for the future if it is to deliver cost-effective technologies. The advent of genetic engineering and nanotechnology in the field of bioconjugates has enabled considerable advances in this process.

Acknowledgements

We thank Dr Ashok K Chauhan, Founder President, Ritanand Balved Educational Foundation, for support. The guidance from Shri. Atul Chauhan, Chancellor, Amity University Uttar Pradesh, and Prof. Dr (Mrs) Balvinder Shukla, Vice Chancellor, Amity University Uttar Pradesh, is greatly appreciated. The publications cited in this chapter provided numerous insights, and we are grateful to their eminent authors.

References

Abraham, C. and Cho, J. (2009) Inflammatory bowel disease. *New England Journal of Medicine* 361(21), 2066–2078.

Alberti, K. and Zimmet, P. (1998) Definition, diagnosis and classification of diabetes mellitus and its complications. Part 1: diagnosis and classification of diabetes mellitus. Provisional report of a WHO Consultation. *Diabetic Medicine* 15(7), 539–553.

Anderson, R., Broadhurst, C., Polansky, M., Schmidt, W., Khan, A., Flanagan, V., Schoene, N. and Graves, D. (2004) Isolation and characterization of polyphenol type-a polymers from cinnamon with insulin-like biological activity. *Journal of Agriculture and Food Chemistry* 52(1), 65–70.

Anderson, R. (2008) Chromium and polyphenols from cinnamon improve insulin sensitivity. *Proceedings of the Nutrition Society*, 67(01), pp.48–53

Andreani, T., Kiill, C., Souza, A., Fangueiro, J., Fernandes, L., Doktorovová, S., Santos, D., Garcia, M., Gremião, M., Souto, E. and Silva, A. (2014) Surface engineering of silica nanoparticles for oral insulin delivery: characterization and cell toxicity studies. *Colloids and Surfaces B: Biointerfaces* 123, 916–923.

Bathe, R., Nimase, P. K., Gali, V. and Suryawanshi, D.M. (2013) Nanotechnology and Diabetes. *International Journal of Advances in Pharmaceutics* 2(4).

Beloqui, A., Coco, R., Alhouayek, M., Solinís, M., Rodríguez-Gascón, A., Muccioli, G. and Préat, V. (2013) Budesonide-loaded nanostructured lipid carriers reduce inflammation in murine DSS-induced colitis. *International Journal of Pharmaceutics* 454(2), 775–783.

Bijsterbosch, M.K. and Van Berkel, T.J. (1990) Uptake of lactosylated low-density lipoprotein by galactose-specific receptors in rat liver. *The Biochemical Journal* 270(1), 233–239.

Bressler, J., Shimmin, L., Boerwinkle, E. and Hixson, J. (2011) Global DNA methylation and risk of subclinical atherosclerosis in young adults: the pathobiological determinants of atherosclerosis in youth (PDAY) study. *Atherosclerosis* 219(2), 958–962.

Brigger, I., Dubernet, C. and Couvreur, P. (2002) Nanoparticles in cancer therapy and diagnosis. *Advanced Drug Delivery Reviews* 54(5), 631–651.

Calder, P. (2012) Mechanisms of action of (n-3) Fatty acids. *Journal of Nutrition* 142(3), 592S–599S.

Carino, G., Jacob, J. and Mathiowitz, E. (2000) Nanosphere based oral insulin delivery. *Journal of Controlled Release* 65(1–2), 261–269.

Chopra, S., Mahdi, S., Kaur, J., Iqbal, Z., Talegaonkar, S. and Ahmad, F. (2006) Advances and potential applications of chitosan derivatives as mucoadhesive biomaterials in modern drug delivery. *Journal of Pharmacy and Pharmacology* 58(8), 1021–1032.

Chourasia, M. and Jain, S. (2004) Polysaccharides for colon targeted drug delivery. *Drug Delivery* 11(2), 129–148.

Elkhodairy, K., Afifi, S. and Zakaria, A. (2014) A promising approach to provide appropriate colon target drug delivery systems of vancomycin HCL: pharmaceutical and microbiological studies. *BioMed Research International* 2014, 1–14.

Engelgau, M., Thompson, T., Herman, W., Boyle, J., Aubert, R., Kenny, S., Badran, A., Sous, E. and Ali, M. (1997) Comparison of fasting and 2-hour glucose and HbA1c levels for diagnosing diabetes: diagnostic criteria and performance revisited. *Diabetes Care* 20(5), 785–791.

Englyst, H. and Macfarlane, G. (1986) Breakdown of resistant and readily digestible starch by human gut bacteria. *Journal of the Science of Food and Agriculture* 37(7), 699–706.

Gappa, H., Baudyš, M., Koh, J., Kim, S. and Bae, Y. (2001) The effect of zinc-crystallized glucagon-like peptide-1 on insulin secretion of macroencapsulated pancreatic islets. *Tissue Engineering* 7(1), 35–44.

Gentschew, L. and Ferguson, L. (2012) Role of nutrition and microbiota in susceptibility to inflammatory bowel diseases. *Molecular Nutrition & Food Research* 56(4), 524–535.

Genuth, S., Alberti, K.G., Bennett, P., Buse, J., Defronzo, R., *et al.* (2003) Follow-up report on the diagnosis of diabetes mellitus. *Diabetes Care* 26(11), 3160–3167.

Han, Y., Zhang, L., Yang, L., Liu, H., Qu, P., Li, W., Jiang, T., Li, S., Jing, Q., Zhang, Q., Xu, B., Li, Y. and Gao, R. (2012) A new generation of biodegradable polymer-coated sirolimus-eluting stents for the treatment of coronary artery disease: final 5-year clinical outcomes from the CREATE study. *EuroIntervention* 8(7), 815–822.

Hua, S., Marks, E., Schneider, J.J. and Keely, S. (2015) Advances in oral nano-delivery systems for colon targeted drug delivery in inflammatory bowel disease: selective targeting to diseased versus healthy tissue. *Nanomedicine* 11(5), 1117–1132.

Jabara, R., Chronos, N. and Robinson, K. (2008) Novel bioabsorbable salicylate-based polymer as a drug-eluting stent coating. *Catheterization and Cardiovascular Interventions* 72(2), 186–194.

Jain, A., Gupta, Y., and Jain, S. (2007) Perspectives of biodegradable natural polysaccharides for site-specific drug delivery to the colon. *Journal of Pharmacy and Pharmaceutical Sciences.* 10(1), pp.86–128.

Jeong, J., Kim, S. and Park, T. (2007) Molecular design of functional polymers for gene therapy. *Progress in Polymer Science* 32(11), 1239–1274.

Kadajji, V.G. and Betageri G.V. (2011) Water soluble polymers for pharmaceutical applications. *Polymers* 3(4), 1972–2009.

Kang, H., Kim, S., Lee, M. and Bae, Y. (2005) Polymeric gene carrier for insulin secreting cells: poly(l-lysine)-g-sulfonylurea for receptor mediated transfection. *Journal of Controlled Release* 105(1–2), 164–176.

Klotz, U. and Schwab, M. (2005) Topical delivery of therapeutic agents in the treatment of inflammatory bowel disease. *Advanced Drug Delivery Reviews* 57(2), 267–279.

Kotla, N., Gulati, M., Singh, S. and Shivapooja, A. (2014) Facts, fallacies and future of dissolution testing of polysaccharide based colon-specific drug delivery. *Journal of Controlled Release* 178, 55–62.

Krauland, A., Guggi, D. and Bernkop-Schnürch, A. (2004) Oral insulin delivery: the potential of thiolated chitosan-insulin tablets on non-diabetic rats. *Journal of Controlled Release* 95(3), 547–555.

Lam, X., Duenas, E., Daugherty, A., Levin, N. and Cleland, J. (2000) Sustained release of recombinant human insulin-like growth factor-I for treatment of diabetes. *Journal of Controlled Release* 67(2–3), 281–292.

Larsen, C. (1989) Macromolecular prodrugs. XII. Kinetics of release of naproxen from various polysaccharide ester prodrugs in neutral and alkaline solution. *International Journal of Pharmaceutics* 51(3), 233–240.

Libby, P. (2002) Inflammation in atherosclerosis. *Nature* 420(6917), 868–874.

Liu, C. (2015) A novel ligand conjugated nanoparticles for oral insulin delivery. Available at: www.ncbi.nlm.nih.gov/pubmed/26203690 (accessed 1 October 2015).

Lloyd-Jones, D., Adams, R., Brown, T., Carnethon, M., Dai, S., De Simone, G., Ferguson, T., Ford, E., Furie, K., Gillespie, C., Go, A., Greenlund, K., Haase, N., Hailpern, S., Ho, P., Howard, V., Kissela, B., Kittner, S., Lackland, D., Lisabeth, L., Marelli, A., McDermott, M., Meigs, J., Mozaffarian, D., Mussolino, M., Nichol, G., Roger, V., Rosamond, W., Sacco, R., Sorlie, P., Stafford, R., Thom, T., Wasserthiel-Smoller, S., Wong, N. and Wylie-Rosett, J. (2010) Executive summary: heart disease and stroke statistics – 2010 update: a report from the American Heart Association. *Circulation* 121(7), 948–954.

Lorenzo-Lamosa, M., Remuñán-López, C., Vila-Jato, J. and Alonso, M. (1998) Design of microencapsulated chitosan microspheres for colonic drug delivery. *Journal of Controlled Release* 52(1–2), 109–118.

Lucas, M., Cooper, B., Lei, F., Johnson, I., Holmes, G., Blair, J. and Cooke, W. (1978) Acid microclimate in coeliac and Crohn's disease: a model for folate malabsorption. *Gut* 19(8), 735–742.

Mailänder, V. and Landfester, K. (2009) Interaction of nanoparticles with cells. *Biomacromolecules* 10(9), 2379–2400.

Marion-Letellier, R., Savoye, G., Beck, P., Panaccione, R. and Ghosh, S. (2013) Polyunsaturated fatty acids in inflammatory bowel diseases. *Inflammatory Bowel Diseases* 19(3), 650–661.

Moon, M. (2010) AHA sets new goals in cardiovascular health. *Internal Medicine News* 43(3), 38.

Panyam, J. and Labhasetwar, V. (2012) Biodegradable nanoparticles for drug and gene delivery to cells and tissue. *Advanced Drug Delivery Reviews* 64, 61–71.

Philip, A. and Philip, B. (2010) Colon targeted drug delivery systems: a review on primary and novel approaches. *Oman Medical Journal* 25(2), 70–78.

Pollman, M. (2009) Engineering a bioresorbable stent: REVA programme update. *EuroIntervention* 5(F), F54–F57.

Rajendran, N. (2010) Role of diet in the management of inflammatory bowel disease. *World Journal of Gastroenterology* 16(12), p.1442.

Rajpurohit, H., Sharma, S., Sharma, P. and Bhandari, A. (2010) Polymers for colon targeted drug delivery. *Indian Journal of Pharmaceutical Sciences* 72(6), 689.

Rana, S., Sharma, S., Malik, A., Kaur, J., Prasad, K., Sinha, S. and Singh, K. (2013) Small intestinal bacterial overgrowth and orocecal transit time in patients of inflammatory bowel disease. *Digestive Diseases and Sciences* 58(9), 2594–2598.

Reddy, S. (1999) Novel oral colon-specific drug delivery systems for pharmacotherapy of peptides and nonpeptide drugs. *Drugs Today* 35(7), 537.

Remaury, A. and Paris, H. (1992) The insulin-secreting cell line, RINm5F, expresses an alpha-2D adrenoceptor and nonadrenergic idazoxan-binding sites. *Journal of Pharmacology and Experimental Therapeutics* 260(1), 417–426.

Rubinstein, A. (1995) Approaches and opportunities in colon-specific drug delivery. *Critical Reviews in Therapeutic Drug Carrier Systems* 12(2–3), 101–149.

Sahal, D., Ramachandran, J. and Fujita-Yamaguchi, Y. (1988) Specificity of tyrosine protein kinases of the structurally related receptors for insulin and insulin-like growth factor I: Tyr-containing synthetic polymers as specific inhibitors or substrates. *Archives of Biochemistry and Biophysics* 260(1), 416–426.

Sartorius, T., Peter, A., Schulz, N., Drescher, A., Bergheim, I., *et al.* (2014) Cinnamon extract improves insulin sensitivity in the brain and lowers liver fat in mouse models of obesity. PLoS ONE, 9(3). DOI: 10.1371/journal.pone.0092358

Sonia, T. and Sharma, C. (2012) An overview of natural polymers for oral insulin delivery. *Drug Discovery Today* 17(13–14), 784–792.

Thia, K., Loftus Jr, E., Sandborn, W. and Yang, S. (2008) An update on the epidemiology of inflammatory bowel disease in Asia. *The American Journal of Gastroenterology* 103(12), 3167–3182.

Van den Mooter, G. and Kinget, R. (1995) Oral colon-specific drug delivery: a review. *Drug Delivery* 2(2), 81–93.

Vftter, K. (1986) Diabetes mellitus. WHO Technical Report Series 727. 113 Seiten. World Health Organization, Geneva 1985. Preis: 9.00 Sw. fr. Nahrung, 30: 700. DOI: 10.1002/food.19860300713.

Wang, N., Adams, G., Buttery, L., Falcone, F. and Stolnik, S. (2009) Alginate encapsulation technology supports embryonic stem cells differentiation into insulin-producing cells. *Journal of Biotechnology* 144(4), 304–312.

Wild, S., Roglic, G., Green, A., Sicree, R. and King, H. (2004) Global prevalence of diabetes: estimates for the year 2000 and projections for 2030. *Diabetes Care* 27(5), 1047–1053.

Witczak, Z. (1997) Polysaccharides in medicinal applications, edited by Severian Dumitriu, reviewed in *Journal of Carbohydrate Chemistry* 16(2), 245–247.

Zinn, A., Felson, S., Fisher, E. and Schwartzbard, A. (2008) Reassessing the cardiovascular risks and benefits of thiazolidinediones. *Clinical Cardiology* 31(9), 397–403.

15 Bioengineered Wound and Burn Healing Substitutes: Novel Design for Biomedical Applications and General Aspects

Erdal Cevher[1], Ali Demir Sezer[2,*] and Ayca Yıldız Peköz[1]
[1]*Department of Pharmaceutical Technology, Faculty of Pharmacy, Istanbul University, Istanbul, Turkey;* [2]*Department of Pharmaceutical Biotechnology, Faculty of Pharmacy, Marmara University, Istanbul, Turkey*

Abstract

Wound healing is the inherent ability of an organism to protect itself against injuries. Cumulative evidence indicates that the healing process patterns in part embryonic morphogenesis and may result in either organ regeneration or scarring, phenomena that are developmental stage- or age-dependent. Tissue regeneration by using biomaterials and skin grafting materials in periapical surgery is an example of tissue engineering technology. Significant progress has been made in the development of *in vitro*-engineered skin substitutes that mimic human skin, either to be used for the replacement of lost skin or for the establishment of *in vitro* skin research models. Full-thickness skin deficits are indications to autologic skin graft. In extensive skin injuries an employment of skin substitutes is sometimes necessary. This review presents the classification of skin substitutes (permanent, temporary, biological, synthetic). The different kinds of skin substitutes approved for commercial production are described (epidermal substitutes, dermal substitutes, composite dermo-epidermal substitutes). The possibilities of clinical applications of skin equivalents and results obtained by many authors after employment of artificial skin are also presented. Important new discoveries in key elements of engineering of tissue-engineered skin including cell sources, biomaterials and growth factors, etc., are summarized. Basic and clinical applications for engineered skin substitutes in cell therapy, tissue engineering and biomedical research continue to drive design improvements premised on these structure- and function-based engineering paradigms.

Structure and Function of the Skin

Skin, responsible for protecting the body from noxious external factors and for keeping the internal systems intact, has a surface area of 1.8 m^2 and makes up about 16% of the body weight. It consists of three layers: epidermis, dermis and subcutis (Junqueria *et al.*, 1992).

Epidermis

The epidermis is the outer layer of stratified squamous epithelium. Its average thickness is about 0.1 mm, but the thickness is greater on the palm and sole (0.8–1.4 mm) (Leeson *et al.*, 1988a). The main function is to protect the body from outer conditions. It further consists of five

*Corresponding author. E-mail: adsezer@marmara.edu.tr

sublayers: stratum basale, stratum spinosum, stratum granulosum, stratum lucidum and stratum corneum. These five also represent the stages of maturation of keratin by keratinocytes (Junqueria *et al.*, 1992).

Basal cell layer (stratum basale)

This layer comprises mostly either dividing or non-dividing keratinocytes. There are keratin tonofibrils in the cells; the latter are attached to the basement membrane by the hemi-desmosomes. Up to 5–10% of basal cell population is made up of melanocytes (Gartner and Hiatt, 2001). Melanin is synthesized in these cells and is transferred via dendritic processes to the neighbouring keratinocytes. Originating from the neutral crest, the melanocyte population is at its most dense on the face and other exposed sites. Some merkel cells can also be found in the basal cell layer. These are closely associated with terminal filaments of cutaneous nerves and thus they are assumed to have a role in sensation. There are neuropeptide granules as well as neurofilaments and keratin in their cytoplasms (Young and Heath, 2000).

Prickle cell layer (stratum spinosum)

This layer of polyhedral cells is formed by daughter basal cells migrating upwards. These are interconnected by desmosomes. A supportive mesh in the cytoplasm of these cells is formed by keratin tonofibrils (Leeson *et al.*, 1988a). Dendritic and immunologically active Langerhans cells are mostly found in this layer (Junqueria *et al.*, 1992).

Granular cell layer (stratum granulosum)

Cells become flattened and lose their nuclei in the granular cell layer. There are keratohyalin granules in the cytoplasm, together with membrane-coating granules. Their lipid contents are expelled into the intercellular spaces (Young and Heath, 2000).

Translucent layer (stratum lucidum)

The skin found in the palms and soles is known as thick skin.

Horny layer (stratum corneum)

The end result of keratinocyte maturation exists in the horny layer, composed of sheets of overlapping polyhedral cornified cells with no nuclei (corneocytes) (Young and Heath, 2000). This layer, consisting of several cells, is thick on the palms and soles, but less thick elsewhere in the body. The corneocyte cell envelope is broadened and keratin tonofibrils in a matrix formed from the keratohyalin granules replace the cytoplasm. Lipid glue, which is partly derived from the membrane-coating granules, sticks the cells together (Young and Heath, 2000).

Dermis

The dermis is defined as a tough supportive connective tissue matrix, containing specialized structures, found immediately below the epidermis and intimately connected. Its thickness varies, being thin (0.6 mm) on the eyelids, but thicker (3 mm or more) on the back, palms and soles (Leeson *et al.*, 1988b). The papillary dermis (the thin upper layer) lies just below, and interdigitates with the epidermal rete ridges. It consists of a kind of loosely interwoven collagen. Coarse and horizontally running bundles of collagen are found in the deeper and thicker reticular dermis. Up to 70% of the dermis is made of collagen fibres and they impart a toughness and strength to the structure. Loosely arranged elastin fibres, running in all directions in the dermis, provide elasticity to the skin. They are numerous near hair follicles and sweat glands, and less so in the papillary dermis. The ground substance of the dermis is a semi-solid matrix of glycosaminoglycans (GAG) (Young and Heath, 2000). GAG allows dermal structures some movement. The dermis contains fibroblasts (which synthesize collagen, elastin, other connective tissue and GAG), dermal dendrocytes (dendritic cells with a probable immune function), mast cells, macrophages and lymphocytes (Young and Heath, 2000).

Subcutaneous layer

The subcutis consists of loose connective tissue and fat (up to 3 cm thick on the abdomen). The subcutaneous tissue layer has a few different functions, one of which is to provide insulation to help maintain body temperature in a cold environment. The last function of the subcutaneous tissue layer is to act as a passageway for the blood vessels and nerves from the dermis of the skin into the muscles.

Wound Type

The term wound, as described in the commonly accepted terminology, usually refers to an acute disruption or mechanical trauma such as a gunshot or a scratch (Kapoor and Appleton, 2005). They could be classified as chronic, acute and post-operative (Mutsaers *et al.*, 1997).

Acute wounds

These are usually traumatic. They may be cuts, abrasions, lacerations, burns, to name a few (Porth, 1998). They have the ability to respond rapidly to treatment and normally heal in an orderly and efficient manner. In the absence of complications, they progress smoothly through the four distinct but overlapping phases of the healing process: haemostasis, inflammation, proliferation and remodelling (Chanson *et al.*, 2005).

Burns

These can be thermal, electrical, chemical or caused by radiation. The depth can be characterized by the degree or description of the thickness of the burn. The most common terms are first degree (epidermis only, e.g. sunburn), second degree (epidermis and partial thickness of the dermis, e.g. scalding injuries) and third degree (full thickness of the dermis, e.g. long exposure to flames; Ruszczak, 2003). First-degree burns require no specific care. Second-degree burns are generally determined by the blistering or denudation of the dermis. Depending on their depth, they vary on moistness, colour and pain. Third-degree burns are defined by their dry, leathery and insensate quality. A combination of different degrees is called indeterminate. Burns with initially partial thickness can evolve into full thickness by either infection or hypoperfusion of the wound (Atiyeh *et al.*, 2005).

Chronic wounds

These are wounds that fail to heal within a period of time. There is not a specific time for this period, but most physicians concede 3–4 months of failure to heal as chronic. The time period of wound healing itself depends on factors such as the size of the wound, its cause, and the patient's general clinical status. Several observations suggest that persistent tissue ischaemia in the vicinity of the wound is an important underlying feature of chronic wounds that severely impair the healing process (Bennett and Schultz, 1993; Schultz, 2000; Whitney and Wickline, 2003). Long-lasting ischaemia exposes wounds to infection, inflammation and necrosis. Diabetes, pressure sores, peripheral vascular disease and venous stasis are also some of the delaying factors. Some topical agents that improve wound healing in the young have been found to fail to work in the aged. Thus, it was concluded that physiological response to wounding is deranged in the elderly (Porth, 1998). In dermatology, the preferred term for a chronic wound is chronic cutaneous ulcer (Yager *et al.*, 1996; Trengove *et al.*, 2000). An ulcer is classified as a depressed lesion in which the epidermis or at least the upper dermis has been damaged (Margolis, 1999). Erosion is a focal loss of the epidermis without the involvement of the dermis. An ulcer does not emerge out of nothing. An antecedent initial pathologic lesion such as a papule or a pustule would cause the wound in the first place (Phillips, 2001).

Post-operative wounds

These are intentional acute wounds. Sutures, clips or tape are used for healing. Some surgical wounds are left open to heal by second intention. This allows drainage of infected material (Cohen *et al.*, 1999; Ennis and Meneses, 2000).

Wound Treatment

Since they have a considerable impact on patient morbidity and mortality, acute or chronic wounds pose a substantial problem in healthcare. Wounds are one of the most serious health considerations in quality of life. Those involved in medical research or the medical care sector need to investigate the causes, classification and treatment of wounds with the scientific rigour that the subject deserves (Mutsaers *et al.*, 1997).

History of wound treatment

Wound care practices have improved tremendously over the years. A range of methods has been used, dating back to ancient times. The earliest known example was described in the Ebers Papyrus, found in Egypt (Cohen *et al.*, 1999). It is known that ancient Egyptians used frog skin and castor oil for wound treatment. Throughout history, people have used many materials of biological origin. Various experiments on animals were conducted, although they had limited benefit to human health (Cohen *et al.*, 1999). The first skin grafting experiments were documented in 1804 by the Italian scientist Guiseppe Baronio, in a sheep model (Flye, 1989). The progenitor of the present-day techniques of skin grafting, Baronio conducted a skin grafting experiment by using epidermal cells (Flye, 1989). Inspired by this work, Swiss medical researcher and surgeon Jacques-Louis Reverdin developed the so-called pinch-type grafts for human skin grafting in the 1860s (Pangarkar and Hutmacher, 2003). Mangoldt, on the other hand, described a technique in 1895, where he scraped off the superficial layer of epithelium from the patient's forearm with a scalpel until light wound secretion occurred (Langer and Vacanti, 1993; Pangarkar and Hutmacher, 2003). At the beginning of the 19th century, various researchers performed experiments with small skin pieces in the form of explant cultures (Flye, 1989). Peter Medawar proved that keratinocytes isolated from dynamically cultured skin fragments could be implanted back to the donor without induction of a strong foreign-body reaction by successfully isolating proliferating epidermal cells (Gibson and Medawar, 1943). In the 1970s, Rheinwald and Green succeeded in reproducibly culturing a large number of non-differentiated keratinocytes on a feeder layer of freshly irradiated mouse fibroblasts (Rheinwald and Green, 1975; Green *et al.*, 1979). The use of cultured epidermal autografts (CEA) started after that application. From the early 1970s many clinical scientists believed that skin coverage in chronic and acute wounds such as burns, chronic ulcers, scar revision surgery and vitiligo could be effectively addressed by CEA, despite several shortcomings (Cuono *et al.*, 1986). In the late 20th century, more clinical data were published and manufacturers began to introduce occlusive dressings designed to preserve and protect the moist environment in the wound. Recent dressings accelerate the re-epithelialization, stimulate collagen synthesis and create a hypoxic environment at the wound bed to promote angiogenesis and to decrease the pH at the surface. This process forms an environment wherein bacteria cannot grow, thus decreasing the infection rate. These dressings are indeed advantageous compared to gauze dressings in terms of patient comfort and compliance, and reduction in scarring (Sarabahi, 2012). Nowadays, researchers are looking for new materials to overcome the disadvantages of traditional dressings such as gauze, paraffin gauze and biological dressings. Methyl cellulose was the first synthetic material used for wound coverage (Shakespeare and Shakespeare, 2002). The need for physical protection from external factors and conditions is the common feature of all these materials. In recent years, many novel materials have been developed in forms of films, sprays, foams and gels, to name a few, for understanding wound-healing mechanisms (Shakespeare and Shakespeare, 2002).

Skin substitutes

Skin substitutes are a heterogeneous group of substances that help temporary or permanent closure of many types of wounds. These kinds of products are not substitutes for adequate surgical debridement or standard surgical therapies (i.e. flap coverage), but are certainly alternatives. They may require a less vascularized wound bed and thus provide reconstructive solutions. They increase the dermal component of the healed wound, reduce or remove inhibitory factors and also decrease the inflammatory response (Shakespeare, 2001). Therefore, they have been proved to be more effective and to provide rapid and safe coverage. The skin substitutes also suppress donor site morbidity, like pain and hypertrophic scar formation. These are very important in massive burns where autografts fail to provide an adequate skin cover (Sheridan *et al.*, 2001). They also enable practitioners to use an approach more analogous to a reconstructive elevator, rather than a ladder, endorsing flexibility within the reconstructive ladder. Practitioners cultivate up and down the reconstructive ladder from one extreme end to the other of the coverage options, skipping between steps if desired (Shakespeare, 2001; Sheridan *et al.*, 2001). There are many skin substitutes in the medical world. However, the claims about which may be an ideal skin substitute are yet to be quantified and verified. These product types have their own applications, strengths and disadvantages and require a head-to-head comparison, but this might not be feasible in practice (Shores *et al.*, 2007; Singh and Shenoy, 2012) (Table 15.1).

Temporary impervious dressing materials

There are many potential benefits of temporary skin substitutes: transient physiologic wound closure, pain control, absorption of wound exudate and prevention of wound desiccation (Quinn *et al.*, 1985). Physiologic wound closure means a degree of protection from mechanical trauma, vapour transmission characteristics similar to skin and a physical barrier to bacteria. These membranes help in maintaining a moist environment with low bacterial density. The four general uses are: (i) as a dressing on donor sites to facilitate pain control and epithelialization from skin appendages; (ii) as a dressing on clean superficial wounds to a similar end; (iii) as a temporary physiologic closure of deep dermal and full-thickness wounds after excision prior to autografting or healing of underlying widely meshed autografts; and (iv) as a test graft in questionable wound beds. There are a large number of such membranes in common use (Quinn *et al.*, 1985; Shakespeare, 2001; Singh and Shenoy, 2012).

Table 15.1. Characteristics of an ideal skin substitute (from Singh and Shenoy, 2012).

Ability to resist infection	Flexible
Ability to withstand wound hypoxia	Conforms to irregular wound surfaces
Cost efficient	Can be used 'off the shelf'
Does not require refrigeration	Cannot transmit viral diseases
Non-toxic and non-antigenic properties	Does not become hypertrophic
Does not incite inflammatory response	Grows with a child
Long-term wound stability	Can be applied in one operation
Epidermal and dermal components	Elasticity and high mechanical strength
Ability to resist shearing forces	Bioadhesiveness to the wound surface
Widely available	Compatibility with topical therapeutic agents
Easy to store and use	Biodegradable
Prevents water loss	Optimum oxygen permeability
Long shelf-life	Could be easily sterilized

Single layer materials (biological skin substitutes)

Biological skin substitutes function temporarily 'like skin', fully fulfilling the functions of the skin. There are many available substitutes and their market costs are relatively low. Synthetic skin substitutes can be synthesized on demand and modulated for specific purposes with increased control over scaffold composition, but biological skin substitutes have a more intact and native structure that allows the construction of a more natural new dermis. They also have superb re-epithelialization characteristics due to the presence of a basement membrane (Halim *et al.*, 2010). However, natural constructs can exhibit problems in terms of slow vascularization of the material. The most widely used biological substitutes worldwide are amnion, amniotic mesenchymal stem cells, potato peelings and boiled banana leaf (Pruitt and Levine, 1984).

HUMAN AMNION. The amnion is a thin, semi-transparent tissue forming the innermost layer of the fetal membrane (Gallin and Hepperle, 1998). The amnion has been used as an epidermal barrier since 1910. Human amniotic membrane is used in many different countries to cover clean superficial wounds, including partial-thickness burns, donor sites and freshly excised burns awaiting donor site availability (Subrahmanyam, 1995; Ramakrishnan and Jayaraman, 1997;

Nanchahal *et al.*, 2002). The epithelium in human amniotic membrane provides a good protection from evaporative loss. It also includes barrier function, while the fibronectin and collagen matrix provide some dermal function. Its transparency allows good wound surveillance. It is minimally adherent and thus dressings need to be changed every 2 days. The practical disadvantages include: hard to obtain, prepare and store; and must be changed frequently. Hence, it has more significant potential for infectious disease transmission than other products (Halim *et al.*, 2010).

AMNIOTIC MESENCHYMAL STEM CELLS. Even though their therapeutic potential for wound healing has been scarcely investigated, amniotic mesenchymal stem cells (AMMs) have been recognized as promising stem cell resources. Through secretion of angiogenic factors and enhanced engraftment/differentiation capabilities, AMMs possess considerable therapeutic potential for chronic wounds (Lobmann *et al.*, 2003; Kim *et al.*, 2012). Transplanted AMMs appear to exhibit high engraftment rates and demonstrate keratinocyte-specific proteins and cytokeratin in the wound area, indicating a direct contribution to cutaneous closure.

OASIS WOUND MATRIX. OASIS wound matrix is produced from porcine small intestine submucosa. It is intended for wound closure stimulation in acute and chronic wounds, and in burns. In order to prevent immunological responses, it is freeze-dried and decellularized. Positive results have been obtained in randomized prospective controlled multicentre trials for chronic leg ulcer treatment with faster healing time and less ulcer recurrence (Mostow *et al.*, 2005). It has also been shown to support *in vitro* epidermal differentiation and basement membrane formation (Lindberg and Badylak, 2001). It was also evaluated *in vivo* in rodent full-thickness wounds as a dressing material. It contributes towards contraction minimization and had no effect on epithelialization (Prevel *et al.*, 1995). However, no results of clinical trials regarding its use in full-thickness wound management have been reported so far.

POTATO PEELINGS. These are used as temporary dressings for burn patients, but are not technically skin substitutes. They are organic materials that are highly available in rural areas. They help in limiting moisture loss from large burns in the body, thus they also help in limiting fluid and electrolyte losses. They have to be changed at regular intervals and pose the risk of leading to infections (Van de Velde *et al.*, 2011). The studies comparing potato peeling dressings with gauze alone have shown that they indeed are better than the gauze during the healing phase. However, there is no direct evidence of an independent antibacterial effect. Van de Velde *et al.* (2011) concluded that potato peelings advance healing, possibly by maintaining moisture and reducing desiccation. The full-thickness skin defects in rats covered with dressings made of boiled potato peelings revealed that wounds could be closed within 14 days and histologically complete repair of epidermis could be found. The cork layer of the potato peel also prevents dehydration of the wound and protects against exogenous agents (Dattatreya *et al.*, 1991).

BOILED BANANA LEAF. The boiled banana leaf as a temporary dressing was first used in 1996 (Gore and Akolekar, 2003). There are several advantages of boiled banana leaf over potato peelings. During dressing change, the pain tolerance for banana leaf over potato peel dressings was greater. The dressing is effective and it is the cheapest dressing for partial-thickness burn wounds. Its preparation is simple and can easily be taught and learnt (Gore and Akolekar, 2003).

> An important factor in the healing of superficial and moderate partial thickness burn is early and effective coverage with a dressing that protects the wound from trauma and dessication and is non-adherent. In our country cost is also a very important factor. Disturbed by the pain and anxiety experienced by the patients during and after dressing changes, search for a new dressing material was begun for partial thickness burn wounds and the banana leaf dressing (BLD) was subsequently developed and optimised by June 1996.
>
> (Gore, M.A., Akolekar, D. (2003))

Bi-layered tissue-engineered materials

TRANSCYTE. This is a bilayer, temporary skin substitute containing human biologically active healing factors in direct contact with the burn wound surface and an external synthetic barrier against environmental factors. A biologic covering, TransCyte (Advanced Tissue Sciences, USA), marketed as 'Dermagraft-Transitional Covering', has been used for the treatment of partial-thickness wounds in both children and adults (Demling and DeSanti, 2002). This material consists of human newborn fibroblasts cultured on the nylon mesh of Biobrane. In a prospective, randomized, comparison study with silver sulfadiazine, the results have shown that burns treated with TransCyte healed more quickly with less hypertrophic scarring without any infection (Noordenbos *et al.*, 1999; Loss *et al.*, 2000).

APLIGRAF. Apligraf (Organogenesis Inc., USA) was the first composite skin graft approved by the US Food and Drug Administration (FDA) (Bell *et al.*, 1981). It consists of a type I collagen gel and allogenic fibroblast mixture on which a confluent layer of allogenic keratinocytes is cultured (Teumer *et al.*, 1998; Trent and Kirsner, 1998). The material can be easily peeled off the supporting mesh and carrier medium. It is resilient, relatively easy to handle and has a shelf-life of several days. It is stated that the cell lines obtained from foreskin are non-immunogenic and therefore there is a rejection possibility. The general consensus is that long-term survival of allogenic keratinocytes and fibroblasts in deep dermal wounds is limited; however, tissue-engineered allografts may still have some benefit. They can act as viable allogenic cells delivering biologic mediators (growth factors such as TGF-beta and IL-1) capable of accelerating the repair process (Falanga, 1998; Falanga and Sabolinski, 1999). Apligraf is also approved for the treatment of chronic ulcers caused by venous insufficiency that have not responded to standard care. It has also been used in a variety of other clinical settings such as acute surgical defects following micrographic surgery, blistering disorders, ulcerations due to inflammatory conditions such as pyoderma gangrenosum, on split-thickness donor sites, pressure ulcers and a number of other skin defects (Falanga, 1998; Kirsner, 1998). Figure 15.1 illustrates the potential mechanism of action during the early and late phase of wound healing in chronic wounds with Apligraf.

POLYACTIVE. PolyActive is a bilayer product based on autologous cultured keratinocytes and fibroblasts seeded into a PolyActive matrix. This porous matrix consists of a biodegradable polymer, a mixture of polyethylene oxide terephthalate and polybutylene terephthalate. This enables the slow release of a drug into the patient's body (El Ghalbzouri *et al.*, 2004). Slow release can decrease the side effects, enhance the therapeutic efficacy and reduce the number of injections during therapy. This polymer is commonly used for bone reconstruction and its use for skin repair has not been extensively studied. Autologous cells are used in the product, so it does not pose the same potential risks as those associated with allogeneic material, such as cross-contamination by infective agents or immune rejection. This feature suggests potential benefits over allogeneic-based bioconstructs (Xiao *et al.*, 1999; El Ghalbzouri *et al.*, 2004).

Single layer durable skin substitutes

CULTURED SKIN SUBSTITUTE. Rheinwald and Green (1975) described a novel way of culturing human keratinocytes, allowing rapid expansion of the original sample by several thousands. One can grow between 1 and 2 cm^2 of human epidermis from 2 to 3 cm^2 of skin. In order to yield a single-cell keratinocyte suspension, the technique involves taking a small skin biopsy. This is then inoculated onto a layer of irradiated 3T3 mouse fibroblasts and incubated in a growth medium with various supplements until the cells reach confluence; the process is repeated several times. The final cultured sheets are enzymatically released from the support surface, and are attached to a petrolatum-impregnated gauze. After that, they are ready to be placed on the wound. These final sheets made clinical applications of cultured epidermal autografts (CEA) possible.

Subsequently many variations on Rheinwald and Green's methodology have been developed to improve its efficiency (Pittlekow and Scott, 1986; Grimwood, 1988). Cultured skin substitute is a CEA version of these techniques, with the addition of a cultured autologous dermal layer. This layer makes it a more anatomically correct skin substitute. The product was developed at the University of Cincinnati and Shriners Hospitals for Children in Cincinnati,

USA (Cuono *et al.*, 1986) and is still one of the most advanced autologous skin substitutes available. Meanwhile, cultured keratinocyte grafting has become established as a treatment option in severe burn injuries and in the clinical management of other wounds such as chronic venous ulcers (Fig. 15.2). Keratinocytes were grown to confluency to form a dense epithelial sheet. By enzymatic treatment, the cultured epithelium was then detached as a coherent sheet from culture flasks. Later, it was transferred on a surgically prepared grafting bed to provide permanent coverage of excised third-degree burn wounds. The detachment of a coherent epidermal sheet is difficult and time-consuming. Enzymatic detachment of the epithelial sheet from plastic surfaces causes the sheet to shrink. Inconsistencies in the graft have the tendency to form blisters; and scars due to exposure to shear forces are other problems. Hence, the CEA technique is improved by growing keratinocytes on a transplantable substrate that can be detached together with the epithelial sheet. Attaching the CEA with surgical clips to a backing of petrolatum gauze was one of the first improvements. Using a suitable substrate to provide mechanical stability allows the application of epidermal grafts on areas subjected to mechanical stress and might also improve the overall outcome of the regenerated tissue (Horch *et al.*, 2005).

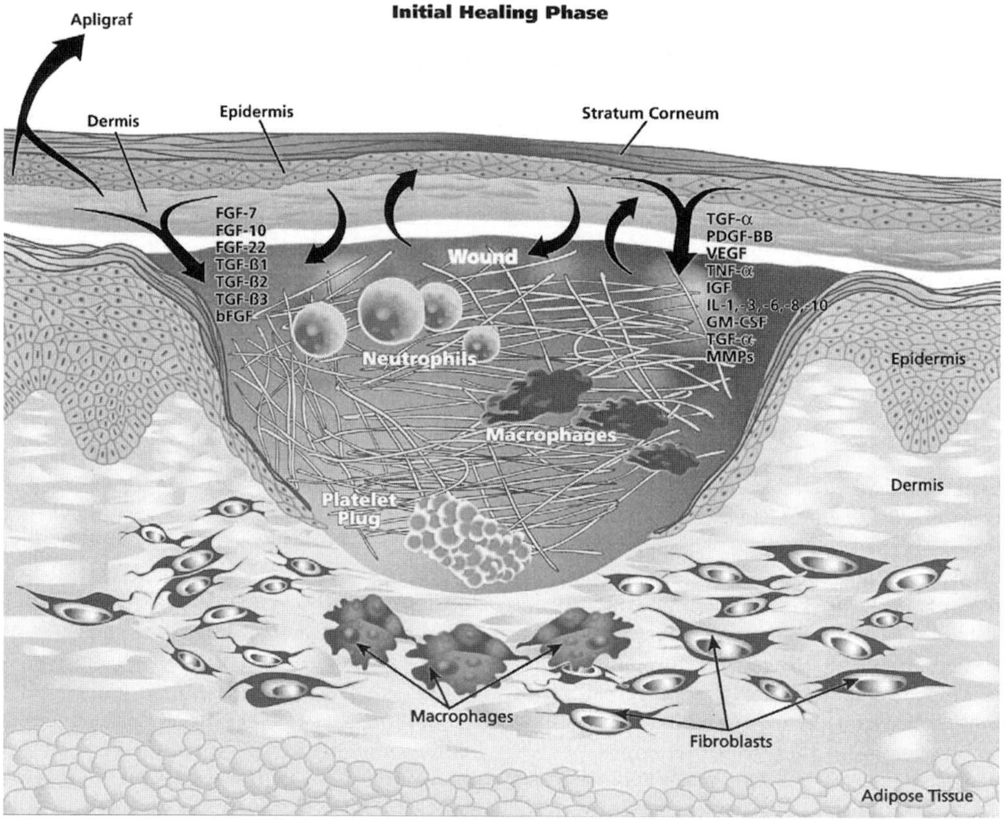

Fig. 15.1. Potential mechanism of action with Apligraf: (a) during initial phase of wound healing in chronic wounds; (b) during late phase of wound healing in chronic wounds. bFGF, basic fibroblast growth factor; FGF, fibroblast growth factor; GM-CSF, granulocyte/monocyte colony-stimulating factor; IGF, insulin-like growth factor; IL, interleukin; MMPs, matrix metalloproteinases; PDGF, platelet-derived growth factor; TGF, transforming growth factor; TNF, tumour necrosis factor; VEGF, vascular endothelial growth factor (from Jimenez and Jimenez, 2004).

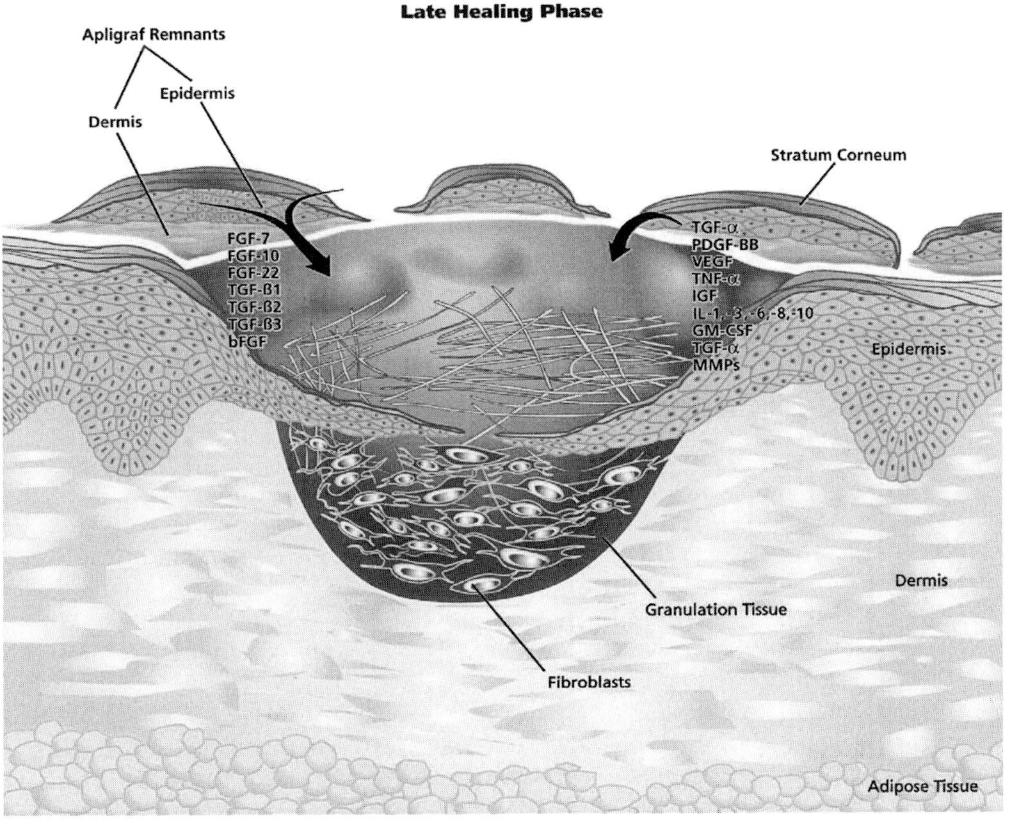

Fig. 15.1. Continued.

ORCEL. OrCel (Ortec International Inc., USA) is a bilayer dressing close to normal skin. It was developed as a tissue-engineered biological dressing, indicated in the treatment of chronic wounds and skin graft donor sites. OrCel consists of allogeneic fibroblasts and keratinocytes seeded on opposite sides of a bi-layered matrix of bovine collagen. It received FDA approval in March 2001 for the treatment of ulcers caused by epidermolysis bullosa. In September 2001, it was approved for the treatment of split-thickness skin graft donor sites and is now available on the market. It is shipped to the clinician frozen in a sealed pouch, then after a brief thawing and rinsing, it can be placed onto the wound. It is then secured with a non-adherent dressing of choice. Still *et al.* (2003) demonstrated that, when grafted onto split-thickness wounds, Orcel accelerated healing speed and reduced scarring compared to standard dressing (Biobrane-L).

ALLODERM. Alloderm is designed to be combined with a thin epithelial autograft during application (Wainwright, 1995). Skin is procured from cadaver donors, properly screened for disease. Using hypertonic saline, the epithelial elements of the grafts are removed and the remaining dermis is treated in a hygienic solution to inactivate any viruses. The material is then freeze-dried. This process results in a non-antigenic complete dermal scaffold with base membrane proteins, including laminin and type IV and VII collagen. With an overlying ultra-thin epithelial autograft, the material is then rehydrated and applied to

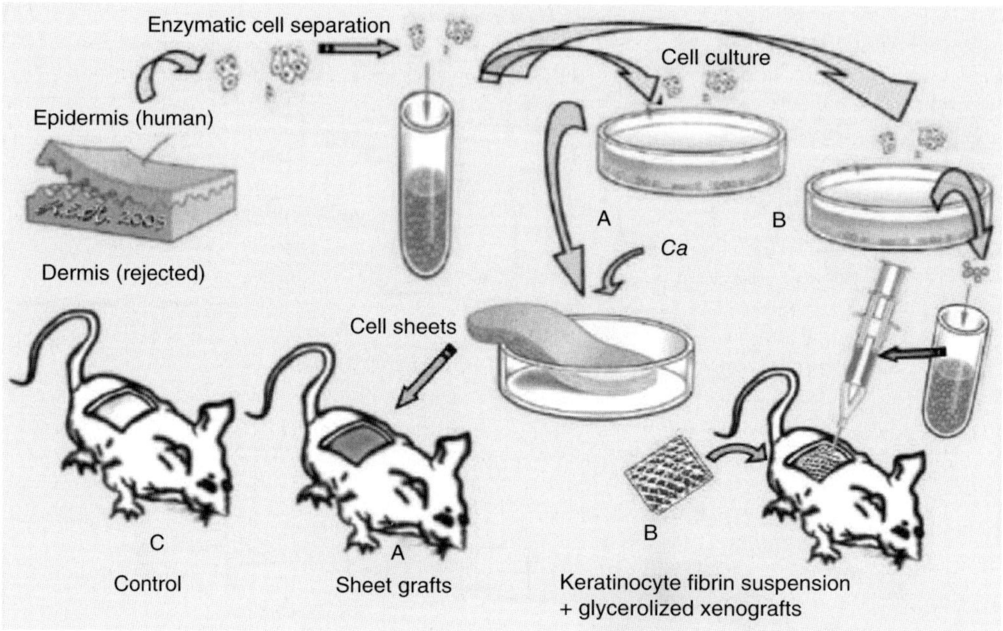

Fig. 15.2. Experimental setup of testing the performance of tissue-engineered skin substitutes in full-thickness nude mouse skin wounds. Comparison of cultured and stratified epithelial sheet grafts after calcium addition (A) versus sub-confluently cultured non-stratified keratinocyte single cell suspensions in fibrin sealant; (B) with xenogenic meshed allograft skin as an overlay versus control wounds; (C) model mimicking the clinical reality in large burn wounds (from Horch *et al.*, 2005).

wounds. Clinical experience on acute and reconstructive burn wounds has been positive (Wainwright, 1995; Wainwright *et al.*, 1996).

Composite skin substitutes

Skin grafts

XENOGRAFT. Xenografts are skin substitutes harvested from animals for temporary graft use in humans (Lloyd *et al.*, 1998; Halim *et al.*, 2010). All these are temporary, as the host immune system will eventually reject them. Thus, they will have to be either replaced by autografts or by other substitutes (Sheridan and Tompkins, 1999; Halim *et al.*, 2010). Frog skin is the earliest known xenograft application for wound coverage, and was used in Ancient Egypt (Cohen *et al.*, 1999; Balasubramani *et al.*, 2001). Essentially the frog skin can be replaced with porcine products. De-epidermized xenografts consist only of varying thickness of dermal tissue. They are generally used for temporary coverage of clean wounds, such as superficial second-degree burns and donor sites (Ruszczak, 2003; Whitney and Wickline, 2003). They also have a wide use in patients with toxic epidermal necrolysis. Porcine xenograft suppresses wound colonization. These xenografts do not vascularize, but will adhere to a clean superficial wound and provide excellent pain control during the healing process (Cohen *et al.*, 1999; Balasubramani *et al.*, 2001).

ALLOGRAFT. The use of skin allograft goes back to World War II. Cadaveric skin allograft application is one of the most commonly used skin substitutes in burn wound treatment. There are also reports on skin allografts from living donors (Jones *et al.*, 2002). The gold standard of skin substitute for coverage of large, full-thickness burns is mainly considered to be fresh cadaver allografts (Balasubramani *et al.*, 2001; Nanchahal *et al.*, 2002). In the same way, their practical use is severely limited due to availability. Human skin stored at 40°C in an appropriate medium maintains its efficacy for several weeks, but its viability decreases severely (DeBono *et al.*, 1998). Skin banks worldwide differ in their codes of conduct and techniques of preservation and storage (Greenleaf and Hansbrough, 1994; Kalter and de By, 1997). There are two main types of cadaveric skin allografts, depending on the methods of processing and storage: (i) cryopreserved allografts; and (ii) glycerol-preserved allografts (GPAs). GPAs are more popular and are commonly used in clinical practices. When the skin allograft is used in a sandwich grafting technique to overlay the autograft, it prevents desiccation of the wound bed in the interstices of widely expanded autografts, as well as reducing bacterial colonization. Thus, the autograft is protected from shear. When a skin allograft is applied, the epithelialization of the wound bed is accelerated as well (Cuono *et al.*, 1987; Langdon *et al.*, 1988).

Several European companies have been studying AEGs and they were commercialized as active wound dressings (CryoCeal, XCELLentis, Belgium) for burns and ulcers. Since allogenic cells survive only for a short period, an allogeneic epidermal graft might be evaluated as a biologically active temporary wound dressing In this period, growth factors capable of accelerating the wound-healing process may be produced by epidermal keratinocytes (Parenteau *et al.*, 2000).

AUTOGRAFTS. Autografts are tissues grafted from the individual concerned and placed in a new position on their body. Autografts are the 'gold standard' for achieving skin coverage (Shakespeare, 2001).

There are three main categories:

- Split-thickness skin grafts
- Full-thickness skin grafts
- Cultured autologous skin

Split-thickness skin grafts. A split-thickness skin graft (STSG) contains epidermis and a predetermined amount of dermis (Quinn *et al.*, 1985; Atiyeh *et al.*, 2005). These are subdivided into thin (0.125–0.275 mm), medium (0.275–0.4 mm) and thick (0.4–0.75 mm) layers depending on the amount of dermis in them (Phillips, 1998). They are often used in surgery to cover post-operative defects or accelerate the epithelialization of chronic wounds such as leg ulcers (Pruitt and Levine, 1984; Falanga *et al.*, 1998). STSG is characteristically prepared from full-thickness skin (FTS) using a dermatome. The dermatome removes much of the dermis from the FTS. For *in vitro* transdermal drug delivery studies, STSGs remain the common skin model. There are three advantages to this technique: (i) the graft assimilates itself better to the edges of the wound with the bevelled edge, as it drapes over the wider edge of soft tissue. This also avoids the poor take in right-angled corners; (ii) since there is less clinical retraction, the bevelled excision margin displays the blood vessels, particularly at the galeal layer. This aids effective haemostasis; and (iii) as the angles are softened, the post-operation aesthetics are improved. The graft can then be secured using sutures or staples and covered by a dressing of choice (Seto *et al.*, 2010).

Full-thickness skin grafts. Full-thickness skin grafts (FTSGs) consist of epidermis and entire dermis, including adnexal structures such as hair follicles, sebaceous glands and sweat glands (Sheridan and Tompkins, 1999). In addition to covering deep skin defects down to the fascia on every part of the body, recovery of the facial defects caused by skin cancers is one of the most important uses of FTSGs. They are especially useful for covering the defects of nasal tip, nasal dorsum, nasal ala, lateral nasal sidewall, lower eyelid and ear. In general, avascular tissue such as exposed bone, cartilage, tendon or nerve devoid of periosteum, perichondrium, peritenon or perineurium, cannot support FTSGs. However, small areas may be grafted with the bridging phenomenon. FTSGs can also provide relatively good colour, texture and thickness that matches properly selected defects. When skin is grafted to the recipient site where no adipose tissue remains, wound contraction is minimized (Yamaguchi *et al.*, 2000). Full-thickness replacements of human skin are the most technologically advanced skin substitutes that are currently available. These materials are a composite of cultured human fibroblasts grown on a collagen sponge. On one side they are covered with a sheet of cultured human keratinocytes. The cells mature normally so that the resulting material histologically resembles the normal skin. However, it is without appendigeal structures and the undulations of the rete ridges, which are characteristic of intact skin (Falabella *et al.*, 1999).

Cultured autologous skin. The autologous skin graft has been the 'gold standard' of wound closure for many years. Unfortunately, in patients with massive and large burns, there is no donor site available. Also, harvesting skin grafts creates secondary wounds that may lead to additional complications and scarring. Since the 1950s scientists have attempted to culture human and mammalian tissues, including human keratinocytes. Unfortunately, these attempts produced only small colonies of cells until 1975 (Rheinwald and Green, 1975; Dioguardi *et al.*, 1990).

MYSKIN. MySkin uses subconfluent autologous living keratinocytes that are grown on a silicone support layer with a specially formulated surface coating (Moustafa *et al.*, 2004). The time needed for cell culture is decreased and handling and application of keratinocyte grafts becomes easier. Proliferatively active keratinocytes can be delivered to the patient with greater time flexibility (Hernon *et al.*, 2006) in a way that cannot be achieved with confluent-cultured epithelial sheet grafts. MySkin is intended for the treatment of diabetic neuropathic ulcers, pressure ulcers, partial-thickness burns and skin graft donor sites (Moustafa *et al.*, 2007). In combination with meshed skin grafts it could be used for the treatment of full-thickness wounds; it cannot be applied alone to deep wounds.

LASERSKIN. Laserskin (Fidia Advanced Biopolymers, Italy) is an epidermal autograft composite. It uses autogenous keratinocytes from the patient that are cultured in the laboratory and seeded onto a membrane (consisting of 100% esterified hyaluronic acid), which is laser micro-perforated. Biopsy of the patients for whom autogenous keratinocytes may be cultured and expanded is required. After 10 days, autogenous keratinocytes are seeded onto the membrane and delivered to the clinical site for engraftment (Lam *et al.*, 1999).

Tissue-engineered skin

INTEGRA. The first dermal substitute clinically used was Integra 'artificial skin' (Integra Life-Sciences Corp., USA). It was developed in the 1980s by a biomaterials research team from the Massachusetts General Hospital and Massachusetts Institute of Technology (Tompkins and Burke, 1990). It is a membrane that provides a temporary vapour and bacterial barrier as well as scaffold functionality for dermal regeneration. The material is designed for use in freshly excised burn wounds, and has been approved for clinical use in patients with life-threatening burns. The inner layer of this material is a 2 mm thick combination of fibres of collagen, isolated from bovine tissue, and the glycosaminoglycan (chondroitin-6-sulfate), with a pore size of 70–200 μm and a structure that allows fibrovascular growth, following which it is designed to biodegrade (Yannas *et al.*, 1981, 1982). Fabrication of the device requires a precipitation of glycosaminoglycan and collagen fibres, later freeze-dried and cross-linked by glutaraldehyde. The outer layer is 0.02286 cm. Polysiloxane polymer with vapour transmission characteristics is analogous to epithelium. It is placed on excised full-thickness burns, and after fibrovascular in-growth has taken place into the inner layers, the outer silicone membrane is interchanged with an ultra-thin epithelial autograft 2–3 weeks later (Tompkins *et al.*, 1989). Integra can be applied to a freshly excised burn wound and in-growth from the underlying wound creates a neodermis. A super-thin split-thickness skin graft must be applied to replace the silastic membrane when this is completely vascularized (~2 weeks). Similar to Alloderm, it allows for immediate coverage of large wounds, and has proven benefits over superior cosmesis. In contrast to the vast experience in favour of Integra (Heimbach *et al.*, 2003), prospective, randomized evidence for its superiority to other methods is lacking. In a study with 106 patients, it had a median take rate of over 80% (Heimbach *et al.*, 2000). However, the artificial skin achieved either a high take rate or a very low take rate in some patients. It was also reported that as surgeons became more experienced in working with the material, the take rate in individual centres finally improved significantly. There are several different techniques in clinical use, and there is no consensus on the best protocol for application of this material (Moiemen *et al.*, 2001). Originally developed as a wound dressing, Integra has recently been manufactured so that fibroblasts and epidermal cells have been seeded into the collagen matrix (Chan *et al.*, 2001). Clinical reports in patients with large burns have been favourable (Hutmacher, 2000). As in any occlusive wound dressing, submembrane purulence must be accounted for and promptly treated (Goodwin *et al.*, 1998).

BIOBRANE. Biobrane (Winthrop Pharmaceuticals, USA) is a biosynthetic skin substitute composed of a bilaminated membrane formed by nylon mesh filled with type I porcine collagen (dermal analogue) and covered by a thin lamina of silicone (epidermal analogue) (Ou *et al.*, 1998). It is considered a semi-impermeable substitute and its small pores allow drainage of the transudate (Fig. 15.3). It enables fibroblasts and capillaries to invade the wound and repair the dermal defect. Re-epithelialization is possible thanks to the presence of keratinocytes at the edge of the wound. It has proven benefits in treatment of superficial and medium clean burns of partial thickness, which are not caused by chemicals or petroleum-based products. It is effective through the temporary coverage of donor areas in partial skin grafts and the protection of autogenous mesh grafts (Quinn *et al.*, 1985).

With the enormous clinical experience since its development in 1979, Biobrane has become a standard for skin substitute coverage of thermal injuries, compared to other products. There are reports about its applications in patients with toxic epidermal necrosis, chronic wounds or following skin resurfacing. It is also used in plastic and reconstructive surgery. There is a fair amount of evidence to support the use of Biobrane in the management of burns, particularly in partial-thickness burns in children (Barret *et al.*, 2000). When dressed with Biobrane, patients with superficial partial-thickness burns experienced less pain as compared to silver sulfadiazine dressing, and wound-healing time and length of hospital stay reduced significantly (Frew *et al.*, 2013). However, older (> than 24 h) and deeper wounds are at risk of infection and may not be suitable for treatment with Biobrane (Ou *et al.*, 1998; Barret *et al.*, 2000).

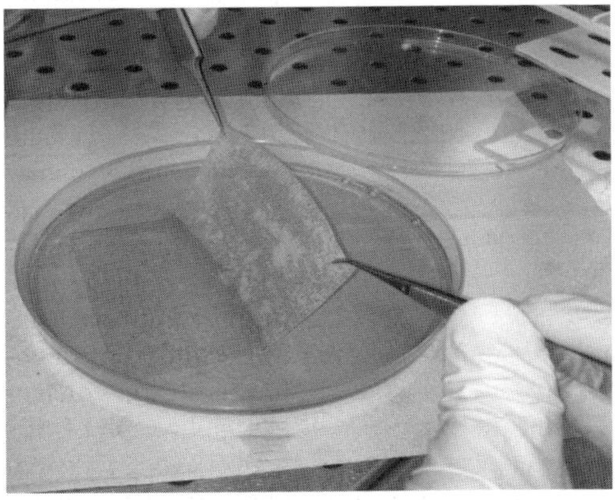

Fig. 15.3. Biobrane in culture medium (from Frew *et al.*, 2013).

EPICEL. This matrix is indicated for use in patients who have deep dermal or full-thickness burns comprising a total body surface of greater than or equal to 30%. In patients for whom split-thickness autografts may not be an option due to the severity and extent of their burns, Epicel may be used alone or in conjunction with split-thickness autografts. The availability of autologous tissue with permanent cover from small amounts of donor tissue is one of its advantages. This is theoretically ideal for patients with high total body surface area injuries with limited or no adequate donor sites (Kapoor and Appleton, 2005). On the other hand, Epicel is relatively expensive, has a long preparation time, the resultant skin is fragile secondary to the thinness of the epidermal grafts and there is only a short window for grafting (Atiyeh *et al.*, 2005).

EPIDEX. EpiDex (Modex Therapeutics Ltd, Switzerland) is cultured from outer root sheet cells of antigen hair follicles plucked from the patient's scalp. The outer root sheet contains the precursor cells for epidermal keratinocytes. These are maintained in the natural wound healing. They have a high proliferative capacity independent of the age of the hair follicle donor. Under tissue culture conditions, the cells proliferate and differentiate, resulting in an epidermal equivalent, which closely resembles human epidermis within 2–3 weeks. After hair follicles are harvested, cells are processed step by step to allow the creation of a cryopreserved cell suspension. The grafts are produced as 1-cm diameter discs and they are rendered onto the silicone membrane for handling purposes. The individual autograft discs are placed in a non-flapping fashion into the wound bed. They are then secured with a proper dressing (Mulder, 1999; Kuroyanagi *et al.*, 2001).

Polymeric-based skin substitutes

Polymeric-based skin substitutes are constructed out of non-biological molecules and polymers that do not exist naturally in human skin. These constructs must be stable and biodegradable, and must provide an adequate environment for the regeneration of tissue. In order to allow ingrowths of blood vessels and fibroblasts, and coverage by epithelial cells, they should maintain their three-dimensional structure for at least 3 weeks. Biodegradation should take place after this period. This process should occur without massive foreign-body reaction as this increases various inflammatory responses that may be associated with profound scarring. These substitutes should be composed of immunocompatible materials to avoid immunoreactive processes (Trumble *et al.*, 2002).

The artificial nature of these skin substitutes has some distinct advantages and disadvantages. The composition and properties of the product can be precisely controlled and additives such as growth factors and matrix components may be added to enhance the skin (Shakespeare, 2001). All synthetic membranes are to some extent occlusive. They must be used cautiously, if wounds are not clean and superficial. Submembrane purulence can occur with potentially disastrous results if placed over devitalized tissue. These products may also avoid complications caused by potential disease transmission. However, the architecture of synthetic skin substitutes does not resemble native skin and they principally lack a basement membrane. The use of non-biological components can be problematic when trying to produce a material that is biologically compatible with the skin (Ou *et al.*, 1998).

Several polymeric-based skin substitutes are proved to be practical for wound coverage. Dermagraft, Hyalomatrix and Renoskin are polymeric-based skin substitutes that are available in the market (Shakespeare, 2001; Hashimoto *et al.*, 2004). There is also a substantial number of synthetic substitutes undergoing *in vitro* or animal testing.

DERMAGRAFT. Dermagraft (Advanced BioHealing, USA) is a commercially available living dermal replacement (Williamson *et al.*, 2002). It consists of a synthetic, knitted PLA (Polylactic acid)/ PGA (Polyglycolic acid) Vicryl mesh, seeded and cultured with foreskin fibroblasts (passaged up to eight times). The skin graft is cryopreserved following the culturation of the scaffold/cell construct for up to 20 days in a closed bioreactor system (Naughton *et al.*, 1997). Dermagraft has been approved by the FDA during October 2001 for treatment of full-thickness diabetic foot ulcers of greater than 6 week duration (Naughton, 2000; Williamson *et al.*, 2002). With enrolment of 314 patients suffering from chronic diabetic foot ulcers, Marston *et al.* (2003) conducted a randomized, controlled, single-blind, multi-centred study. The efficacy of Dermagraft was compared with that of a moist saline gauze covering. Both groups received standard care such as pressure-reducing footwear. After 12 weeks of treatment, 30% of patients treated with Dermagraft achieved wound closure compared with 18.3% of patients in the control group. Gentzkow *et al.* (1996) performed another unique randomized, controlled, single-blind, clinical trial. With respect to the control group, the rate of complete healing of diabetic foot ulcers proved higher in patients treated with Dermagraft. This study also investigated the frequency of Dermagraft application and suggested that weekly usage was better. Dermagraft is also used for burns and chronic wounds. It is used as a temporary or permanent covering to support the take of meshed STSG on excised burn wounds (Noordenbos *et al.*, 1999).

In terms of wound infection, wound exudate, wound healing time, wound closure and graft take, Dermagraft appears to produce results as good as allograft. It needs a simple thawing and rinsing procedure, and is later cut and secured in the wound bed with dressings (Gentzkow *et al.*, 1996). It is removed more easily than allograft and thus increases patient satisfaction. There have been no reports of adverse reactions to Dermagraft, no evidence of rejection, early deterioration or separation from the wound. So far, there have been no reported safety issues. The success of this product depends on its use in a clear, non-infected, viable wound bed that has been debrided. The treatment is done with pressure-relieving footwear in tandem. The fibroblasts used in the cell culture have been widely tested and have shown no transmission of any infectious agents (Finch and Hyder, 1999; Gentzkow *et al.*, 1999).

HYALOMATRIX. Hyalomatrix (Fidia Advanced Biopolymers, Italy) is a bilayer hyaluronan base scaffold with autologous fibroblast. It has an outer silicone membrane. The silicone membrane acts as a temporary epidermal barrier, and the scaffold delivers hyaluronan to the wound bed (Milella *et al.*, 2002). It is commonly used in burn care for full or deep partial-thickness wound temporary repair. Alternative indications are difficult-to-heal wounds, post-traumatic or complicated surgical wounds as a good option for temporary coverage (Radomsky *et al.*, 1999). To assess the clinical and histological long-term outcomes of autologous skin grafting of fresh surgical wounds following previous repair with the hyaluronic acid three-dimensional scaffold Hyalomatrix, an observational study was carried out by Faga *et al.* (2012). Histological observations showed that it promoted the growth of a connective tissue histochemically similar to normal dermis within the first 3 weeks of application. The clinical problem of secondary contracture is still unsolved with a good dermis and no signs of scarring, even though the histological features would suggest regeneration of a skin-like tissue (Faga *et al.*, 2012).

RENOSKIN. Renoskin is a new bilayer dermal matrix for dermal reconstruction composed of a reinforced silicon film and a porous matrix of pure crosslinked bovine collagen. Preclinical assays consisting in *in vitro* fibroblasts and keratinocytes co-culture have been carried out, together with a comparison study with another bilayer dermal substitute on full-thickness wounds on pigs (Fabienne *et al.*, 2007). *In vitro* reconstructed dermis showed more important cell proliferation and a lower contraction rate for Renoskin. Immunohistological studies have shown comparable results for collagen type I synthesis and collagen type I and elastin genic expression. With a differentiated epidermis and a functional basal membrane, reconstructed epidermis was equivalent in both cases. *In vivo*, Renoskin showed no tearing around the staples and a lower rate of detachment and infection. Dermal matrices take rates were 100% for both products and take of the epidermal grafts were 75% equivalent. The percentage of wound retraction of Renoskin was similar to that of the reference material at around 35% (Fabienne *et al.*, 2007).

ACTICEL. Acticel provides superior compressive support and elasticity, is weatherproof and offers significant advantages over other neoprene and elastic support materials. Acticel, a non-autologous version of Epicel, was developed as an over-the-counter product.

OTHER POLYMERIC SKIN SUBSTITUTES. Some new generation biopolymers are used for burn and wound healing treatment, such as thiolated silicone oil (Partenhauser *et al.*, 2016); modified chitosan derivative; chitosan as a modifying component of an artificial scaffold (Romanova *et al.*, 2015); and chitosan in artificial extracellular matrix (aECM) proteins. Over the last decade, many artificial proteins that have incorporated functional domains derived from nature (or created *de novo*) have been reported. They are used for tissue repairing and wound healing (Low *et al.*, 2015).

Conclusion

In summary, the current state of play in the development of skin substitutes is as follows. Products formed from allogeneic cultured cells are commercially available in combination with biomaterials as treatments to provide wound cover and stimulate wound closure. Following on from developments in the 1980s involving the use of cultured keratinocyte grafts in wound healing, the first decade of the century saw great progress in the fabrication of composite grafts. The momentum in this new scientific field built up rapidly between 2005 and 2010. This has been part of a general movement towards *in vitro* fabrication of bioactive or living tissue replacement materials that has been termed *tissue engineering*. The momentum in this new scientific field has built up rapidly during the latter half of the decade. Limited resources have made the need for comparisons of clinical effectiveness and cost ever more important. In the future, skin-substitute materials may be able to stimulate regeneration rather than repair, and tissue-engineered skin may match the quality of split-skin autografts, our present gold standard. There is little doubt that the volume of research and development in tissue replacements for bone, cartilage, cardiovascular tissue and even nerve replacements – together with fundamental studies on the isolation and exploitation of mesenchymal stem cells and other pluripotent cells (e.g. embryonic stem cells) or even totipotent stem cells – will cross-fertilize with skin research. It is, therefore, perhaps not too optimistic to expect tissue-engineered skin grafts with a clinical performance equivalent to the current gold standard skin autografts, in the not-too-distant future.

References

Atiyeh, B.S., Hayek, S.N. and Gunn, S.W. (2005) New technologies for burn wound closure and healing-review of the literature. *Burns* 31, 944–956.

Balasubramani, M., Kumar, T.R. and Babu, M. (2001) Skin substitutes: a review. *Burns* 27, 534–544.

Barret, J.P., Dziewulski, P., Ramzy, P.I., Wolf, S.E., Desai, M.H. and Herndon, D.N. (2000) Biobrane versus 1% silver sulfadiazine in second-degree pediatric burns. *Plastic and Reconstructive Surgery* 105, 62–65.

Bell, E., iv., Sher, S., Merrill, C., Sarber, R., Hull, B., Nakatsuji, T., Church, D. and Buttle, D.J. (1981) Development and use of a living skin equivalent. *Plastic and Reconstructive Surgery* 67, 386–392.

Bennett, N.T. and Schultz, G.S. (1993) Growth factors and wound healing: part II. Role in normal and chronic wound healing. *American Journal of Surgery* 166, 74–81.

Chan, E.S., Lam, P.K., Liew, C.T., Lau, H.C., Yen, R.S. and King, W.W. (2001) A new technique to resurface wounds with composite biocompatible epidermal graft and artificial skin. *Journal of Trauma and Acute Care Surgery* 50, 358–362.

Chanson, M., Derouette, J., Roth, I., Foglia, B., Scerri, I., Dudez, T. and Kwak, B.R. (2005) Gap junctional communication in tissue inflammation and repair. *Biochimica et Biophysica Acta* 1711, 197–207.

Cohen, I.K., Diegelmann, R.F., Yager, D.R., Wornum, I.L., Graham, M.F. and Crossland, M.C. (1999) Wound care and wound healing. In: Schwartz, S.I. (ed.) *Principles of Surgery*. Vol. 1. McGraw-Hill, New York, pp. 263–295.

Cuono, C.B., Langdon, R. and McGuire, J. (1986) Use of cultured epidermal autografts and dermal allografts as skin replacement after burn injury. *Lancet* 1, 1123–1124.

Cuono, C.B., Langdon, R., Birchall, N., Barttelbort, S. and McGuire, J. (1987) Composite autologous-allogeneic skin replacement: development and clinical application. *Plastic and Reconstructive Surgery* 80(4), 626–637.

Dattatreya, R.M., Nuijen, S., van Swaaij, A.C. and Klopper, P.J. (1991) Evaluation of boiled potato peel as a wound dressing. *Burns* 17(4), 323–328.

DeBono, R., Rao, G.S. and Berry, R.B. (1998) The survival of human skin stored by refrigeration at 4°C in McCoy's 5A medium: does oxygenation of the medium improve storage time? *Plastic and Reconstructive Surgery* 102(1), 78–83.

Demling, R.H. and DeSanti, L. (2002) Closure of partial-thickness facial burns with a bioactive skin substitute in the major burn population decreases the cost of care and improves outcome. *Wounds* 14, 230–234.

Dioguardi, D., Brienza, E., De Robertis, M. and Di Lonardo, A. (1990) Skin substitutes in burn treatment-our experience. *Annals of the MBC* 3, 1–7.

El Ghalbzouri, A., Lamme, E.N., van Blitterswijk, C., Koopman, J. and Ponec, M. (2004) The use of PEGT/PBT as a dermal scaffold for skin tissue engineering. *Biomaterials* 25, 2987–2996.

Ennis, W.J. and Meneses, P. (2000) Issues impacting wound healing at a local level: the stunned wound. *Ostomy Wound Management* 46(Suppl 1A), 39S–48S.

Fabienne, B., Daniel, H., Odile, D., Michel, M., Benjamin, H. and Hanane, C. (2007) Development of a new bilayer dermal matrix RENOSKIN : preclinical data. *Burns* 33S, S1–S172.

Faga, A., Nicoletti, G., Brenta, F., Scevola, S., Abatangelo, G. and Brun, P. (2012) Hyaluronic acid three-dimensional scaffold for surgical revision of retracting scars: a human experimental study. *International Wound Journal* 10, 329–335.

Falabella, A.F., Schachner, L.A., Valencia, I.C. and Eaglstein, W.H. (1999) The use of tissue-engineered skin (Apligraf) to treat a newborn with epidermolysis bullosa. *Archives of Dermatology* 135, 1219–1222.

Falanga, V. (1998) Apligraf treatment of venous ulcers and other chronic wounds. *Journal of Dermatology* 25, 812–820.

Falanga, V. and Sabolinski, M. (1999) A bilayered living skin construct (APLIGRAF®) accelerates complete closure of hard-to-heal venous ulcers. *Wound Repair and Regeneration* 7, 201–207.

Falanga, V., Margolis, D., Alvarez, O., Auletta, M., Maggiacomo, F., Altman, M., Jensen, J., Sabolinski, M. and Hardin-Young, J. (1998) Rapid healing of venous ulcers and lack of clinical rejection with an allogeneic cultured human skin equivalent. Human skin equivalent investigators group. *Archives of Dermatology* 134, 293–300.

Finch, P.M. and Hyder, E. (1999) Treatment of diabetic ulceration using dermagraft. *The International Journal of Clinical Foot Science* 9, 156–163.

Flye, M.W. (1989) *Principles of Organ Transplantation*. W.B. Saunders Company, Harcourt Brace Jovanovich, Inc., Richmond, Texas, USA, pp. 87–95.

Frew, Q., Philp, B., Shelley, O., Myers, S., Navsaria, H. and Dziewulski, P. (2013) The use of Biobrane as a delivery method for cultured epithelial autograft in burn patients. *Burns* 39, 876–880.

Gallin, W.J. and Hepperle, B. (1998) Burn healing in organ cultures of embryonic chicken skin: a model system. *Burns* 24, 613–620.

Gartner, L.P. and Hiatt, J.L. (2001) *Color Textbook of Histology*. Vol. 14. W.B. Saunders Company, Philadelphia, Pennsylvania, USA, pp. 325–342.

Gentzkow, G.D., Iwasaki, S., Hershon, K., Mengel, M., Prendergast, J.J., Ricotta, J., Steed, D. and Lipkin, S. (1996) Use of dermagraft, a cultured human dermis, to treat diabetic foot ulcers. *Diabetes Care* 19, 350–354.

Gentzkow, G.D., Jensen, J., Pollak, R., Kroeker, R., Lerner, J., Lerner, M. and Iwasaki, S. (1999) Improved healing of diabetic foot ulcers after grafting with dermagraft, a living human dermal replacement. *Wounds* 11, 77–84.

Gibson, T. and Medawar, P.F. (1943) Fate of skin homografts in man. *Journal of Anatomy* 77, 299–310.

Goodwin, C.J., Braden, M., Downes, S. and Marshall, N.J. (1998) Release of bioactive human growth hormone from a biodegradable material: poly(e-caprolactone). *Journal of Biomedical Materials Research* 40, 204–213.

Gore, M.A. and Akolekar, D. (2003) Evaluation of banana leaf dressing for partial thickness burn wound. *Burn* 29, 487–492.

Green, H., Kehinde, O. and Thomas, J. (1979) Growth of cultured human epidermal cells into multiple epithelia suitable for grafting. *Proceedings of the National Academy of Sciences of the USA* 76, 5665–5668.

Greenleaf, G. and Hansbrough, J.F. (1994) Current trends in the use of allograft skin for patients with burns and reflections on the future of skin banking in the United States. *Journal of Burn Care & Rehabilitation* 15, 428–431.

Grimwood, R.E. (1988) Keratinocyte culture techniques: methods leading to the successful grafting of cultured epidermal cells. *Journal of the Association of Military Dermatologists* 14, 4–7.

Halim, A.S., Khoo, T.L. and Yussof, S.J.M. (2010) Biological and synthetic skin substitutes: an overview. *Indian Journal of Plastic Surgery* 43(Suppl), 23–28.

Hashimoto, T., Suzuki, Y., Tanihara, M., Kakimaru, Y. and Suzuki, K. (2004) Development of alginate wound dressings linked with hybrid peptides derived from laminin and elastin. *Biomaterials* 25, 1407–1414.

Heimbach, D., Luterman, A., Burke, J., Cram, A., Herndon, D., Hunt, J., Jordan, M., McManus, W., Solem, L. and Warden, G. (2000) Artificial dermis for major burns. A multi-center randomised clinical trial. *Annals of Surgery* 208, 313–320.

Heimbach, D.M., Warden, G.D., Luterman, A., Jordan, M.H., Ozobia, N. *et al.* (2003) Multicenter post approval clinical trial of integra(r) dermal regeneration template for burn treatment. *Journal of Burn Care & Rehabilitation* 24, 42–48.

Hernon, C.A., Dawson, R.A., Freedlander, E., Short, R., Haddow, D.B., Brotherston, M. and MacNeil, S. (2006) Clinical experience using cultured epithelial autografts leads to an alternative methodology for transferring skin cells from the laboratory to the patient. *Regenerative Medicine* 1, 809–821.

Horch, R.E., Kopp, J., Kneser, U., Beier, J. and Bach, A.D. (2005) Tissue engineering of cultured skin substitutes. *Journal of Cellular and Molecular Medicine* 9, 592–608.

Hutmacher, D.W. (2000) Scaffolds in tissue engineering bone and cartilage. *Biomaterials* 21, 2529–2543.

Jimenez, P.A. and Jimenez, E.S. (2004) Tissue and cellular approaches to wound repair. *The American Journal of Surgery* 187, 56–64.

Jones, I., Currie, L. and Martin, R. (2002) A guide to biological skin substitutes. *British Journal of Plastic Surgery* 55, 185–193.

Junqueria, L.C., Carneiro, J. and Kelley, O.R. (eds) (1992) *Basic Histology, Skin*. Vol. 18. Prentice-Hall International Inc., Rio de Janeiro, pp. 357–370.

Kalter, E.S. and de By, T.M. (1997) Tissue banking programmes in Europe. *British Medical Bulletin* 53, 798–816.

Kapoor, M. and Appleton, I. (2005) Wound healing: abnormalities and future therapeutic targets. *Current Anaesthesia & Critical Care* 16, 88–93.

Kim, S.W., Zhang, H.Z., Guo, L., Kim, J.M. and Kim, M.H. (2012) Amniotic mesenchymal stem cells enhance wound healing in diabetic NOD/SCID mice through high angiogenic and engraftment capabilities. *PloS One* 7, 1–11.

Kirsner, R.S. (1998) The use of Apligraf in acute wounds. *The Journal of Dermatology* 25, 805–811.

Kuroyanagi, Y., Yamada, N., Yamashita, R. and Uchinuma, E. (2001) Tissue-engineered product: allogeneic cultured dermal substitute composed of spongy collagen with fibroblasts. *Artificial Organs* 25, 180–186.

Lam, P.K., Chan, E.S., To, E.W., Lau, C.H., Yen, S.C. and King, W.W. (1999) Development and evaluation of a new composite laserskin graft. *Journal of Trauma and Acute Care Surgery* 47, 918–922.

Langdon, R.C., Cuono, C.B., Birchall, N., Madri, J.A., Kuklinska, E., McGuire, J. and Moellmann, G.E. (1988) Reconstitution of structure and cell function in human skin grafts derived from cryopreserved allogeneic dermis and autologous cultured keratinocytes. *Journal of Investigative Dermatology* 91, 478–485.

Langer, R. and Vacanti, J.P. (1993) Tissue engineering. *Science* 260, 920–926.

Leeson, T.S., Leeson, C.R. and Paparo, A.A. (1988a) The skin and its appendages (the integument). *Text/Atlas of Histology*. Vol. 10. W.B. Saunders Company, Philadelphia, Pennsylvania, USA, pp. 308–327.

Leeson, T.S., Leeson, C.R. and Paparo, A.A. (1988b) The skin and its appendages (the integument). *Text/Atlas of Histology*. Vol. 10. W.B. Saunders Company, Philadelphia, Pennsylvania, USA, pp. 362–393.

Lindberg, K. and Badylak, S.F. (2001) Porcine small intestinal submucosa (SIS): a bioscaffold supporting in vitro primary human epidermal cell differentiation and synthesis of basement membrane proteins. *Burns* 27, 254–266.

Lloyd, L.L., Kennedy, J.F., Methacanon, P., Paterson, M. and Knill, C.J. (1998) Carbohydrate polymers as wound management aids. *Carbohydrate Polymers* 37, 315–322.

Lobmann, R., Pittasch, D., Mühlen, I. and Lehnert, H. (2003) Autologus human keratinocytes cultured on membranes composed of benzyl ester of hyaluronic acid for grafting in nonhealing diabetic foot lesions. *Journal of Diabetes and its Complications* 17, 199–204.

Loss, M., Wedler, V., Kunzi, W., Meuli-Simmen, C. and Meyer, V.E. (2000) Artificial skin, split-thickness autograft and cultured autologous keratinocytes combined to treat a severe burn injury of 93% of TBSA. *Burns* 26, 644–652.

Low, P.S., Tjin, M.S. and Fong, E. (2015) Design and construction of artificial extracellular matrix (aECM) proteins from *Escherichia coli* for skin tissue engineering. *Journal of Visualized Experiments* 11(100), e52845.

Margolis, D. (1999) Risk factors associated with the failure of a venous leg ulcer to heal. *Archive of Dermatology* 135, 920–926.

Marston, W.A., Hanft, J., Norwood, P. and Pollak, R. (2003) The efficacy and safety of dermagraft in improving the healing of chronic diabetic foot ulcers. *Diabetes Care* 26, 1701–1705.

Milella, E., Brescia, E., Massaro, C., Ramires, P.A., Miglietta, M.R., Fiori, V. and Aversa, P. (2002) Physico-chemical properties and degradability of non-woven hyaluronan benzylic esters as tissue engineering scaffolds. *Biomaterials* 23, 1053–1063.

Moiemen, N.S., Staiano, J.J., Ojeh, N.O., Thway, Y. and Frame, J.D. (2001) Reconstructive surgery with a dermal regeneration template: clinical and histologic study. *Plastic and Reconstructive Surgery* 108, 93–103.

Mostow, E.N., Haraway, G.D., Dalsing, M., Hodde, J.P. and King, D. (2005) Effectiveness of an extracellular matrix graft (OASIS Wound Matrix) in the treatment of chronic leg ulcers: a randomized clinical trial. *Journal of Vascular Surgery* 41, 837–843.

Moustafa, M., Simpson, C., Glover, M., Dawson, R.A., Tesfaye, S., Creagh, F.M., Haddow, D., Short, R., Heller, S. and MacNeil, S. (2004) A new autologous keratinocyte dressing treatment for non-healing diabetic neuropathic foot ulcers. *Diabetic Medicine* 21, 786–789.

Moustafa, M., Bullock, A.J., Creagh, F.M., Heller, S., Jeffcoate, W., Game, F., Amery, C., Tesfaye, S., Ince, Z., Haddow, D.B. and MacNeil, S. (2007) Randomized, controlled, single-blind study on use of autologous keratinocytes on a transfer dressing to treat nonhealing diabetic ulcers. *Regenerative Medicine* 2, 887–902.

Mulder, G.T. (1999) The role of tissue engineering in wound care. An update on the most recent developments and treatments in tissue engineering and recommendations and considerations for practice. *Journal of Wound Care* 8, 21–24.

Mutsaers, S.E., Bishop, J.E., McGrouther, G. and Laurent, G.J. (1997) Mechanisms of tissue repair: from wound healing to fibrosis. *The International Journal of Biochemistry & Cell Biology* 29, 5–17.

Nanchahal, J., Dover, R. and Otto, W.R. (2002) Allogeneic skin substitutes applied to burns patients. *Burns* 28, 254–257.

Naughton, G., Mansbridge, J. and Gentzkow, G. (1997) A metabolically active human dermal replacement for the treatment of diabetic foot ulcers. *International Journal of Artificial Organs* 11, 1203–1210.

Naughton, G.K. (2000) Dermal equivalents. In: Lanza, R.P., Langer, R. and Vacanti, J. (eds) *Principles of Tissue Engineering*. 2nd edn. Academic Press, San Diego, California, USA, pp. 891–901.

Noordenbos, J., Dore, C. and Hansbrough, J.F. (1999) Safety and efficacy of TransCyte for the treatment of partial-thickness burns. *Journal of Burn Care & Rehabilitation* 20, 275–281.

Ou, L.F., Lee, S.Y., Chen, Y.C., Yang, R.S. and Tang, Y.W. (1998) Use of biobrane in pediatric scald burns-experience in 106 children. *Burns* 24, 49–53.

Pangarkar, N. and Hutmacher, D.W. (2003) Invention and business performance in the tissue engineering industry. *Tissue Engineering* 9, 1313–1322.

Parenteau, N.L., Harding-Young, J. and Ross, R.N. (2000) Skin. In: Lanza, R.P., Langer, R. and Vacanti, J. (eds) *Principles of Tissue Engineering*. 2nd edn. Academic Press, San Diego, California, USA, pp. 879–887.

Partenhauser, A., Zupančič, O., Rohrer, J., Bonengel, S. and Bernkop-Schnürch, A. (2016) Thiolated silicone oils as adhesive skin protectants for improved barrier function. *International Journal of Cosmetic Science* 38(3), 257–65.

Phillips, T.J. (1998) New skin for old: developments in biological skin substitutes. *Archive of Dermatology* 134, 344–349.

Phillips, T.J. (2001) Current approaches to venous ulcers and compression. *Dermatologic Surgery* 27, 611–621.

Pittlekow, M.R. and Scott, R.E. (1986) New techniques for the in vitro culture of human skin keratinocytes and perspectives on their use for grafting of patients with extensive burns. *Mayo Clinical Proceedings* 61, 771–777.

Porth, C.M. (ed.) (1998) Tissue repair and wound healing. *Pathophysiology: Concepts of Altered Health States*. 5th edition. Lippincott, Williams and Wilkins, Philadelphia, USA.

Prevel, C.D., Eppley, B.L., Summerlin, D.J., Sidner, R., Jackson, J.R., McCarty, M. and Badylak, S.F. (1995) Small intestinal submucosa: utilization as a wound dressing in full-thickness rodent wounds. *Annals of Plastic Surgery* 35, 381–388.

Pruitt, B.A. and Levine, N.S. (1984) Characteristics and uses of biologic dressings and skin substitutes. *Archive of Surgery* 119, 312–322.

Quinn, K.J., Courtney, J.M., Evans, J.H. and Gaylor, J.D.S. (1985) Principles of burn dressings. *Biomaterials* 6, 369–377.

Radomsky, M.L., Aufdemorte, T.B., Swain, L.D., Fox, W.C., Spiro, R.C. and Poser, J.W. (1999) Novel formulation of fibroblast growth factor-2 in a hyaluronan gel accelerates fracture healing in nonhuman primates. *Journal of Orthopaedic Research* 17, 607–614.

Ramakrishnan, K.M. and Jayaraman, V. (1997) Management of partial-thickness burn wounds by amniotic membrane: a cost effective treatment in developing countries. *Burns* 23(Suppl. 1), S33–S36.

Rheinwald, J.G. and Green, H. (1975) Serial cultivation of strains of human epidermal keratinocytes: the formation of keratinizing colonies from single cells. *Cell* 6, 331–343.

Romanova, O.A., Grigor'ev, T.E., Goncharov, M.E., Rudyak, S.G., Solov'yova, E.V., Krasheninnikov, S.T., Saprykin, V.P., Sytina, E.V., Chvalun, S.N., Pal'tsev, M.A. and Panteleev, A.A. (2015) Chitosan as a modifying component of artificial scaffold for human skin tissue engineering. *Bulletin of Experimental Biology and Medicine* 159(4), 557–566.

Ruszczak, Z. (2003) Effect of collagen matrices on dermal wound healing. *Advanced Drug Delivery Reviews* 55, 1595–1611.

Sarabahi, S. (2012) Recent advances in topical wound care. *Indian Journal of Plastic Surgery* 45, 379–387.

Schultz, G.S. (2000) Molecular regulation of wound healing. In: Bryant, R.A. (ed.) *Acute and Chronic Wounds: Nursing Management*. Vol. 2. Mosby, Philadelphia, Pennsylvania, USA, pp. 413–429.

Seto, J.E., Polat, B.E., Lopez, R.F.V., Blankschtein, D. and Langer, R. (2010) Effects of ultrasound and sodium lauryl sulfate on the transdermal delivery of hydrophilic permeants: comparative in vitro studies with full-thickness and split-thickness pig and human skin. *Journal of Controlled Release* 145, 26–32.
Shakespeare, P. (2001) Burn wound healing and skin substitutes. *Burns* 27, 517–522.
Shakespeare, P. and Shakespeare, V. (2002) Survey: use of skin substitute materials in UK burn treatment centres. *Burns* 28, 295–297.
Sheridan, R.L. and Tompkins, R.G. (1999) Skin substitutes in burns. *Burns* 25, 97–103.
Sheridan, R.L., Morgan, J.R. and Mohamed, R. (2001) Biomaterials in burn and wound dressings. In: Dumitriu, S. (ed.) *Polymeric Biomaterials*. Marcel Dekker, New York, USA, pp. 451–458.
Shores, J.T., Gabriel, A. and Gupta, S. (2007) Skin substitutes and alternatives: a review. *Advanced Skin Wound Care* 20, 493–508.
Singh, A.K. and Shenoy, Y.R. (2012) Skin substitutes: an Indian perspective. *Indian Journal of Plastic Surgery* 45, 388–395.
Still, J., Glat, P., Silverstein, P., Griswold, J. and Mozingo, D. (2003) The use of a collagen sponge/living cell composite material to treat donor sites in burn patients. *Burns* 29, 837–841.
Subrahmanyam, M. (1995) Amniotic membrane as a cover for microskin grafts. *British Journal of Plastic Surgery* 48, 477–478.
Teumer, J., Hardin-Young, J. and Parenteau, N.L. (1998) Tissue engineered skin. In: Patrick Jr, C.W., Mikos, A.G. and Mcintire, L.V. (eds) *Frontiers in Tissue Engineering*. Elsevier Science Ltd., Oxford, UK, pp. 664–677.
Tompkins, R.G. and Burke, J.F. (1990) Progress in burn treatment and the use of artificial skin. *World Journal of Surgery* 14, 819–824.
Tompkins, R.G., Hilton, J.F., Burke, J.F., Schoenfeld, D.A., Hegarty, M.T., Bondoc, C.C., Quinby Jr, W.C., Behringer, G.E. and Ackroyd, F.W. (1989) Increased survival after massive thermal injuries in adults: preliminary report using artificial skin. *Critical Care Medicine* 17, 734–740.
Trengove, N.J., Bielefeldt-Ohmann, H. and Stacey, M.C. (2000) Mitogenic activity and cytokine levels in non-healing and healing chronic leg ulcers. *Wound Repair and Regeneration* 8, 13–25.
Trent, J.F. and Kirsner, R.S. (1998) Tissue-engineered skin: apligraf, a bi-layered living skin equivalent. *International Journal of Clinical Practice* 52, 408.
Trumble, D.R., McGregor, W.E. and Magovern, J.A. (2002) Validation of a bone analog model for studies of sternal closure. *Annals of Thoracic Surgery* 74, 739–745.
Van de Velde, S., De Buck, E., Dieltjens, T. and Aertgeerts, B. (2011) Medicinal use of potato-derived products: conclusions of a rapid versus full systematic review. *Phytotherapy Research* 25, 787–788.
Wainwright, D., Madden, M., Luterman, A., Hunt, J., Monafo, W., Heimbach, D., Kagan, R., Sittig, K., Dimick, A. and Herndon, D. (1996) Clinical evaluation of an acellular allograft dermal matrix in full thickness burns. *Journal of Burn Care & Rehabilitation* 17, 124–136.
Wainwright, D.J. (1995) Use of an acellular allograft dermal matrix (Alloderm) in the management of full-thickness burns. *Burns* 21, 243–248.
Whitney, J.D. and Wickline, M.M. (2003) Treating chronic and acute wounds with warming: review of the science and practice implications. *Journal of Wound Care* 30, 199–209.
Williamson, D., Couts, P. and Sibbald, R.G. (2002) The role of dermal skin substitutes in the management of 'hard to heal', unusual wounds. *Canadian Journal of Plastic Surgery* 10(suppl A), 27A–30A.
Xiao, Y.L., Riesle, J. and Van Blitterswijk, C.A. (1999) Static and dynamic fibroblast seeding and cultivation in porous PEO/PBT scaffolds. *Journal of Materials Science: Materials in Medicine* 10, 773–777.
Yager, D.R., Zhang, L.Y., Liang, H.X., Diegelmann, R.F. and Cohen, I.K. (1996) Wound fluids from human pressure ulcers contain elevated matrix metalloproteinase levels and activity compared to surgical wound fluids. *Journal of Investigative Dermatology* 107, 743–748.
Yamaguchi, Y., Hosokawa, K., Kawai, K., Inoue, K., Mizuno, K., Takagi, S., Ohyama, T., Haramoto, U., Yoshikawa, K. and Itami, S. (2000) Involvement of keratinocyte activation phase in cutaneous graft healing: comparison of full-thickness and split-thickness skin grafts. *Dermatologic Surgery* 26, 463–469.
Yannas, I.V., Burke, J.F., Warpehoski, M., Stasikelis, P., Skrabut, E.M., Orgill, D. and Giard, D.J. (1981) Prompt, long-term functional replacement of skin. *Transactions – American Society for Artificial Internal Organs* 27, 19–23.
Yannas, I.V., Burke, J.F., Orgill, D.P. and Skrabut, E.M. (1982) Wound tissue can utilize a polymeric template to synthesize a functional extension of skin. *Science* 215, 174–176.
Young, B. and Heath, J.W. (2000) *Wheater's functional histology*. 4th edition. Churchill Livingstone, Philadelphia, Pennsylvania, USA.

Index